Laser Communications in Space

For a complete listing of the *Artech House Optoelectronics Library*,
turn to the back of this book.

Laser Communications in Space

Stephen G. Lambert and William L. Casey

Artech House
Boston • London

Library of Congress Cataloging-in-Publication Data
Lambert, Stephen G.
Laser communications in space/ Stephen G. Lambert and William L. Casey.
p. cm.
Includes bibliographical references and index.
ISBN 0-89006-722-8
1. Laser communication systems. 2. Astonautics–Optical communication systems.
3. Artificial satellites in telecommunication. I. Casey, William L. (William Lee) II. Title.
TK5103.6.L36 1995 95-7139
621.382'7–dc20 CIP

British Library Cataloguing in Publication Data
Lambert, Stephen G.
Laser Communications in Space
I. Title II. Casey. William L.
621.3827

ISBN 0-89006-722-8

685 Canton Street
Norwood, MA 02062

International Standard Book Number: 0-89006-722-8
Library of Congress Catalog Card Number: 95-7139

10 9 8 7 6 5 4 3 2 1

For those who sacrificed the most: my understanding and wonderful wife, Mary and my remarkable children, Lindsay, Greg and Brian.

S. L.

To my wonderful wife LaVerne and our family: Mike, Gail & Sarah, Steve, Kathleen, Huck & Hallie, Kathy, Danny, Eli, Isaac & Evan, Jeff, Lori & Caitlin and Kelly, Noelle, Dominic & Courtney.

W. C.

Contents

Preface

The authors have had the desire for many years to put into a concise volume the basic fundamentals of laser communication system design. A book which provides the necessary depth for the technically pure as well as provides the top level aspects in a manner understandable by the nontechnical manager was the goal. While survey texts have been written, a simple, concise text from a designers point of view, had never been attempted on the subject of laser communications.

With the breadth of knowledge and understanding brought by the over 35 years of combined experience on the subject matter and with the ease of accessibility to the technical experts in all related fields (many of whom resided internal to the authors' parent company—McDonnell Douglas Aerospace) we felt this text would give an unparalleled view of the field of laser communications.

This book has been organized into 12 chapters. Chapter 1 provides the history and background of laser communications at the time of this writing. Chapters 2–7 identify architectures, design methodologies for systems, optical design guides and a detailed description of the acquisition, tracking and communication processes. Chapter 8 deals with the state-of-the-art technology and Chapter 9 is a primer on how to design a system. Chapter 10 identifies weight and power trends and Chapter 11 provides a comparison to RF systems and identifies key parametrics for decision making. Finally, Chapter 12 overviews potential applications for laser communication systems.

The appendix provides useful equations and rules of thumb and is intended as a quick reference guide for the designer.

The text postulates many systems applications for laser communications, including several military applications. However, the views expressed in this text are those of the authors and do reflect the official policy or position of the Department of Defense or the U.S. Government. The same can be said of the authors' parent company, McDonnell Douglas Corporation.

Steve Lambert
Bill Casey

Acknowledgments

Many individuals and organizations have contributed indirectly to the success of this text. The works of many technical experts have made this text possible. The early works of Dr. Mike Fitzmaurice and his colleagues at NASA-Goddard Space Flight Center along with the contributions of Frank Goodwin and his co-workers at Hughes are given recognition for beginning laser communication technology development and for the development of the early theoretical understanding of this then new technology. Much of the theory, still used today, was developed by these early pioneers.

Monte Ross and his co-workers at McDonnell Douglas began the first development of laser communications space flight hardware. Supported by visionaries in the U.S. Air Force, such as Lt. Col. Paul Freedman, with both requirements and funding, the McDonnell Douglas team demonstrated transfer of one gigabit of information via an optical link from an aircraft to the ground over 15 years ago! This demonstration paved the way for additional space flight development work.

At about the same time, Dr. Vincent Chan and his colleagues at the Massachussetts Institute of Technology Lincoln Laboratory perceived the need for even greater data rates from laser communication systems and began in earnest work that advanced the state-of-the-art in coherent laser communications technology. The early foundations established by these technology leaders continues today.

Acknowledgement is also deserved of Dr. Jim Lesh, who with his group at NASA's Jet Propulsion Laboratory, has led the effortsassociated with deep space interplanetary optical communications. Their illumination of the Galileo probe from the Earth at six million kilometers was an outstanding achievement.

Equally outstanding work has been performed across both oceans by the Europeans and Japanese whose thrust has been to develop, space qualify and demonstrate on-orbit working laser communications payloads.

In addition to the primary developers and technology leaders, many individuals have contributed in many ways, both directly and indirectly, to the preparation of this manuscript. The authors would like to acknowledge many of their colleagues and peers whose hardwork and dedication have contributed to the success of this text and to the field of laser communications. Each deserves special recognition, and as such, are mentioned herein.

Abshire, Jim
Allen, Lou
Anderson, Richard
Araki, Kenichi
Arikawa, H.
Arima, Y.
Arimoto, Y.
Armentrout, Ben
Aruga, T.
Baker, Bob
Barry, Dale
Bear, Phil
Begley, Dave
Berner, Charlie
Bondurant, Roy
Boronson, Don
Bott, Eric
Brashears, Mel
Breault, R.
Bristow, Tom
Brixey, Harley
Brookman, Jim
Brosen, Pierre
Brown, Bill
Burke, Rusty
Busch, Tim
Carlson, Bob
Carson, Don
Carter, Nick
Casey, Rob
Caudill, Lou
Chen, Chien
Chenoweth, Amos
Chow, Ken
Clarke, Ernie
Cornwell, D.
Cosgrove, M.
Cotton, John
Craig, Richard
Crofts, Duane
Deadrick, Rob
Deckelman, Bill
DePaula, Ramon
Dick, Jim
Doughty, Glen
Driemeyer, Dan
Driesewerd, Doug
Dyer, Buddy
Eagen, Jerry
Ebben, Tom
Elliott, Gary
Fancher, Scott
Feldmann, Bob
Fender, Pete
Fenner, Wayne
Fette', Randy
Fox, Alan
Freidell, Jim
Fries, Wally
Fujiise, M.
Furuhama, Yoji
Gagliardi, Robert
Gass, Karl
Gnojewski, Lenny
Gordinier, Mike
Green, Sam
Greenwell, Rick
Hammond, Sam
Hayden, Bill
Hemmati, Hamid
Henkin, Ernie
Hinrichs, Carl
Hodges, Geoff
Hulsey, Dan
Huther, Kevin
Inagaki, K.
Jackson, John
Jeromin, Lori
Johann, Ulrich
Johnson, Clyde
Johnson, Edward
Johnson, Kevin
Jones, Robert
Kalb, Austin
Karp, Sherman
Katzman, Morris
Kauffman, Bob
Kaufmann, John
Kelsey, John
Kennedy, Chan
Kobylinski, Rich
Kokal, Jim
Korevaar, Eric
Krainak, Mike
Laurent, B.
Leeb, Walter
Levy, Joe
Linford, Rod
Liu, C. S.
Lopez, J.
Marshalek, Bob
Mattassov, George

Matthews, N.
Maynard, John
McGrew, Jimmy
McIntyre, Robert
Mecherle, Steve
Merritt, Jim
Meyer, Danny
Miller, John
Miller, Ted
Mingo, Joe
Minnott, Peter
Mooradian, Greg
Morris, Tim
Muhonen, John
Nicholson, Don
Oleski, Paul
Oppenhauser, Gotthard
Paddon, Earl
Palmer, Karen
Palmer, Les
Paul, Dilip
Pautler, Jim
Pearce, Herb
Pekkarinen, Pat
Phinney, Damon
Pillsbury, Alan
Popescu, Alexander
Pratt, William
Price, Tracy
Prior, Ray
Radecki, Dan
Rice, Bob
Roessler, Norris
Rose, Gary
Ruggieri, Neil
Scholtz, Arpad
Scifres, Don
Seery, Bernie
Shepherd, Jim
Shields, Dale
Shikatani, M.
Shimizu, M.
Simmons, Shawn
Smith, Jim
Smith, Warren
Sontag, Heinz
Spatscheck, T.
Spence, Rodney
Starbuck, Chris
Stark, Richard
Stukel, Steve
Sun, Xiaoli
Swanson, Eric
Taylor, John
Teague, Ralph
Testaguzza, Lou
Titterton, Paul
Tribula, Ed
Ulrich, George
Utrecht, Fred
Van Moui, Tran
Vetter, Hans
Vukabratovich, Dan
Waldo, Dale
Ward, Connel
Webb, Paul
Whited, Chuck
Wichman, Darrel
Wilson, Keith
Windham, Joe
Wolf, John
Yasukawa, K.
Yong, K.
Yu, Y.
Zino, Joe

Special thanks is also given to Richard Stark for his review and suggestions on Chapter 10 of this text. And finally, a very special thanks is given to Dr. Tom Hartwick for his tireless editing of the original manuscript. Dr. Hartwick's suggestions and comments have contributed to the clarity and accuracy of the text.

Chapter 1

Past, Present, and Future

1.1 INTRODUCTION

Remember looking at black and white television with three local network channels; listening to *amplitude modulated* (AM) radio, because *frequency modulated* (FM) radio wasn't invented yet; taking vacuum tubes to the hardware store to be tested; changing speeds on your phonograph between 45, 33 1/3, and 78; watching Neil Armstrong take his historic step onto the surface of the moon; and owning a 16-KB computer? Now you may be sitting in your family room watching a large screen color television with over 100 channels available at your fingertips beamed directly into your home via direct broadcast satellite; listening to FM and even AM stereo in your car or while you jog; explaining to your children what long-playing phonograph albums and vacuum tubes were; having only slight interest in astronauts being launched into space so routinely that it's second page news; or pondering whether 4, 8, 16, or 64 MB are needed in your laptop personal computer. In the past 30 years, technology has led the way to improved living standards. One can't help but think that this technology explosion, that was the sixties, seventies, and eighties, won't continue on indefinitely. There are movements in this direction. the most notable being the appearance of the beginnings of the "information superhighway."

With this explosion in technology comes an equal and complementary explosion in information. Information is the commodity of the future. The transfer of this commodity around the globe in a timely manner, real time if possible, is driving the information business to higher and higher data rates. What technology will be the backbone of this data transfer? Already fiber optics, the foundation of many electro-optic technologies, is playing an important role in terrestrial information transfer. Fiber networks are currently transferring data at over two gigabits per second. Space to space crosslinks will afford information transfer to even the most remote sites on Earth without the expense of intermediate ground relay stations.

Laser crosslinks will enable this transfer of data between satellites at rates compatible with the ground fiber networks.

This is an exciting era for space laser communications. Not only is information transfer driving the requirements to higher data rates, but laser crosslink technology explosions, global development activity, and increased hardware/design maturity are all contributing to interest in space laser communication.

Space laser communication technology advancements have erupted in the past 15 years. Advances have been made in all areas of laser communications technology. Receiver configurations have improved with advanced techniques in the area of silicon *avalanche photodiodes* (APDs) and fiber preamplifiers. Beam steering has evolved into two-axis configurations that replace single-axis devices, which had to be used in pairs. Optical and structural material advances have given rise to radiation hardened glasses and optical coatings as well as lightweight structures. Where once cumbersome laboratory lasers were a part of the system design, laser communication technology has now evolved into the use of semiconductor laser diodes that could, if required, operate on flashlight batteries. In addition to the evolution of semiconductor laser diodes, the evolution of solid state, diode-pumped lasers has steadily improved with advances in numerous host materials and diode pumping schemes.

While many of the improvements are noteworthy, perhaps the single most important to space laser communications, certainly in the last five years, has been the development of a reliable, high-power, single-mode laser diode as a directly modulatable laser source. Where output powers of less than 50 mW were the only single-mode sources available five years ago, (even these with questionable reliability), commercial products are available today with powers of 200 mW, CW and lifetimes greater than ten years. Even more exciting has been the introduction of one watt, *master oscillator*, *power amplifier* (MOPA) devices. The MOPA devices should replace the single-mode sources in the not so distant future for high data rate and long-range applications.

These technology advances offer the space laser communication system designer the flexibility to design very lightweight, high-bandwidth, low-cost communication payloads for satellites whose launch costs are a very strong function of launch weight.

We are seeing a surge in the space laser communication development activities around the globe. Until recently, the United States government was funding the development of an operational space laser crosslink system employing solid-state laser technology [1]. Other system designs and component technologies are under development. The *National Aeronautics and Space Administration* (NASA) is developing technology and studying the applicability of space laser communication to NASA's tracking and data relay network both as crosslinks and for user relay links [2, 3]. NASA's *Jet Propulsion Laboratory* (JPL) is studying the development of large space and ground-based receiving stations and payload designs for optical data transfer from interplanetary spacecraft [4]. A U.S.-backed space exploration initiative is postulating optical data transfer from the Moon and Mars back to Earth-orbiting stations [5]. In addition to the work being performed by the United States government, the *European Space Agency* (ESA) and its Japanese equivalent, are currently developing experimental systems for launch in the middle to late 1990s [6, 7].

1.2 HISTORY

Space laser communications is not new. Laser communication's humble beginnings date back to the early sixties. Figure 1.1 illustrates the major milestones in the history of laser communications to date. Early systems looked at very high-data-rate applications. This was natural so as to leapfrog the more established *radio frequency* (RF) technology. Laser communication needed a niche of its own and high data rates appeared to fit that niche nicely. After all, a strong RF communication contingent was going to have to be convinced that lasers could indeed transfer information as efficiently as the older, more established RF technology with which they were so comfortable. And laser communications couldn't be a threat to the security of the current RF systems if it was going to have a chance of being accepted. Hence, a data rate of one gigabit per second was set as a requirement for the first systems.

In the beginning, two technologies were pursued by the U.S. government; a coherent *carbon dioxide* (CO_2) system by NASA [8] and a *neodymium yttrium aluminum garnet* (Nd:YAG) direct detection system by the U.S. Air Force [9]. The CO_2 system development program (undertaken by Hughes for NASA) was pursued for five years and subsequently terminated. The CO_2 system work, however, continues to this day in Europe.

After several years of study of a Nd:YAG direct detection system, the U.S. Air Force began an initial competition, in the early seventies, for a laser crosslink technology demonstration in space. McDonnell Douglas, St. Louis, Missouri, and Lockheed Missiles and Space Company, Sunnyvale, California, were the initial developers of the technology. McDonnell was awarded the contract to develop and build a space laser communication payload that would communicate with a ground station at an Air Force facility at White Sands, New Mexico [10]. This program was known as the *Space Flight Test System* (SFTS), or the Air Force 405B program, and was funded out of Wright Patterson Air Force Base, Dayton, Ohio. Its primary goal was the development of space-qualified laser communications hardware.

Unfortunately for space laser communications, funding cuts in the 405B program resulted in a downscoping of the technology demonstration program. The funding cuts eliminated the space segment from the budget, and the program was reconfigured as an aircraft to ground station experiment. At that time the name was

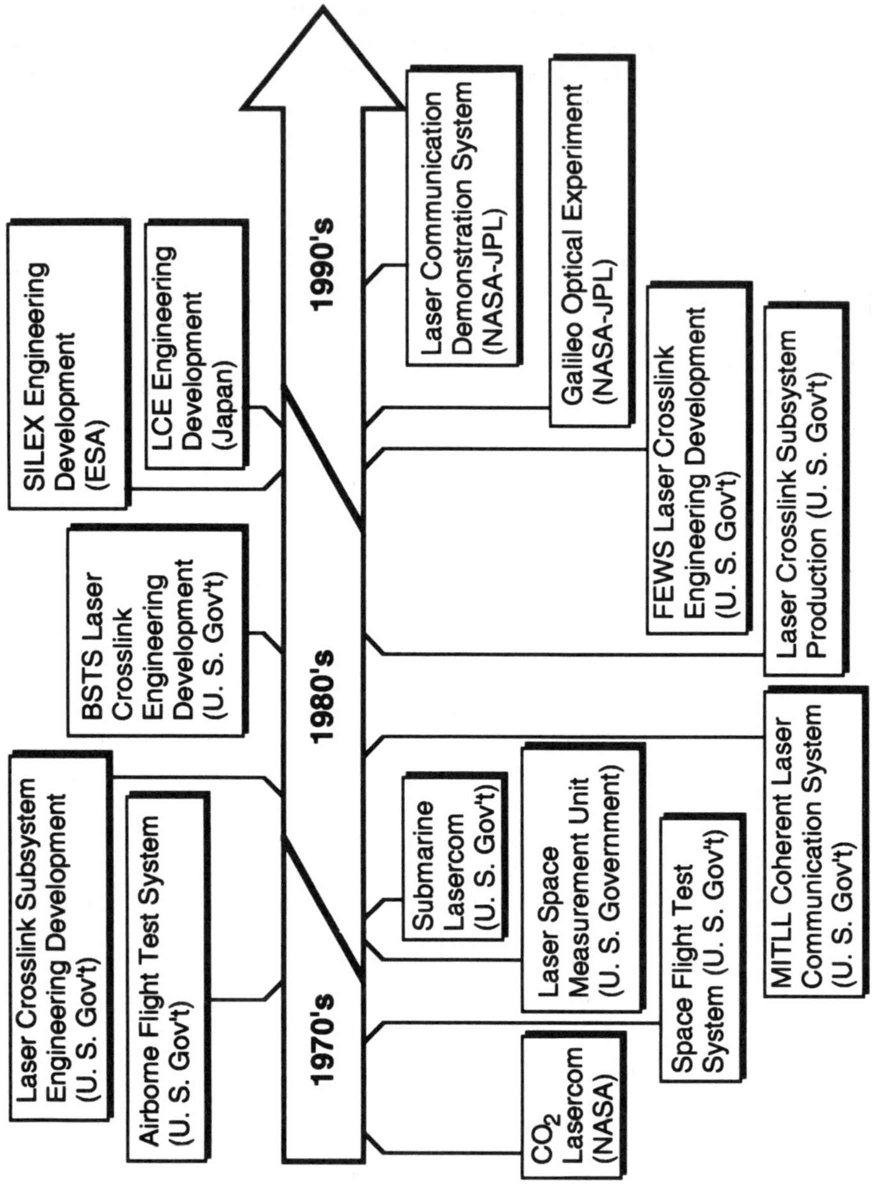

Figure 1.1 Laser communications historical perspective.

changed from SFTS to the *Airborne Flight Test System* (AFTS) [10]. McDonnell finished the development of the two terminals, ground and aircraft, and flew several experiments in an EC-135 test aircraft in the late seventies. These experiments using solid-state laser technology, specifically Nd:YAG, proved the feasibility of transferring one gigabit per second of data from an aircraft to a ground station.

The field tests were broken into six sorties. The first and second proved the feasibility of transmitting one gigabit of information over the atmospheric turbulent channel. The third demonstrated the acquisition of the ground station from an airborne EC-135 aircraft. The fourth sortie verified complete end to end mode logic sequencing. The fifth verified the quality of the return 20-Kbps voice channel and the 1-Gbps data channel. And the final sortie demonstrated communication from the aircraft to the ground at slant ranges of up to 100 km [10].

The success of the AFTS program became immediately apparent when McDonnell Douglas was awarded a contract in the early eighties to develop and build a space to space laser crosslink for a geosynchronous satellite system. This development became known as the *laser crosslink subsystem* (LCS) and was to be an upgrade payload to an existing satellite system. The proposed system [1] used Nd:YAG solid-state laser technology similar to that flown on the AFTS demonstration. The laser reliability was enhanced using diode pumping of the laser medium as opposed to the flashlamp pumping used on AFTS.

The LCS, like its predecessor AFTS, was a direct-detection laser communication system. That is to say, the information was transferred at baseband and directly demodulated at the receiver back into the transmitted signal waveform. (This differs from coherent detection, which imparts the message onto a carrier signal and uses a local oscillator at the receiver to downconvert the carrier to baseband or to an intermediate frequency carrier.) The LCS used a pulsed Nd:YAG laser and imparted five bits of information on every pulse using an M-ary signal set and pulse interval modulation. The LCS system used a common optical path and both transmit and receive wavelengths were at the Nd:YAG fundamental lasing frequency, namely 1.064 μm. Polarization and temporal isolation were used to separate the transmit and receive signals. The LCS transferred data at a modest 1.3 Mbps with a return rate of four kilobits per second. The LCS was chosen for this application primarily because of it ability to be integrated into the satellite and not interfere with existing payload functions, not because of its data rate capacity. The LCS incurred a series of technology setbacks and the program was terminated in late 1993 after several units were space qualified and delivered. The official reason for termination was the removal of the need for crosslinks on the satellite system, but one has to wonder if the technology setbacks and late deliveries contributed to the decision to terminate the development short of completion.

While the LCS development was riddled with problems, both technical and programmatic, it had achieved many firsts and advanced the understanding and technology in many areas. Included in the technology advances were avalanche photodiode detectors, laser-diode pump sources, radiation hardening techniques and optical coatings for optical communication system applications. Additionally, integration, mechanical tolerancing, optical alignment, and design for spaceflight techniques were developed. The LCS was also to be checked out on-orbit using a ground station designed and built for this purpose [11].

At the same time the LCS was being developed, work was being performed at the *Massachusetts Institute of Technology's Lincoln Laboratory* (MITLL) on advanced technology laser communication systems [12–14]. The personnel at MITLL embarked on the development of a coherent optical communication system. This system was to use a quaternary *frequency shift keyed* (FSK) signal set and was to use a local oscillator to mix the received optical signal to an intermediate carrier frequency for eventual detection using standard techniques. This heterodyne detection approach used gallium arsenide laser-diode technology and offered data rates at 220 Mbps. Developmental hardware was built with anticipation of fielding an experiment onboard NASA's *Advanced Communication Technology Satellite* (ACTS). Unfortunately, the NASA funding was trimmed and the experiment was downgraded to a laboratory demonstration. MITLL has developed many of the components and subassemblies for this type of system and has qualified hardware in several areas.

In the middle eighties, a rigorous crosslink trade study was conducted comparing direct-detection laser communications versus conventional 60-GHz RF technology for use on the then *Strategic Defense Initiative Organization's* (SDIO) *Boost Surveillance and Tracking System* (BSTS). This study was conducted by one of the prime contractor teams and compared the mission requirements of the satellite and the complementary enhancements each technology brought to the system. When the results were tabulated, the laser communication approach was chosen, primarily because of cost, weight, and power savings over the state-of-the-art RF approach. Once again, as with the LCS previously, the data rate was not a factor for the selection of optical communications.

Development of the BSTS laser crosslink system, which used gallium arsenide laser-diode technology, began in earnest in 1987 with the demonstration and validation phase. However, in the early nineties with the world threat situation changing, the BSTS satellite system was terminated. Once again funding became a stumbling block and, as a result a laser crosslink was thwarted in its development and deployment. The program, with changes, was to reappear as the *Follow-on Early Warning System* (FEWS) program.

Perhaps the most successful demonstration of laser communication, and a record-breaking one at that, occurred in December, 1992. The scientists at the JPL successfully illuminated the camera on the Galileo spacecraft with an Earth-based pulsed green laser uplink. Detection of the pulses by the *charged coupled device* (CCD) camera was accomplished and the data retransmitted to the ground via the spacecraft RF link. The spacecraft distance was six million kilometers from Earth, making this the longest range laser communication link ever recorded. The proof of the detection of the laser energy now paves the way for future work on deep-space optical data transmission from planetary exploration craft to Earth at much higher data rates, with much smaller antennas and much lower payload weights than with conventional RF approaches [4].

The FEWS laser crosslink was a multiterminal design and was in the demonstration and validation phase of a competitive satellite contract when the satellite program was canceled. Again, a laser crosslink development program was terminated prematurely due to lack of funding.

Laser communications has made significant inroads into the communications community in the last five years. Successful demonstrations like the *Galileo Optical Experiment* (GOPEX) and the successful delivery and launch of the ESA payload will further enhance its image in this community. Additional head-to-head competitions with RF approaches need to be conducted by satellite builders who need to understand the advantages that laser communication brings them and the flexibility it offers their satellite designs. Only through persistence by the laser communication community will laser communication be accepted by the communication community for inclusion into the majority of design tradeoffs. The authors expect that laser communication will dominate satellite crosslink applications in the future, including commercial applications.

1.3 PRESENT ONGOING EFFORTS

Space laser communication is beginning to be accepted as a viable and reliable means of transferring data between satellites. Engineers and scientists working in this area have the distinction of being on the forefront of an emerging technology.

Presently, ongoing hardware development efforts include ESA's *Space Intersatellite Link Experiment* (SILEX) and the Japanese's *Laser Communication Experiment* (LCE). The United States development programs ended with the termination of both the production of the laser crosslink subsystem and the FEWS satellite program.

The SILEX program involves the transmission of up to 50 Mbps of information from a French SPOT IV satellite in low Earth orbit to the *Advanced Relay and Technology Mission* (ARTEMIS) satellite in geosynchronous orbit. This system uses single-mode gallium arsenide laser diode technology and is scheduled for launch in the mid to late 1990s. The system employs an avalanche photodiode detector for communication with *nonreturn to zero* (NRZ) modulation. Acquisition and tracking is accomplished using CCD based array detectors [6].

The Japanese LCE will be integrated onto the Japanese Engineering Test Satellite IV. Initially, tests will be conducted against a ground station from the satellite. Eventually, the experiment will be expanded to include a low Earth to geosynchronous satellite link similar to the SILEX experiment [7]. Discussions between the Japanese and Europeans have begun on a possible joint linkage of the two experimental systems, transferring data from one to the other. Both political as well as technical issues still need to be worked to make this joint venture a reality.

Ongoing work does exist in the United States as a recent study was initiated by NASA-JPL for a high-data-rate link transfer (> 750 Mbps) space demonstration. Anticipated follow-on efforts for hardware development are subject to funding constraints, but could be awarded as early as 1995.

1.4 FUTURE OUTLOOK

The technology advances, which have been the basis for the growth in the industry in the last five years, are anticipated to continue. Standardization of crosslink data rates, formats, and designs is the beginning to interoperable system designs. The international nature of the technology certainly speaks well for the interest that is now and will continue. The successful demonstrations of SILEX and LCE will pave the way for future system designs and operational systems in the years beyond 2000. The United States, which once was the leader in laser crosslink technology has managed to yield the technological leadership to the Europeans and Japanese. The authors only hope that the technological developments and personal sacrifice of those many engineers and scientists involved in the early U.S. development efforts will not be forgotten.

Many future applications exist. Relays from *low Earth orbit* (LEO) to *geosynchronous Earth orbit* (GEO) satellites, LEO networks with point-to-point connectivity, LEO to ground, GEO to ground, LEO and GEO to aircraft, and aircraft to aircraft links are all possible in the near-Earth realm. Satellite to submarine communications is technically feasible and future operational needs may require its development. Deep space development will be enhanced by the deployment of large Earth orbiting and Earth-based receivers and will enable high-data-rate transfer from planetary mission spacecraft.

Future advances in the areas of laser technology and receiver performance as well as the introduction of coherent systems will drive the data rate capacities well beyond the gigabit per second regime. Laboratory demonstrations today have shown multigigabit capabilities.

References

[1] Deadrick, R. B., and Deckelman, W. F. "Laser Crosslink Subsystem—An Overview," *SPIE OE-LASE* '92, Vol. 1635, Los Angeles, CA, Jan. 23–24, 1992.

[2] Hayden, W. L., Krainak, M. A., Cornwell, Yu, A. W., and Sun, X. "Overview of Laser Communication Technology at NASA Goddard Space Flight Center," *SPIE OE-LASE* '93, Vol. 1866, Los Angeles, CA, Jan. 20–21, 1993.

[3] Fitzmaurice, M. W., and Hayden, W. L. "Space Station Laser Communication Transceiver," *SPIE-OE-LASE* '91, Vol. 1417, Los Angeles, CA, Jan. 21–22, 1991.

[4] Lesh, J. R. "Recent Progress in Deep Space Optical Communications," *SPIE OE-LASE* '93, Vol. 1866, Los Angeles, CA, Jan. 20–21, 1993.

[5] Hall, J. R., and Hastrup, R. C. "Telecommunications and Navigation Systems Design for Manned Mars Exploration Missions," *SPIE Proceedings*, Vol. 1059, Los Angeles, CA, Jan. 16–18, 1989.

[6] Popescu, A., F., and Furch, B. "Status of the European Developments for Intersatellite Communications," *SPIE OE-LASE* '93, Vol. 1866, Los Angeles, CA, Jan. 20–21, 1993.

[7] Shikatani, M., Toyoda, M., Tukami, H., Araki, K., Isogai, M., Suzuki, V., and Aruga, T. "Ground System Development for the ETS-VI/LCE Laser Communications Experiment.," *SPIE OE-LASE* '93, Vol. 1866, Los Angeles, CA, Jan. 20–21, 1993.

[8] McElroy, J. H., et al. "CO_2 Laser Communications Systems for Near-Earth Space Applications," *IRE Proc.*, Vol., 65(2), Feb. 1977.

[9] Barry, J. D., et al. "1000-Mbits/s Intersatellite Laser Communications System Technology," *IEEE Trans. on Communications*, April 1976.

[10] Katzman, M. *Laser Satellite Communications*, Englewood Cliffs, NJ: Prentice Hall, 1987, pp. 56–67.

[11] Davis, D. J., Deadrick, R. B., and Stahlman, J. R. "ALGS: Design and Testing of an Earth to Satellite Optical Transceiver," *SPIE OE-LASE* '93, Vol. 1866, Los Angeles, CA, Jan. 20–21, 1993.

[12] Boronson, D. M. "An Overview of Lincoln Laboratory Development of Lasercom Technologies for Space," *SPIE OE-LASE* '93, Vol. 1866, Los Angeles, CA, Jan. 20–21, 1993.

[13] Pillsbury, A. D. "LITE Engineering Model-II: the Flight Qualified Subsystem," *SPIE OE-LASE* '93, Vol. 1866, Los Angeles, CA, Jan. 20–21, 1993.

[14] Boronson, D. M. "LITE Engineering Model-I: Operation and Performance of the Communications and Beam Control Subsystems," *SPIE OE-LASE* '93, Vol. 1866, Los Angeles, CA, Jan. 20–21, 1993.

Chapter 2

Architecture and Applications

2.1 WHY LASER COMMUNICATIONS?

Laser communication systems offer many advantages over *radio frequency* (RF) systems. Before proceeding to the task of understanding the technologies and the methodology for designing a laser communication system, it is worthwhile to initially explore these advantages, many of which relate directly to the architecture and applications [1].

Most of the differences between laser communications (sometimes referred to as lasercom) and RF arise from the very large difference in the wavelengths. RF wavelengths are thousands of times longer than those at optical frequencies. This high ratio of wavelengths leads to some interesting differences in the two systems. First, the beamwidth attainable with the laser communication system is narrower than that of the RF system by the same ratio at the same antenna diameters (the telescope of the laser communication system is frequently referred to as an antenna). For a given transmitter power level, the laser beam is brighter at the receiver by the square of this ratio due to the very narrow beam that exits the transmit telescope. Taking advantage of this brighter beam or higher gain, as it may quantitatively be stated, permits the laser communication designer to come up with a system that has a much smaller antenna than the RF system and further, need transmit much less power than the RF system for the same received power. The narrow laser beam is a two-edged sword, however, since it is much harder to point and acquisition of the other satellite terminal is more difficult.

System comparisons reveal these advantages of laser communications over RF:

- Smaller antenna size;
- Lower weight, usually significant;
- Lower power;
- Minimal integration impact on the satellite.

Last (but very important), laser communications is capable of much higher data rates than RF, again by virtue of that same wavelength (frequency) ratio.

2.2 ARCHITECTURE

Communications architecture is the overall structure or network of satellites and ground stations tied together by communication links. The term ground station denotes Earth-located stations including airborne, land-mobile, land-fixed, shipborne, and submarine terminals. The architecture defines plans, needs, and capabilities as they relate to the present or the future of military or commercial satellite communications. For military satellite communications architecture, the architecture must provide for secure, survivable communications in support of our troops at all levels of conflict.

Military communications planners use a methodology in coming up with a near-term or future satellite communications architecture. Starting with the full spectrum of user requirements or needs and developing a threat estimate for the time period, a baseline architecture or structure is established. Of course, the baseline capabilities will probably fall short of the needs, so these shortcomings must be addressed and rectified. Trades are performed to include and analyze alternate approaches or solutions and the results evaluated in detail. Important factors are included in these detail evaluations, such as risk, the hardware state-of-the-art, and of course, the cost. Out of these detailed studies, a military satellite architecture emerges that is tailored to the task and fits into the budget allocated.

Commercial architecture follows the same procedure in developing an architecture except that jamming resistance and immunity from nuclear weapons scintillation is not required. Privacy is a driving requirement. Of course, commercial satellite communications are concerned with future needs based on projections, with the goal in mind of satisfying future customer needs and making a profit from the operations. Interestingly enough, commercial satellite communications are used by the military on a demand basis under conditions of stress, such as conflict. Desert Storm was a good example where commercial satellite communication carried some of the military communications traffic, freeing up critical military communication assets.

Some of the major architectural considerations include what frequencies to use for each link, the arrangement of the satellites to be placed in orbit, the size of the satellites, the lifetime required of the satellites, the survivability, the matching of the users to the systems, the cost (which is becoming increasingly more important with shrinking budgets), the data rates required to transmit the data, the amount of data rate growth that must be included, and a transition plan that details how and when the new system will be phased into the overall communications architecture. Laser communications must conform to the same considerations as RF systems as applied to an architecture.

Today, interoperability and multiuse satellites are major considerations for those who operate commercial and military communications systems. Satellites are becoming lighter by necessity since light weight can be equated to smaller launch vehicles and consequently, lower overall cost. There is a tremendous difference in launch costs as we move from a given launch vehicle to the next larger size. This can be as much as several hundred million dollars over the life (lifecycle cost) of a communications satellite system with even a small number of satellites. Of course, this is where laser communications shines since laser communications terminals are significantly lighter than those of RF. This weight difference is even more pronounced at the higher data rates projected for many future systems.

An overall military satellite communication architectures is shown in Figure 2.1, which illustrates the users and links that form the architecture. The individual links and the flow of data for each application segment of the overall architecture illustrated will be discussed briefly in Section 2.4. Note in Figure 2.1 that satellite to satellite links (crosslinks) as well as satellite to ground/aircraft/ship/submarine links are included.

As part of commercial or military architecture, air to air, air to ground, and ground to ground communications are necessary but are not considered in this book. The next section will address the satellite orbits that comprise the architectural configurations.

2.3 SATELLITE ORBITS

There are several basic orbits that are used in satellite communication architectures. The most common communications orbit is the *geostationary Earth orbit* (GEO) illustrated in Figure 2.2.

This is the orbit used by virtually all communication relay satellites as well as meteorological satellites. The satellite is placed in a near-zero inclination orbit (equatorial orbit) at 35,786-km altitude. The period is exactly equal to the rotation of Earth so that the satellite appears to be stationary when viewed from the ground. If the inclination (angle the orbit makes with the equator) is not essentially zero, the satellite describes a figure eight over a day's time as viewed from Earth and is commonly termed a geosynchronous orbit. The inclination could be as great as 90 deg. Very little antenna pointing or tracking is required with a geostationary orbit so this satellite orbit is a favorite for ground stations. The pickup of television with a fixed satellite dish is an example of receiving from a geostationary satellite. A network or architecture using geostationary satellites is easy to set up and control compared to the dynamic network resulting from using satellites in nonstationary orbits. There is a lack of coverage above 70-deg latitude due to the equatorial location of the satellite. Also, the launch costs are higher than for nonstationary satellites since larger, more energetic launch vehicles are required to boost the

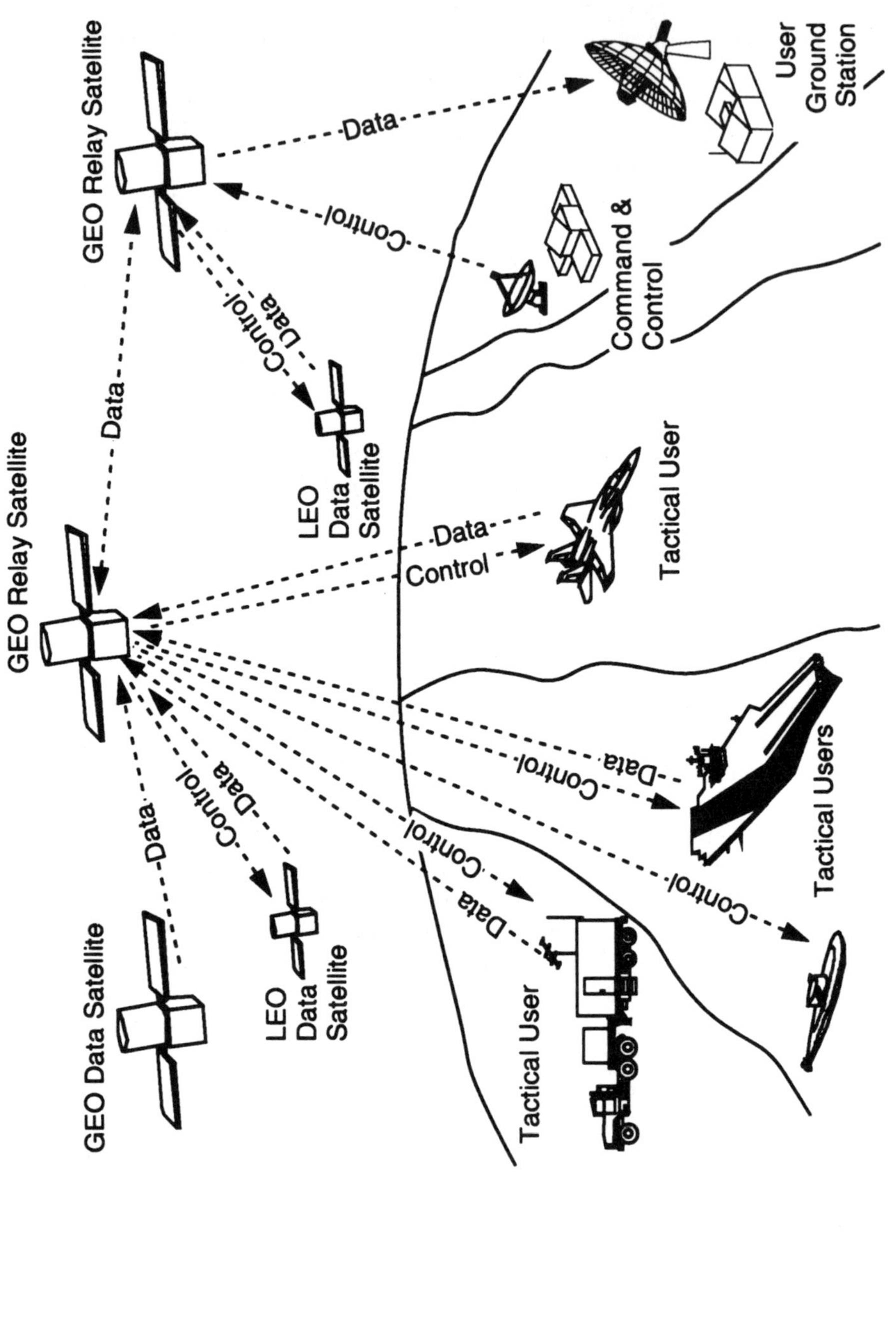

Figure 2.1 A military satellite communications architecture.

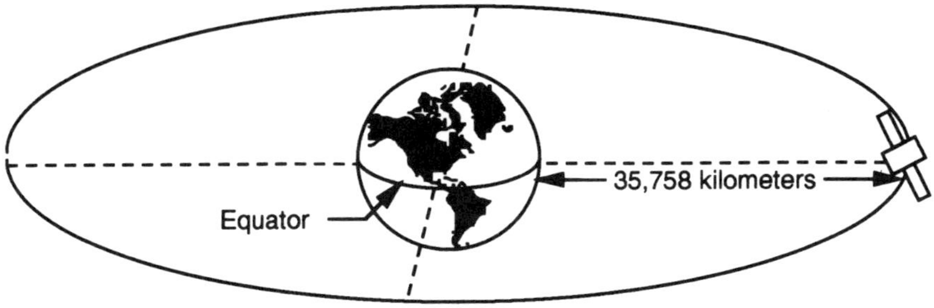

Figure 2.2 Geostationary orbit.

satellite into this high orbit. Orbital slots are assigned for these satellites since with the proliferation of communications satellites and other military satellites it is easy to see that clutter is possible to such an extent that the uplinks to the satellites could interfere with each other.

To overcome the shortcoming of the geostationary orbit in providing northern polar coverage, the former Soviet Union used a Molniya orbit. The satellites are placed in highly elliptical orbits with an apogee (farthest location from Earth) of 40,000 km, a perigee (closest location from Earth) of 500 km, and an inclination angle of 63.4 deg. The apogee is over the North Pole to cover the northern latitudes. The period of the satellite is 12 hr, but because it is highly elliptical, the satellite dwells for some 8 hr of the orbit over the northern hemisphere.

Two or more Molniya satellites placed in different planes as illustrated in Figure 2.3 and phased properly in their orbits can continuously observe the northern regions of Earth. Due to the dynamics of the satellite orbits, the antenna pointing must be continuously changed at the ground stations and the links switched between satellites as they move in and out of sight.

Low Earth orbit (LEO) satellites are used for data gathering by the military, NASA, and international users. Military and commercial Earth mapping and other missions require that the data gathered be sent to the ground for use in agriculture, geological exploration, sociogeographic studies, and so forth. The period of a LEO orbit at an altitude of several hundred kilometers is about an hour and a half. Orbits can be circular or elliptical. The shuttle and forthcoming space station are in LEO orbits (see Figure 2.4).

Frequently the data are gathered and recorded onboard the satellite and dumped to the ground when the satellite passes over a ground station (store and forward). See Figure 2.5.

If so desired, the data may be sent by crosslink to a geostationary relay satellite, which in turn may relay it to another geostationary satellite, which in turn will send it to the ground. This relay application is of particular importance to laser

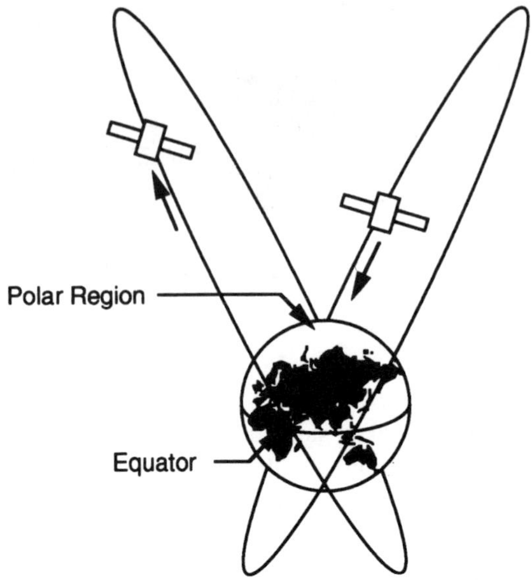

Figure 2.3 Molniya orbits illustrating continuous north latitude coverage.

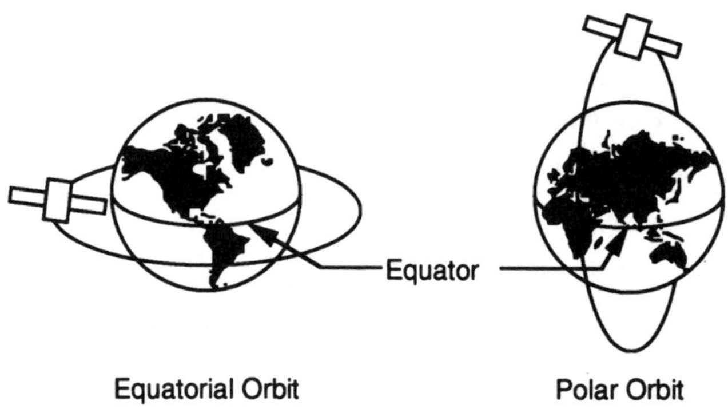

Figure 2.4 Low-Earth orbits.

Figure 2.5 LEO store-and-forward configuration.

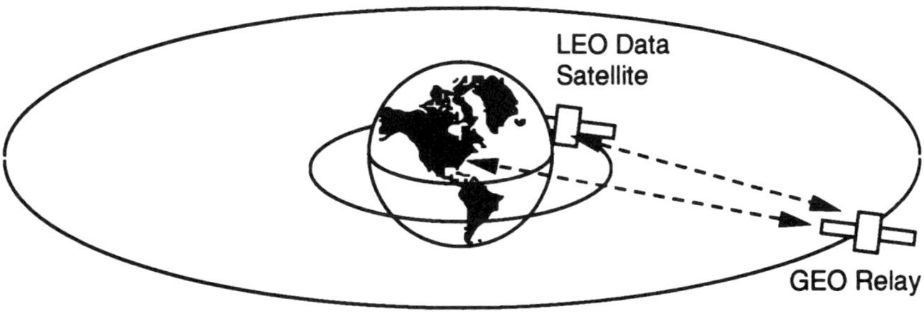

Figure 2.6 Geostationary relay.

communications since many benefits accrue over RF systems. Figure 2.6 illustrates use of a geostationary relay crosslink used to relay data gathered by a LEO satellite.

Data which are of a time-critical nature may be handled this way but it is obvious this is a more complex and costly approach than the store-and-forward approach. There are limitations, however, on what quantity and data rate can be recorded on board a satellite.

There is another LEO architecture, termed internetting. If a number of LEO satellites are placed in low altitude orbits of 500 to 3,000 km and interconnected with crosslinks, a highly survivable constellation architecture results. There are so many possible paths that the loss of several of the satellites does not disable the communications capability. Jamming the network from the ground is difficult since each satellite sees only a small portion of Earth. The shortcoming of the approach is of course cost, along with complexity of networking, rerouting the data, sorting out the individual packet messages, and delivering them to the proper users (see Figure 2.7).

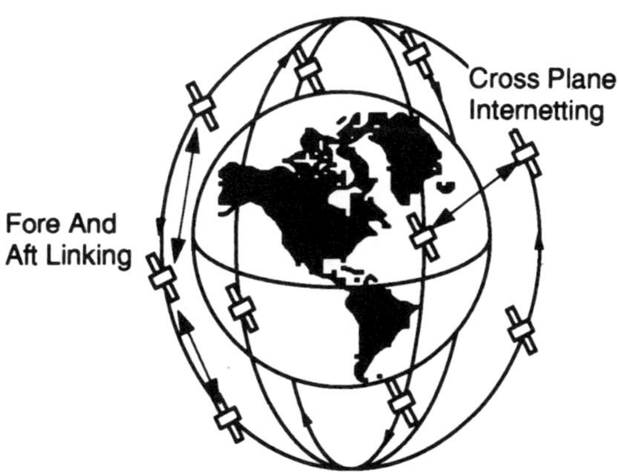

Figure 2.7 Internetting LEO constellation.

There are also hybrid systems of satellites used to configure certain architectures. For example, geostationary satellites may be combined with Molniya satellites to provide near-global communications coverage. Another approach is to combine high and low-altitude satellites realizing the advantages of both. It may be seen that each architecture embodies both advantages and disadvantages and must be studied in depth if an efficient and useful architecture is to result.

2.4 LASER COMMUNICATION APPLICATIONS OVERVIEW

There are a number of applications for which laser communications is well suited. Laser communications will never totally replace RF systems for space to ground communications where the message must be sent from space to ground or ground to space. The reason for this is of course that laser signals do not readily pass through clouds. When we say readily there is one application that exists for laser communication where even a laser's poor penetration through clouds can be used and that is in submarine laser communications. In Chapter 12, the reasons for this are explained. Suffice it to say that, in general, laser communications from space to Earth is subject to the vagaries of cloud cover. There are mitigating approaches, however, that permit us to get around the problem of cloud cover and these are discussed at some length in Chapter 12.

The range of space applications for laser communications was illustrated in Figure 2.1. These included the following:

- Satellite crosslinks;
- Satellite to aircraft;
- Satellite to ground;
- Satellite to submarine.

Satellite crosslinks are communication links in space and may be from LEO to LEO, LEO to GEO, or from GEO to GEO. Of course, these links are full duplex; that is, data flows both directions simultaneously. For laser links, there is no broadcast capability; links are point to point and cooperative effort is required between the terminals to close the links and transmit data.

LEO to GEO crosslinks are frequently used to transmit data from a data gathering LEO to a GEO where the data in turn will be transmitted to a user on the ground. In this case, the link is asymmetric; that is, high data rate is sent from the LEO to the GEO while low data rate for satellite command and control is passed in the opposite direction from the GEO to the LEO. The alternative to transmitting the data as it is taken is to record the data (store and forward mentioned previously). If the data gatherer is capable of extremely high data rates, the data must be transmitted if the onboard recorder is not capable of recording at the required rate or is unable to store the very large amount of data required.

GEO to GEO links are useful. For a military satellite system, for example, a GEO relay may be used to avoid the use of a vulnerable ground station located on foreign soil. In times of conflict, the security of the ground station may be in question and if the link is of strategic importance, a GEO relay may be used. This situation is illustrated in Figure 2.8.

A satellite to aircraft link application can involve data being gathered by an aircraft (for example, a reconnaissance aircraft), and sent to a satellite, or the opposite where the aircraft receives the data for end-user use. It is easy to envision the relaying of command data to a satellite and thence to an airborne command post for force direction. These two applications are illustrated in Figures 2.9 and 2.10. The reconnaissance aircraft data rate may be much higher than that of the command post.

A satellite to ground link may be an excellent application for laser communications since the data rate is virtually unlimited. RF systems have difficulty transmitting very high data rates to the ground from synchronous orbit. RF must transmit a wavelength, which passes through the atmosphere as does laser communications. The windows for RF are at the lower frequencies (longer wavelengths), which means that RF antennas tend to be extremely large (to reduce beam divergence) and large power levels are required. In addition, crowded frequency assignments are increasingly causing problems for RF downlinks.

Laser communications, on the other hand, possesses wavelengths that pass through the atmosphere readily. This coupled with the need for relatively small antennas offers laser communications a big advantage over RF systems provided the cloud cover problem can be overcome or mitigated. Numerous studies have indicated that cloud cover can be mitigated by the expedient of providing several ground stations at different locations so that the transmission can be sent to the one

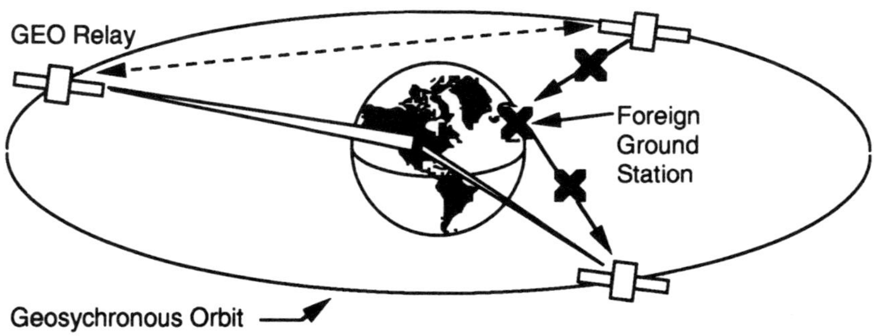

Figure 2.8 Use of a GEO relay to avoid a foreign ground station.

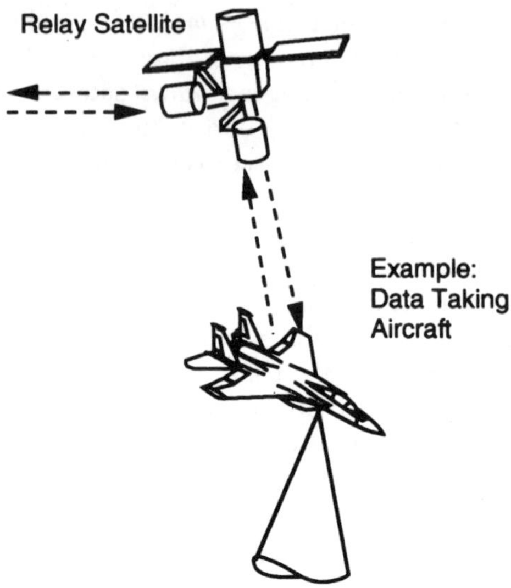

Figure 2.9 Aircraft to satellite data link.

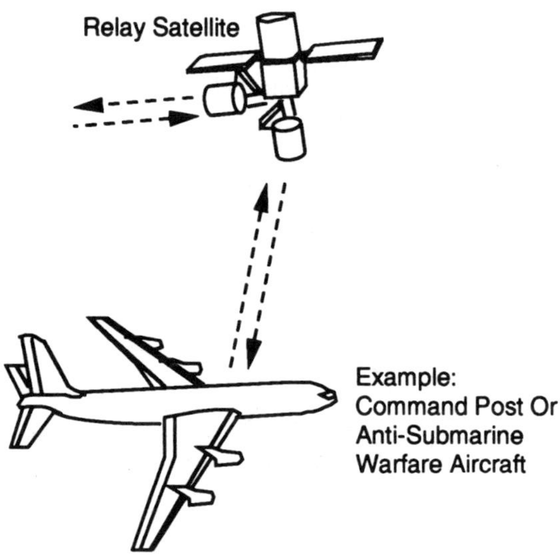

Figure 2.10 Satellite to aircraft link.

that is cloud-free. This is termed cloud-cover diversity and four stations can provide an availability of over 97%. If a single aircraft is flown at 35,000 foot altitude, an availability of over 95% is obtained. As well, the aircraft can simply fly around the cloud cover for nearly 100% availability. The data pass through the atmosphere relatively intact even at low elevation angles, although not as low as with RF. This was demonstrated during the *Airborne Flight Test System* (AFTS) flights where 1-Gbps data rate was transmitted to the ground at an average bit error rate of 10^{-6} from ranges of up to 100 km and at elevation angles as low as 7 deg [2]. Privacy/*low probability of intercept* (LPI) is also an issue since an RF transmission may make a footprint several hundred kilometers in diameter on the ground while a laser communication system would produce a footprint of only several hundred meters in diameter.

Satellite to submarine, commonly known as *submarine laser communications* (SLC), is an exceedingly interesting application for laser communication. Seawater transmits bluish-green light well enough to permit communication to a significant depth. The United States government has for a number of years investigated the technology required to transmit data from a satellite to a submerged submarine. Tests have been performed where an aircraft, carrying a spaceborne-type laser has successfully transmitted data to a submarine at depth through thick cloud cover. Earlier it was mentioned that clouds can be penetrated. There is a caveat, however, in that only data rates to a few hundred bits per second can be received. This is due to the fact that as the beam passes through the cloud, it scatters and arrives at the bottom of the cloud from a number of different directions and at different times. This multipath effect stretches the communications pulse so much that only a relatively low data rate can be received. This is not much of a problem if low data rate command data are being transmitted. There has even been some investigation and testing of a two-way laser link to subs. The satellite to submarine link application is illustrated in Figure 2.11.

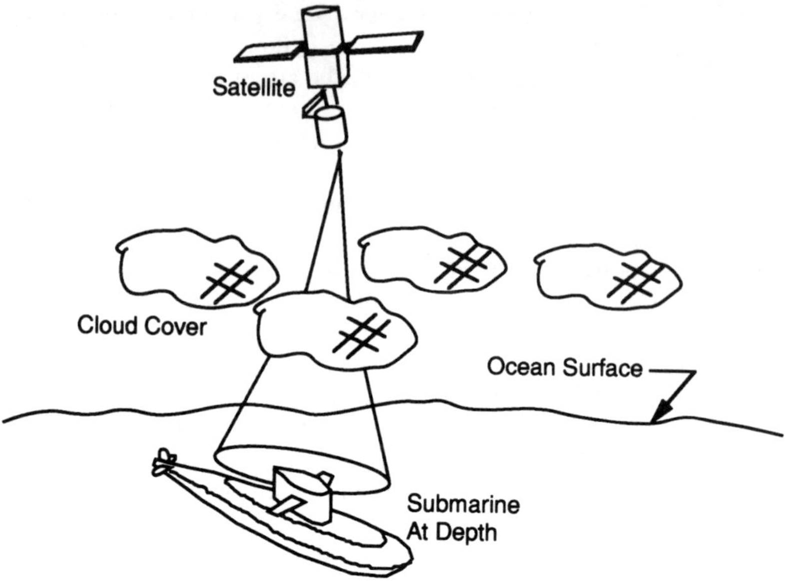

Figure 2.11 Satellite to submarine application.

References

[1] Sebacher, K. S., Lambert, S. G., Pautler, J. A. and Carter, J. P. "Laser Crosslink Configurations for Satellite Communications Systems," *Conf. Proc. of the Military Communications Conf. - MILCOM* '85.

[2] Katzman, M., Ed. *Laser Satellite Communications*, Englewood Cliffs, NJ, Prentice-Hall, 1987, pp. 56–67.

Chapter 3

Systems Design Methodology

The systems engineering design methodology plays an important part in any systems implementation. Requirements must be clearly understood; properly analyzed; and allocated to the functional elements, which are the building blocks of the system. Laser communication system design is not an exception to this disciplined approach. In fact, because of some unique interactions between subsystem elements, laser communication is more apt to benefit from a disciplined approach to system design.

Understanding requirements is perhaps the most important element in system design, for without a proper understanding of customer needs, the system cannot cost effectively be designed and manufactured. Once requirements (data rate, range, bit error rate, etc.) are understood and iterated with the customer, the process of design implementation takes place. Each requirement is analyzed and allocated to a functional entity (tracking system, data transfer, etc.). Finally, the functional entities are examined for commonality and system synthesis occurs in the form of physically implemented hardware/software (telescope, laser source, detector, etc.). Of course, iteration occurs at each level, until the "best" solution that fits the requirements set is identified. Best in this context refers to the overall system. As an example, for a satellite system, the best solution may include a slightly heavier communication system that realizes a lighter weight overall satellite solution.

For any space-based communication system, *radio frequency* (RF) or laser communication, the driving system requirements almost always come down to the basic need to conserve weight and power on the satellite. For any satellite system, weight and power are directly related to lifecycle cost as they affect launch costs, on-orbit life, and payload capabilities. Minimizing these critical system parameters within the performance requirements established for data transfer is usually the number one priority for any communication payload.

With this in mind, this chapter begins a disciplined approach to laser communication system design. First, the generic functions of a laser communication system are presented. Then, the system is identified in terms of link analysis on the three critical links of concern—namely, link establishment, link maintenance, and data transfer. Next, design and technology options and implementation techniques are identified. The pros and cons of each are outlined. And finally, the details of system level link analysis are discussed.

3.1 GENERIC SYSTEM DESIGN

The basic functional components of any point-to-point space laser communication system are given in Figure 3.1. A basic space-based laser communication system consists of an optical power source, relay and steering optics, a transmit-to-receive diplexer, an optical transmit aperture, an optical collection aperture, a detector, a coarse pointing mechanism, a control function, and a satellite interface. Sometimes these functions are shared with the host satellite. Most often though, the laser communication system is required to act as a standalone entity and will include these functions. In addition, power conditioning and regulation, while not shown in Figure 3.1, are typically part of the laser communication functional responsibility.

By describing the typical operation of a space-based laser communication system the functionality of the above-mentioned elements are identified. The system is required first to establish the link between two satellites. Starting from ephemeris and attitude data, typically provided by the satellite, the system begins the process of acquisition. A signal is generated by the controller to point the optical antennas (transmit and receive) via the coarse pointing mechanism along the line of sight between the two satellites. A pilot signal is generated in the controller function and modulates the optical source. The optical energy is generated, then transferred and steered, by the optical elements in the relay and steering optics to the transmit aperture. There, the light exits the system and traverses the link distance to the far satellite. A transmit-to-receive diplexer is used to isolate the transmit signal from the receive signal.

Optical energy impinging on the receive aperture is collected and relayed via the relay and steering optics to the optical detector, which has been simultaneously looking for a similar signal from the far satellite. Once the signal is received, controller logic begins the process of narrowing the acquisition uncertainty until both systems have locked on to each other's signal. Controller logic then commands optical beamsteering elements to keep the received signal boresighted on the optical detector (usually a quadrant device). The links are now tracking and link maintenance keeps the systems dynamically engaged through the satellite disturbance environment. Finally, the mission data is accepted via the interface with the satellite. At the optical source the electrical signal is converted to an information carrying, modulated optical signal and relayed to the exit aperture via optical elements. This signal is then transferred in a similar manner as the acquisition and

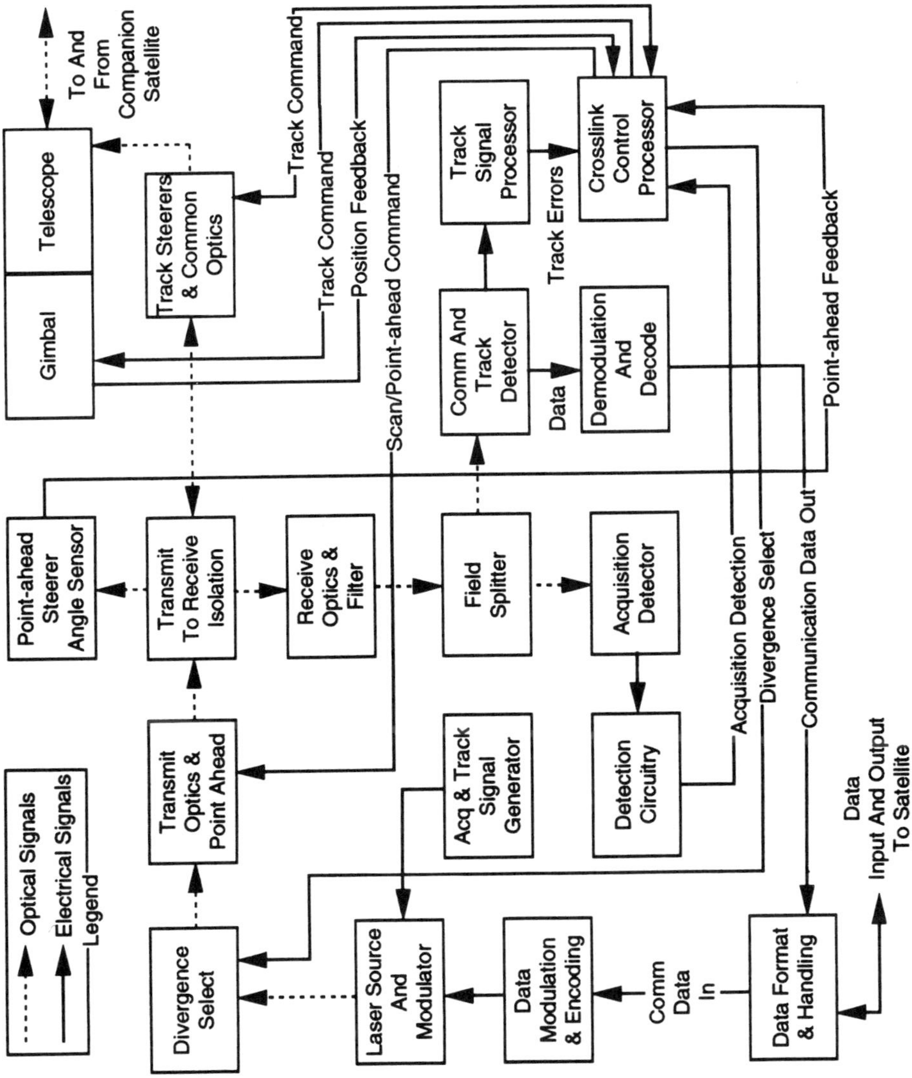

Figure 3.1 Basic laser communication system functional block diagram.

tracking signals to the far satellite, where it is collected and focused onto the optical detector. There, it is converted to an electrical signal that is demodulated back into the original signal and transferred to the satellite.

As can be seen, having these basic building blocks allows the definition of a space based laser communication system. Some systems, with lesser requirements, may require less functionality. For example, short-range systems may have sufficient laser power to utilize larger transmit beams, thereby negating the need for steering optics. Others might rely on the satellite to provide coarse pointing. Each system design will have a unique solution built upon the functional foundation identified by Figure 3.1.

3.2 END-TO-END LINK DESIGN

The basic building blocks to any laser communication system are the optical propagation models or links. It is with these models that the system design is identified, implementations traded, and sizing performed. Link establishment, link maintenance, and data transfer are the three basic functions of any laser communication system model. This section focuses on the fundamental requirements of each model and the interactions that exist between the three link types.

3.2.1 Link Establishment

Due to the physics of laser beams, namely narrow beamwidths, the perception has been that spatial acquisition of a laser communication link was almost impossible, or at least time consuming. History has shown this not to be true [1]. Using an orderly approach to searching the uncertainty area, acquisition can occur very quickly, usually faster than what is required by the user.

The uncertainty in the acquisition process evolves from the knowledge of where the companion satellites are, relative to each other, and knowledge, locally, of attitude as referenced to the laser communication system line of sight. While many satellites have horizon sensors and some even star trackers to determine their attitudes, transfer of attitude information across the satellite structure to the laser communication system line of sight is still the single largest error in determining the actual line of sight. Other error contributors include the laser communication system gimbal absolute knowledge. Figure 3.2 depicts the general uncertainty regions associated with this acquisition scenario. Techniques, such as an integrated star tracking device or rate gyros, can be employed to reduce the initial uncertainty if attitude knowledge transfer by the satellite exceeds acceptable laser communication system limits.

Given these error sources, the uncertainty region is typically larger than the beam divergences required for detectability. Hence, scanning of the transmit beam

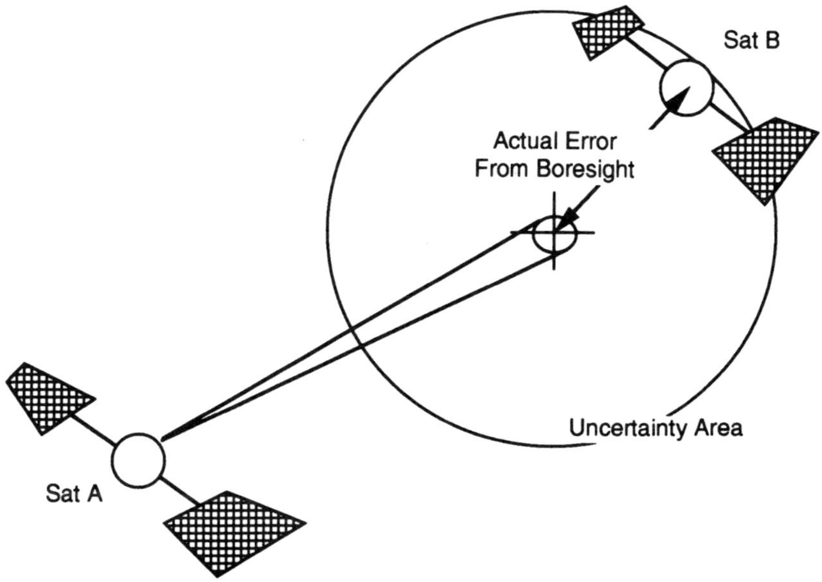

Figure 3.2 Acquisition uncertainty regions.

in a search-type mode is required. Note, in most laser communication systems, the transmit path can be independently scanned without affecting the receive path. For those system designs that only have coarse pointing common to both transmit and receive paths, other techniques can be used.

3.2.1.1 Systems With Coarse and Fine Pointing

A large uncertainty area and a narrow beam divergence can lead to unreasonably long acquisition scan times. It is for this reason that, typically, beam forming is employed on the acquisition beam to yield a divergence that is larger than the system communication or tracking beam. A simple mechanism is employed in the optical path, diverting the beam to additional optical elements that minify the beam, thereby yielding a larger far-field divergence. Once acquisition has occurred, the mechanism is removed and the beam restored to it original communication divergence size.

On the receiving end, the uncertainty is defined similarly. Usually, the receiver boresight is the laser communication system boresight reference. Recall that the receiver viewfield is independent of the transmit beam line of sight in most designs. If the receiver viewfield is smaller than the uncertainty area, scanning the receiver viewfield is required. This, however, is not a desirable situation. In situations where the transmitter and receiver have to be scanned, acquisition times increase as the ratio of uncertainty area to receiver field of view (FOV). Scanning of the receiver is almost never employed if other methods are available to the designer. In most situations, the acquisition receiver viewfield is sized to the uncertainty area. Thus, staring of the receiver viewfield is facilitated. In the rare situations where background is of concern to the acquisition receiver, scanning the receiver viewfield is typically mitigated by a segmented viewfield receiver such as a charge coupled device.

Contributors to the overall acquisition link model are identified in Figure 3.3. Acquisition normally is accomplished using a pilot signal, easily recognizable and detectable at the receiver. In some systems this may be a pulse of known duration. In others, this may be a pure tone at some distinct frequency. Still other approaches may only detect the *direct current* (dc) portion of the signal. The figure of merit for acquisition is always detectability within some timeframe. It includes both spatial and temporal detection (and frequency acquisition in heterodyne systems [2]). The receiving system must detect the signal within a reasonable amount of time and return its beam along the detected line of sight. This takes a few iterations of detections and a reasonable bandwidth of received signal to allow angular control of the return beam.

The acquisition link is affected by background noise and internally generated system noise. The acquisition FOV governs the amount of ambient background

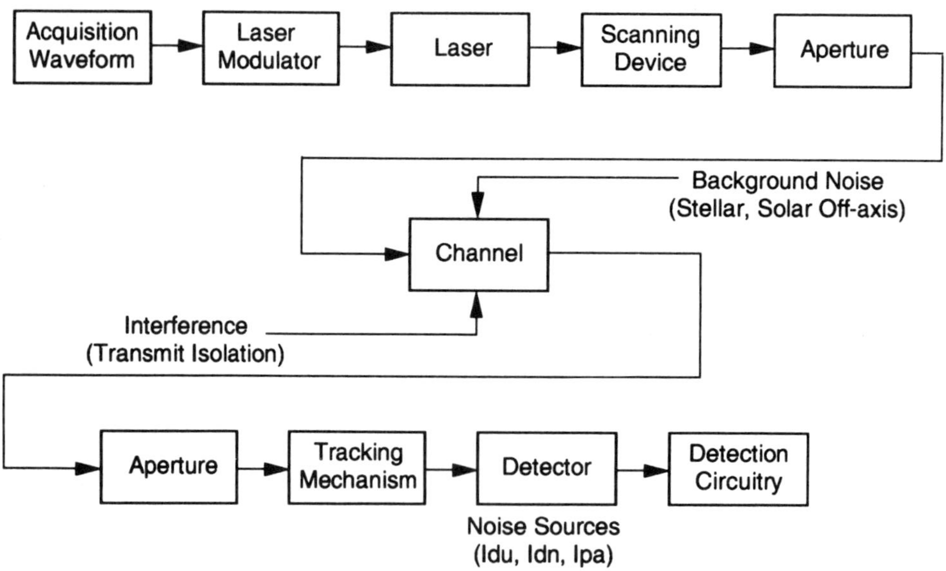

Figure 3.3 Acquisition link model.

light the acquisition system can tolerate. For large background scenarios, segmented viewfields are recommended. Typically, however, acquisition nominally does not have to occur under stressing background conditions.

3.2.1.2 Systems With Only Coarse Pointing

In situations where only coarse pointing is available, as is the case with many short-range links, the acquisition process involves a slightly different approach. If the beam divergence is large enough to cover the uncertainty area, then staring both the transmitter and receiver is recommended. If the situation arises whereby the transmit beam divergence is still smaller than the uncertainty area and scanning is required to cover the uncertainty, the receiver viewfield must be made larger to cover the uncertainty and the additional scan range of the transmit beam. This guarantees that the receiver is not scanned and acquisition times remain reasonable.

3.2.2 Link Maintenance

The transition from acquisition to tracking, before communication commences, is facilitated by using the same type of signal for tracking as was used for acquisition. If a pulsed system was used for acquisition, then using a pulsed tracking system makes sense. If a single optical source is used for acquisition, tracking, and communicating, then a common approach that allows tracking of the communication signal would be desirable. However, even more desirable would be a system that acquires, tracks, and communicates using similar signals.

Maintaining the optical links describes the operation required to keep the links locked onto the partner satellite transmissions through a myriad of disturbances. Some disturbances are self-generated, such as the gimbal jitter. Other disturbances include those from spacecraft momentum wheels, mechanisms, thrusters, solar array drives, and other payload active systems. Very low frequency disturbances such as thermal-induced deformations and line-of-sight tangential velocity must also be corrected or designed against.

Typically, laser communication systems employ active systems to track out the disturbance environment. Gimbal control loops remove the disturbances that are low-frequency high-amplitude and the low-amplitude higher frequency disturbances are removed by beamsteerers actively close-loop controlled about the receiver line of sight. Any excess disturbances must be accommodated by system signal margin. For most satellite systems, gimbal loops of 5 to 10 Hz are adequate, while beamsteerers are typically 100 to 1,000 Hz in closed-loop bandwidth.

For narrow-beam systems the tangential velocity is corrected using a point-ahead system. The transmit beam is pointed ahead to compensate for the propagation delay over the link range. Satellite ephemeris and attitude information are used to determine the point-ahead lead/lag angle. Thermal-induced deformations are accommodated using a periodic alignment of the transmit and receive lines of sight and by implementing, if necessary, a closed-loop dither of the transmit source detectable at the far receiver and closed loop via the communication link overhead data.

Contributors to the overall end-to-end tracking link model are shown in Figure 3.4. The critical figure of merit for tracking links is the *noise equivalent angle* (NEA). The NEA is defined as the residual tracking error along the line-of-sight vector to the companion satellite. The NEA is a function of the received signal to noise ratio (SNR) in the tracking bandwidth, the optical spot size on the detector, and the gain of the tracking system. The amount of residual tracking error controls directly the ability to point back accurately along the line of sight. Additional laser power on the communication and tracking links is required for larger NEAs. In addition, in coherent detection systems the tracking error relates directly to the efficiency of the mixing of the local oscillator with the received optical wavefront. For poorer tracking systems, the mixing efficiency is worse and more signal is required to close the links.

For direct-detection tracking systems, the link is affected by background conditions and internal noise sources. Since the NEA is directly affected by the SNR in the track loop bandwidth, operation of the tracking system in a wide viewfield mode is mostly prohibited in a direct-detection implementation. Normally, however, the tracking system is viewfield-limited to the same level as the communication system.

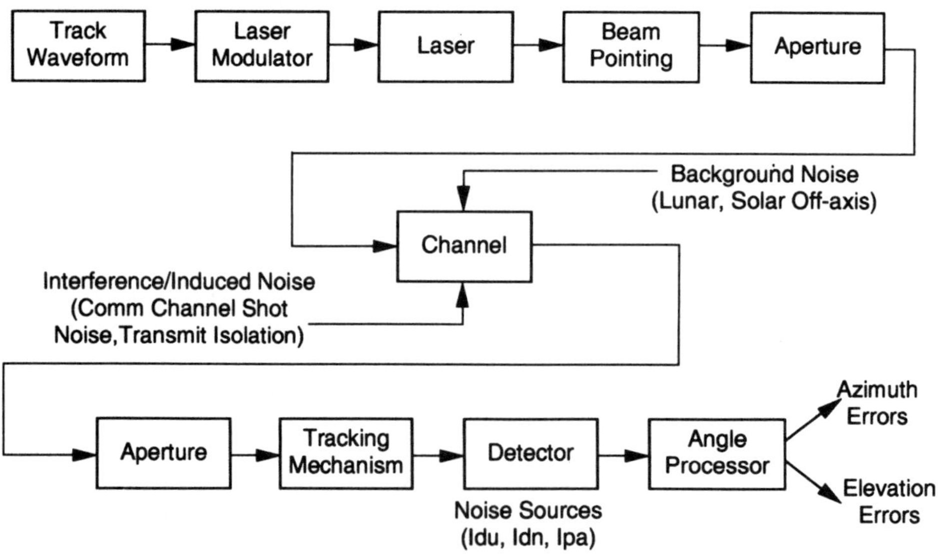

Figure 3.4 Track link model.

Coherent-detection tracking systems are local oscillator shot-noise limited, but typically require tighter tracking errors to ensure highly efficient mixing of the local oscillator and received wavefronts.

3.2.3 Data Transfer

Perhaps the simplest link to understand for the communication engineer is the transfer of data from one satellite to its companion. In most cases, established classical modulation schemes are used. Transfer is accommodated using a variety of noncoherent and coherent schemes. Clock regeneration at the receiver is facilitated by the modulation schemes and classical synchronization techniques. Optical source and modulation technology must be carefully examined to determine the best modulation approach for each source. Not all modulation schemes are possible with all source technology. Limitations exist due to the physics of the source technology. Also, receivers must be matched to the type of modulation used. Once again, not all receivers lend themselves well to all modulation schemes.

Contributors to the overall end-to-end data transfer link model are shown in Figure 3.5. As with the other systems, background energy, in the direct-detection case, is a contributor to the system sensitivity. However, in most cases employing medium to high data rates (megabits to gigabits), except in operation near or through, the solar disk, the signal shot noise and preamplifier noise contributions overwhelm the noise contributions present from background. In the coherent-detection case, the local oscillator shot noise is the major contributor to system noise.

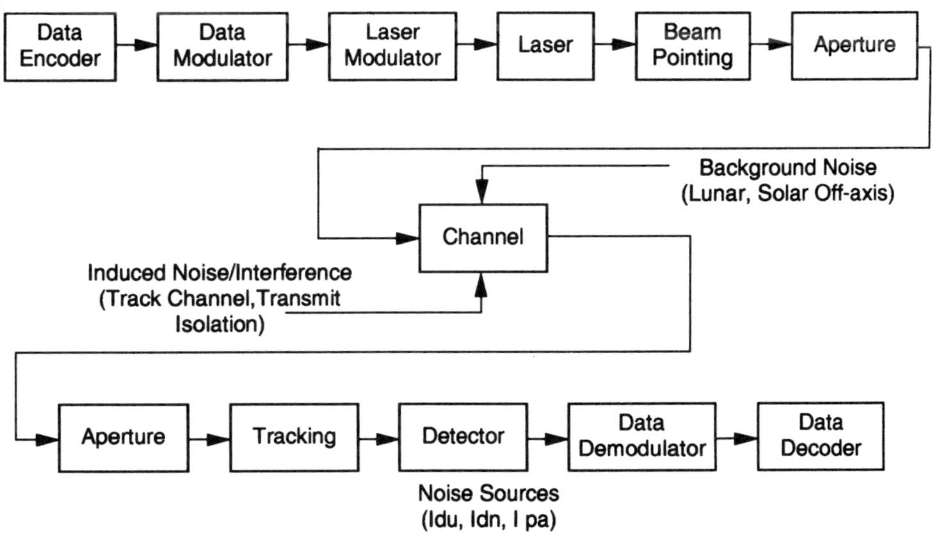

Figure 3.5 Data transfer link model.

3.3 DESIGN APPROACHES

With a top-level overview of the link models completed, discussion now turns towards identifying the system and technology approaches available to the link designer. Many types of laser communication links can, and have been envisioned. This section will identify the pros and cons associated with each technology approach. While classically any approach is feasible, design and technology limitations warrant the use of certain classes of systems over others. This text will identify the current limitations of some systems along with the advantages and disadvantages of each approach. Table 3.1 is a guide to the design approaches identified.

The designer must identify the type of link he or she is considering. Many current satellite systems today use "bent-pipe" or relay technology. That is to say, the information is transmitted from an Earth station via a specific carrier frequency, downconverted to an intermediate frequency at the satellite, and upconverted to another frequency for retransmission. No demodulation of the signal is accomplished onboard the satellites. Source modulation and sink demodulation is the norm for these systems. The current *National Aeronautics and Space Administration* (NASA) *Tracking and Data Relay Satellite System* (TDRSS), the *United States Defense Satellite Communication System* (DSCS), and most all commercial relay satellite systems for television, voice, and data operate in this manner.

Many new systems, however, are beginning to use baseband switching technology. These systems offer superior data quality. The NASA *Advanced Communication Technology Satellite* (ACTS) and the U.S. government's *Military Strategic Tracking and Relay* (MILSTAR) satellite both use the demodulation and remodulation approach to data transfer. Other systems envisioned, such as Motorola's Iridium™, may also use demodulation and remodulation techniques.

These types of links represent two different sets of requirements to the laser crosslink designer. Most classical relay systems cannot afford to do on-orbit demodulation and remodulation, hence laser communication system approaches must be considered that preserve the analog nature of these systems. For the digital baseband system approaches, digital modulation schemes will provide the best

Table 3.1
System Design Approach Choices

	System Approach	*System Type*	*Modulation*
Single channel	Coherent	Homodyne	Analog
or			Digital
Multiplexed channels		Heterodyne	Analog
Wavelength			Digital
Temporal	Direct detection	Pulsed	Digital
Spatial		Quasi-CW	Digital
Code		CW	Analog

performance and fit to the satellite design. When considering the type of laser communication crosslink to field, the designer is limited in his or her trade space by the type of systems interfacing with the satellite.

3.3.1 Coherent-Detection Systems

This class of system identifies the transmission of a modulated signal that utilizes mixing of the optical carrier down to baseband (homodyne detection), or to an intermediate frequency (heterodyne detection), and subsequent demodulation to baseband via conventional RF downconverting. These type of systems are analogous to the classical coherent RF systems and are described below.

In coherent-detection systems, the electric field of the received signal is summed with the electric field of a *local oscillator* (LO) using an optical beam combiner. The resultant signal, ecoh(t), at the combiner output can be expressed as

$$e_{\text{coh}}(t) = E_c \cos[\omega_c t + \theta_c(t)] + E_{\text{lo}} \cos[\omega_{lo} t + \phi_{\text{lo}}(t)] \tag{3.1}$$

where E_c and E_{lo} are the peak amplitudes of the signal and LO electric fields, respectively. The values $\theta_c(t)$ and $\phi_{\text{lo}}(t)$ are the phases of the signal and local oscillator components and ω_c and ω_{lo} are the carrier and LO field angular frequencies.

The electric field is then directed onto a photodiode detector, which can be described as a square law device mathematically. As such, the output current from the detector is proportional to the input electric field squared times the detector responsivity, R_d. Simply stated,

$$i_d(t) = R_d(e_{\text{coh}}(t))^2 \tag{3.2}$$

Now, combining (3.1) and (3.2), and assuming the second-order harmonics and sum frequencies are filtered out, the output current can be written as

$$i_d(t) = R_d\left\{\frac{E_c^2}{2} + \frac{E_{\text{lo}}^2}{2} + E_c E_{\text{lo}} \cos[\omega_{\text{if}} t + \delta(t)]\right\} \tag{3.3}$$

where $\omega_{\text{if}} = \omega_c - \omega_{\text{lo}}$ and $\delta(t) = \theta(t) - \phi_{\text{lo}}(t)$.

Optical power can be expressed in terms of the electric field as

$$P = \frac{E^2}{2} \tag{3.4}$$

Using (3.4) and substituting into (3.3) yields the familiar expression for coherent detection

$$i_d(t) = 2R_d\sqrt{P_cP_{lo}}\cos[\omega_{if} + \delta(t)] \tag{3.5}$$

where P_c and P_{lo} are the coherent signal and LO powers, respectively.

For homodyne, both phase and frequency of the optical carrier are required to be locked for downconversion to occur. Thus, for homodyne detection, (3.5) reduces to

$$i_{dho}(t) = 2R_d\sqrt{P_cP_{lo}} \tag{3.6}$$

In the heterodyne case, only frequency is required to be locked. The phase offset yields an *intermediate frequency* (IF) that is tracked and downconverted in a later stage of processing. Equation (3.5) is reduced for heterodyne to

$$i_{dhe}(t) = 2R_d\sqrt{P_cP_{lo}}\cos[\delta(t)] \tag{3.7}$$

These expressions for homodyne and heterodyne detection both assume the electric fields illuminate the detector uniformly. When this is not the case, mixing efficiencies less than unity degrade the mixing performance. These effects will be discussed in further detail in the chapter on communication receivers.

Generally, classical coherent-detection system designs offer an advantage in data transfer signal margin (7 to 10 dB) over classical direct-detection systems. In addition, the classical modulation schemes employed in RF designs, both analog and digital, can be employed with a coherent approach. However, when system design considerations are taken into account, the weight and power of similar data rate designs below 1 Gbps don't favor either system approach. Typically, producibility of coherent systems is a concern. Optical quality, source stability, and pointing and tracking system requirements offset the advantages in signal margin for the coherent system data links. Ultimately, however, coherent systems offer the potential to be the technology of choice for very high data rates.

3.3.2 Direct-Detection Systems

Direct-detection systems are described by the technique of directly operating on the intensity of the transmitted signal. No mixing occurs at the receiver. The optical signal is converted to an electrical signal directly by the detection process. This process is described by

$$e_{dd}(t) = E_{dd}\cos[\omega_d t + \theta_d(t)] \tag{3.8}$$

where E_{dd} is the direct-detection electrical signal amplitude and ω_d and θ_d are the radian frequency and phase of the incident electric field. Using (3.2) and (3.8) yields

$$i_{dd}(t) = R_d\frac{E_{dd}^2}{2} \tag{3.9}$$

which also can be written as

$$i_{dd}(t) = R_d P_{dd} \tag{3.10}$$

where P_{dd} is the optical power.

Several types of direct-detection systems can be envisioned. The simplest and most often used is digital baseband signaling. Herein, the amplitude (or polarization) of the optical energy is discretely modulated over time to represent binary signals. For example, during a bit period the amplitude could be high for a binary "one" or low for a binary "zero." Several different types of modulation schemes can be used in this digital baseband signaling approach.

Another direct-detection approach involves analog modulation of the optical carrier. Here, the carrier amplitude is modulated linearly by the message signal amplitude. Typically, this is an inefficient modulation technique and usually puts linearity constraints upon the laser source and modulator that are difficult to achieve with present state-of-the-art equipment. However, it is one of the few direct-detection approaches that allows bent-pipe operation to occur.

A final approach that is a subset of the analog approach is identified as subcarrier modulation. This technique uses an RF carrier frequency to modulate the optical carrier amplitude. At the receiver, the direct-detection process strips the RF carrier off and then classical RF processing can occur in the following receiver stages. This technique suffers from many of the same linearity concerns of analog modulation and is useful for only low-frequency analog signals. One advantage this approach has, however, is that several low-frequency signals can be frequency multiplexed onto an RF carrier and transmitted over the optical link. The limitations occur with the frequency requirement of the RF carrier. Typically, RF carriers are 5 to 10 times the highest message frequency to eliminate aliasing at the receiver. RF carriers exceeding 1 GHz would be challenging for a current state-of-the-art direct-detection system.

3.3.3 Multiplexing Approaches

Constant striving to increase the data rates and accommodate multiple users has led to the study of multiplexing approaches using laser communication links. Fortunately the classical techniques for multiplexing several signal channels also applies to laser communication links. In the coherent systems as well as the direct-detection systems, *frequency division multiplexing* (FDM), or in this case, *wavelength, division multiplexing* (WDM), is possible. Several optical frequency sources can be independently modulated and combined to yield multiple channels of data, which are decombined at the receiver. *Time division multiplexing* (TDM) can also be employed where several channels share timeslots at the receiver. Normally this requires bursting data during time windows and synchronizing receiver timing accordingly.

The uniqueness of laser communication, namely narrow beams and narrow viewfields, offers the potential for *spatial division multiplexing* (SDM) as well. Segmented receivers covering separate viewfields could operate on several user channels simultaneously as long as the user channels are spatially separated [3]. Even *code division multiplexing* (CDM) has been studied for use in multiple-channel laser communication links [4].

3.4 DESIGN TRADE OPTIONS

The designer is faced with many options that must be traded to achieve the desired light weight, low-power, low-cost solution. Some of these options are technology related, such as laser choice, detector choice, beamsteerer choices, and isolation approach. Other trade options are related to the techniques that are used to implement the system design. Examples of techniques include acquisition, tracking, communication, optical background mitigation, and coalignment approaches. These trade options are discussed in this section. Approaches are identified and the pros and cons of each approach or technology are outlined as a guide to the system designer.

3.4.1 Laser Technology Base

While the amount of optical power out of laser sources has increased rapidly in recent years, it is generally considered that laser source power is at a premium in any space laser communication link design. For that reason, the optical source power is critical in defining the size of the system. Raw power is not the only requirement of most laser communication sources. The spectral and spatial qualities of the optical source are also important. Because lasers have multiple spatial and spectral modes of operation, the suppression of the undesired modes is crucial to a good laser communication source. A good source will have a narrow, stable spectral linewidth and have a nearly diffraction limited single-mode spatial profile. Thus, brawn and beauty are both important in defining a laser communication optical source.

3.4.1.1 Lasers for Space Communication

Several types of laser devices can be used in laser communications. Table 3.2 outlines the several types of laser devices that can be used in a laser communication system design. Most laser communication work being undertaken today operates in the 0.8 to 1-μm regime. Some work is also being done to piggyback off the fiber-optic technology advancements at 1.3 and 1.5 μm. The available technology base has exploded rapidly in recent years, so this list may further be expanded as new devices are developed. In the early days of laser communication, *carbon*

Table 3.2
Practical Laser Types for Space Communication

Laser Type	*System Type*	*Material*
Solid state	Q-switched	Nd:YAG, Nd:YLF, Nd:YAP
	Cavity-dumped	Nd:YAG, Nd:YLF, Nd:YAP
	Mode-locked	Nd:YAG, Nd:YLF, Nd:YAP
	CW	Nd:YAG, Nd:YLF, Nd:YAP
Semiconductor	Pulsed	GaAlAs, InGaAs, InGaAsP
	Quasi-CW	GaAlAs, InGaAs, InGaAsP
	CW	GaAlAs, InGaAs, InGaAsP

dioxide (CO_2) lasers operating at 10.6 μm were used because of their stable features, their availability, and because they were less sensitive to atmospheric effects than shorter wavelength laser devices. Today, because of the compactness of solid state and semiconductor lasers, CO_2 lasers are rarely considered for use in laser communication applications, especially space-based systems. Gas lasers, such as CO_2, are considered too bulky and too unreliable to be considered for long-term spaceborne communications.

Solid-state lasers became the next devices of choice for space laser communications. Initially, however, the excitation of the solid-state lasing medium (known as "pumping") was performed using flash lamps. Lamp pumping of these solid-state devices became a life-limiting issue for space-based systems. How could a long-life space system be configured that involved flash lamps? The lamp lifetimes were only a few hundred to a few thousand hours long, far too short for space-based satellite systems. Much research went into long-life flash lamps for laser pumping applications. Terrestrial laser systems, used in manufacturing and tactical military applications, benefited directly from this research.

For space-based applications, however, the flash lamp was eventually relegated to second-class status by use of semiconductor laser diodes as pump sources. The laser diodes could be more efficiently matched to the absorption bands of the solid-state medium, thereby improving efficiency and reducing heat loss. Additionally, the diodes, as production processes improved, were shown to have extremely long lifetimes. In some life tests, mean time to failures of greater than 10 to 20 years were not uncommon [5].

The most common solid-state laser medium for space-based laser communication application has been the *neodymium yttrium aluminum garnet* (Nd:YAG) laser. This device has a fundamental lasing wavelength of 1.064 μm and has been doubled using nonlinear crystals to 532 nm. Other close relatives of the Nd:YAG device include *neodymium yttrium aluminum phosphate* (Nd:YAP) and *neodymium yttrium lithium fluoride* (Nd:YLF). These lase at wavelengths of 1.080 for Nd:YAP and either 1.047 or 1.053 μm for Nd:YLF. An advantage of the solid-state devices has been their stable and narrow spectral linewidth and the ability to be designed in either a pulsed or a continuous-wave configuration. In the pulsed configuration, the Nd:YAG device can be operated with very narrow pulsewidths (< 10 ns) with high peak powers (kilowatts or more).

Recently, solid-state lasers have been challenged by the single-mode semiconductor laser diode. Different from their pumping cousins, these devices exhibit single frequency and single spatial-mode performance. Examples of these types of devices are the *gallium arsenide* (GaAs), the *gallium aluminum arsenide* (GaAlAs), the *indium gallium arsenide* (InGaAs), and *indium gallium arsenide phosphate* (InGaAsP) semiconductor devices. These devices are compact, lightweight, easy to operate, directly modulatable, and have become quite powerful. Where single-mode output powers of only a few tens of milliwatts were the only devices available, today devices of 200 mW are commercially available [6]. Issues concerning wavelength stability during modulation have also been alleviated with the use of feedback in the form of external etalons or monolithically imbedded Bragg reflectors.

Because of their compactness, several of these sources can be designed into a single space-based system. Up to 16 semiconductor laser diodes have been successfully combined into a single spatial source using wavelength and polarization combining techniques [7], yielding a combined beam that is only 1.1 to 1.2 times diffraction limited. Where narrow beams are not required, as in the case of short-range links, the compactness of the laser diodes allows the devices to be aperture-shared, yielding more power in the far field. Still, even more power can be achieved with the use of amplification techniques.

Several types of amplification have been considered. One technique is to use solid-state devices, such as the Nd:YAG material, as an external amplifier for the laser diodes. In this technique, the laser diode would act as a master oscillator and the Nd:YAG medium, diode pumped, would amplify the diode output to yield greater output power. Similar techniques have also been developed using erbium-doped fiber-optic material. Here, the master oscillator would be injected into a fiber and amplified by a pumping source also traveling along the same fiber length. The output source would be an amplified version of the input. Several studies and lab experiments [8] have been conducted on the erbium fiber amplifier, which can also be used as a receiver amplifier.

Another type of amplifier is a purely monolithic approach. This technique employs an on-chip laser diode master oscillator injected into an amplifier section. The amount of current applied to the amplifier section produces an amplification of the master oscillator source. The output is a single monolithic device that produces greater than 1W of power [9].

3.4.1.2 Limitations on Laser Technology Choices

There are a large number of laser technologies—and for that matter, a large subset of the technologies—to choose from. Which does the designer choose? How does the designer make a choice? Fortunately, not all lasers are created equal. As discussed earlier, the choice of laser source does not solely depend upon output power. The system designer must look at a complex trade between laser power, far-field gain, spectral width, output wavelength, system range, optical background, link noise sources, data rate, and modulation technique to determine the best laser source for the system.

While this appears to be a single equation with too many unknowns, the criteria can be reduced by experience to a few simple rules of thumb. Notice, for example, that coherent systems put constraints on the laser in the areas of frequency stability and linewidth. Wavelength power combining of sources could not be used in a coherent approach. Also, data rate plays an important role. Intercavity modulation of solid-state devices is limited such that only about 10 Mbps of data can be transmitted via a cavity-dumped, pulsed Nd:YAG system. Single-mode laser diodes are limited in data modulation bandwidth by the relaxation length of the cavity. For many, this means that only a few gigabits per second could be efficiently transmitted by this type of source via direct modulation. Other limitations are outlined in Table 3.3. As can be seen, the choices are still there for the systems designer, but have been narrowed by the applicable technology limitations.

3.4.2 Detector Technology Base

Just like lasers, all detectors are not created equal. The use of the detector determines its design configuration. Each type of detector technology can be optimized for its role. The different types of detectors and common usage are illustrated in Table 3.4. For optical communication systems, the detector technology can be categorized into three distinct groups. First and foremost, detectors are used for communication. Secondly, detectors are used for spatial acquisition of the laser beams. And finally, detectors are used to track the incident optical energy so a return beam can be pointed back toward the transmitting satellite.

3.4.2.1 Communication Detectors

The most common type of detector used for direct detection optical communications has been the *avalanche photodiode* (APD). The APD provides moderate front-end gain (50 to 200), essential for good receiver performance, low noise, high bandwidth, and, depending upon wavelength, good quantum efficiency. The most common APDs are grown from silicon and are responsive from the visible through about 1 μm with varying quantum efficiency (the measure of optical to electrical conversion). Most silicon devices suffer from low quantum efficiency above 1 μm. However, novel techniques have been employed that increase quantum efficiency through the use of surface "dimpling" to about 40% at 1.064 μm. Other APD devices have been fabricated out of InGaAs and InGaAsP. These devices exhibit

Table 3.3
Laser Attributes for Space Laser Communications

Laser Type	*Material*	*Wavelength(nm)*	*Modulation Type**	*Data Rate*	*Peak Power*	*Coherent System*	*Direct-Detection System*
Solid state Pulsed	Nd:YAG Nd:YLF ND:YAP	1064 1047 or 1053 1080	Intracavity PPM, PIM	< 10 MBPS	Very high 10–100 W	No	Yes
			EO or AO External cavity PDBM, PPBM, PQM	< 10 MBPS	Very high 10–100 W	No	Yes
Solid state Modelocked	Nd:YAG Nd:YLF ND:YAP	1064 1047 or 10953 1080	EO or AO External cavity PDBM, PPBM, PQM	Up to 1GBPS	High 10's of W	No	Yes
Solid state CW	Nd:YAG Nd:YLF ND:YAP	1064 1047 or 1053 080	EO or AO External ASK, PSK and analog	> GBPS	1–5 Watts	Yes	Yes
Semiconductor Pulsed	GaAlAs	780–890	Direct drive ASK, PSK, bi-phase, PPM	1–2 GBPS	200 mW	Yes	Yes
	InGaAs	890–980	Direct drive ASK, PSK, bi-phase, PPM	1–2 GBPS	1 W MOPA**	Yes	Yes
Semiconductor CW	GaAlAs	780–890	Direct drive ASK, PSK, FSK, bi-phase, PPM and analog	1–2 GBPS	200 mW	Yes	Yes
	InGaAs	890–980	Direct drive ASK, PSK, FSK, bi-phase, PPM, and analog	1–2 GBPS	1 W MOPA**	Yes	Yes
	InGaAsP	1300 and 1500	Direct drive ASK, PSK, FSK and analog	Multi- GBPS	< 50 mW	Yes	Yes

*ASK = Amplitude shift keying, PSK = phase shift keying, FSK = frequency shift keying, PPM = pulse position modulation, PIM = pulse internal modulation, PDBM = pulse delay binary modulation, PPBM = pulse polarization binary modulation, PQM = pulse quaternary modulation, EO = electro-optic, AO = acousto-optic.
**MOPAs not currently avaialble for all wavelengths in range.

Table 3.4
Practical Detector Types for Space Communication

Application	*Detector Type*	*Materials*
Communication	APD	Silicon, InGaAs, InGaAsP
	PIN	Silicon, InGaAs, InGaAsP
	CCD	Silicon
	PMT	Solid-state silicon photocathode
Tracking	CCD	Silicon
	CID	Silicon
	QAPD	Silicon
	QPIN	Silicon, InGaAs
	APD quadrants	Silicon, InGaAs, InGaAsP
Acquisition	CCD	Silicon
	CID	Silicon
	QAPD	Silicon
	QPIN	Silicon, InGaAs
	APD quadrants	Silicon, InGaAs, InGaAsP

high quantum efficiency at the wavelengths beyond 1 μm and are commonly used at 1.3 and 1.5 μm. These long wavelength detectors have high quantum efficiency at 1 μm also, but their noise performance is substantiallypoorer than their silicon counterparts. As a result, even with low quantum efficiency, the silicon devices normally outperform the longer wavelength devices near 1 μm.

Just as the silicon APD has been the most common device for direct-detection optical communications, the silicon p-intrinsic (PIN) photodiode has been the detector of choice for coherent-system design. PIN diodes exhibit all the features of APD except the avalanche gain. The avalanche gain mechanism found in an APD is not present in the PIN device. Effectively they have a gain of unity. However, in coherent applications, gain is provided by the receiver local oscillator and use of the PIN device is a very cost-effective solution for the coherent receiver. At longer wavelengths, the InGaAs and InGaAsP PINs are used because of their high quantum efficiency. The noise performance burden of these devices, present when used for direct detection, is offset by the local oscillator shot noise in coherent applications.

Photomultiplier tubes (PMTs) were commonly used in the early days of laser communication research. They afforded extremely high front-end gain ($> 10^5$) and allowed very sensitive measurements to be made. Unfortunately, the early PMTs were tube devices and required permanent magnets to establish the proper quantum configuration. The devices were also bulky and heavy. Recently, monolithic PMTs have been developed that may provide good performance in a lightweight package. They should be considered in any design trade space for laser communication receivers.

Other types of detectors have been considered for various applications. For example, one application has postulated the use of a *charge coupled device* (CCD) as a low-data rate receiver, limited by the readout speed of the CCD. Data rates less than 10 Kbps could be expected from devices of this nature. Readout rates of CCD, while increasing, are still well below any high-speed data rate capability. CCDs are commonly constructed in silicon and are signal integrating devices. They make excellent spatial acquisition detectors and star trackers, but communication is better left to the nonintegrating type devices.

3.4.2.2 Acquisition Detectors

As stated earlier, communication is only one part of a laser communication system. Spatial acquisition of the laser source at the receiving end is equally vital. Without the acquisition process, communication could not occur. Detectors, which provide effective spatial acquisition, fall into two categories. The first are the integrating detectors. These are represented by the CCD and the *charge injection device* (CID). Each of these devices is represented by a large array of detectors (individually called pixels). The second category of devices is the N by N detector arrays. These are APD or PIN devices which are arranged in an N by N configuration. The most widely used is the most simplistic of these, the 2 by 2 array, which is commonly called a quadrant detector.

Spatial acquisition most often requires searching over a large uncertainty area. Unless this area is reduced by satellite ephemeris and attitude knowledge transfer to the laser communication system, the area is usually larger than required, or desired, for communication. With large areas comes two solutions. Either small viewfield detectors with long search patterns or large viewfield detectors that stare over the uncertaintly can be used. As stated earlier, staring is the most preferred technique. It affords quick acquisition times. Covering large uncertainty areas leads to large viewfields. The large viewfields allow a large amount of background energy to be collected and focused onto the acquisition detector's focal plane. Hence, background energy, an undesirable noise source, can limit the system design if not effectively reduced.

The CCD and the CID are both segmented focal-plane receivers. Each pixel element effectively is an individual detector looking at only a small portion of the background at any one time. The pixels can number into the millions on a single CCD device. The background energy represented at each pixel is therefore less than 10^{-6} than that incident onto the entire focal plane. Each pixel location represents a specific location in the FOV and is used to direct the return beam in the direction of the received energy. CCDs and CIDs are similar in operation but different in construction. The CID, unlike the CCD, can read its pixel information in a nondestructive way. CCDs always destroy the information during the read process. The CID, however, has been limited in its ability to achieve high pixel density and is currently dwarfed by the CCD in this category. CCD technology has been the choice for many imaging systems and the use of addressable CCD arrays for acquisition and tracking applications make it the detector of choice for these applications.

CCD devices have also been produced with as few as 14 by 14 pixels for acquisition and tracking applications [10]. This allows for faster readout, necessary if using a CCD as a high-speed tracking detector.

With the CCD typically comes excess baggage. While the focal-plane array itself is very small, the readout electronics associated with the CCD have typically taken a large amount of electronics and power. These are precious commodities to any space-based system. Use of N by N APD or PIN arrays require none of the readout electronics associated with the CCD or CID devices.

System designers not willing to burden themselves with these additional electronics have typically used the simple quadrant device as an acquisition detector. Yielding to background noise is usually circumvented by acquiring under low background conditions (i.e., starfield). Quadrant APD and PIN diodes can easily be used as acquisition detectors. The viewfields of these devices are sized to allow staring to occur and detections are identified by quadrant to the detection algorithm. While some linearity is sacrificed, acquisition is still accomplished in a timely manner.

In most systems using quadrant devices, *quadrant APDs*(QAPDs) are used. The avalanche gain is desirable in the detection process for sensitive receiver performance. QAPDs can be produced in a variety of ways. Typically, monolithic devices have limitations in the spacing between quadrants, the so-called "dead zone." Ideally, a device should have an infinitely small dead zone; however, in practice, the dead zones are typically 50 to 100 μm for a good QAPD. For a large FOV device, the spot size incident onto the detector could become lost in the dead zone. Several techniques have been employed to convert the standard QAPD into one with a small (2 to 5 μm) dead zone acceptable for use in a laser communication system.

One technique employs the use of a sharing transition region instead of a finite dead zone between the quadrants. This device has been successfully fabricated and integrated into a laser communication system. The device shares signal from each quadrant. Figure 3.6 illustrates the sharing transition region approach. One drawback of this device is the crosstalk that occurs because the transition is not abrupt.

Another technique employs an external image divider mounted on top of the QAPD dead zone. This device, call the etched silicon divider, uses a crystal-plane etching of silicon to form an image splitter. The silicon divider is reflectively coated and any light incident in the old dead zone is reflected down to the surface of the silicon QAPD, depending upon which side of the divider it impinges upon. A back

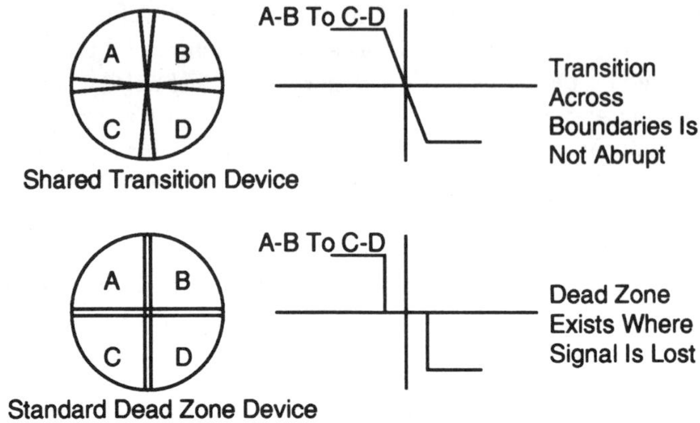

Figure 3.6 Sharing transition quadrant avalanche photodiode.

reflector is also used to capture those light rays that initially hit the surface at angles too shallow to be absorbed. This device has been fabricated and performs extremely well with an effective dead zone of 1 to 2 μm.

Still another technique uses a binary optical fabrication technique. Binary optical techniques allow use of a stepped gradient to control the phase, the amplitude, and the polarization of an optical wavefront. Many nonconventional optical elements can be fabricated in this way. What's attractive in this application is the abrupt junctions, which can be fabricated into a single device. In the case of a quadrant splitter, these abrupt junctions are fabricated at the image-dividing focal plane and the quartered energy bundles are then split and imaged onto four separate APDs monolithically mounted onto the same substrate. This technique has been postulated but not used in any system design of the authors' knowledge.

Finally a technique, which isn't preferred unless absolutely necessary, involves using four separate APDs. The focal plane is split by a machined pyramid whose sides are reflectively coated. The focal-plane reflections are then reimaged onto the four APDs, representing the four quadrants of the focal plane. While this yields a very small dead zone, the additional optics required to image the light onto the detectors and the problems with quantum efficiency matching make this a difficult and heavy way to implement the design.

3.4.2.3 Tracking Detectors

Spatial tracking is the third piece of the laser communication puzzle. Tracking detectors are very similar to acquisition detectors. Any detector previously mentioned as an acquisition detector can also be employed as a tracking detector. Typically, tracking detectors are quadrant devices, although arrays have also been postulated as tracking devices. Tracking detectors usually have a smaller viewfield than acquisition detectors because tracking is usually required through higher background conditions. The optical spot incident on the quadrant device is ratiometrically processed and quadrant amplitude information is provided to the tracking system.

An important feature of the tracking detector is a small dead zone. Tracking sensitivity is a function of the slope of the transfer function. To achieve a high-gain system, small optical image spots are necessary. When using small optical image spots, the dead-zone-to-spot size ratio becomes a critical system signal strength parameter.

3.4.2.4 Hybrid Detectors

In pursuit of the ultimate laser communication system, the system designer is always dreaming of new ways to configure the system with as few detectors as possible. If combining of the three major functions onto a single detector is possible, then that would be the preferred solution. In many system designs some combining of detector functions has occurred.

One such technique has employed the use of a single detector for tracking and communication. Using the same QAPD device, the electronics designers have fabricated a receiver that sums the quadrants on-chip and passes this information to a high-bandpass filter and preamplifier for communication processing. The quadrant information is taken from the opposite side of the QAPD and each quadrant is lowpass filtered and independently amplified [11]. A pictorial schematic of this approach is shown in Figure 3.7. Crosstalk is minimal unless the tracking

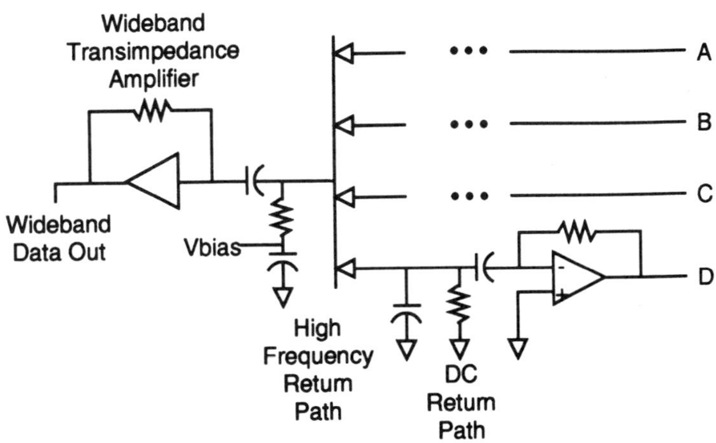

Figure 3.7 Communication and track signal splitting from single-quadrant APD.

bandwidth gets very large or if the data bandwidth becomes low. This technique of using a QAPD for communication and tracking is limited in bandwidth to about 200 MHz by the physical dimensions of the QAPD.

Other techniques have focused on combining the acquisition and tracking functions onto a single device. Ignoring the obvious (i.e., using the same viewfields for each) has led to approaches aimed at multiple viewfield detectors. Using a single QAPD and binary or holographic optics, the designers believe that a receiver that has both a wide and narrow viewfield may be possible. This approach is still in the development stage.

One final technique for combining the viewfields of the acquisition and tracking detectors has been around for many years. Using a bifurcating mirror (i.e., a mirror with a hole in the middle) has been used to successfully split the outer and inner FOVs of two separate detectors. This technique is depicted in Figure 3.8. While it doesn't limit the number of detectors, it simplifies the field splitting and handover from acquisition to tracking detector transition.

3.4.3 Isolation Technology

In most laser communication system designs, receiver sensitivity levels range from hundreds of picowatts to tens of nanowatts. At the same time, transmitter power

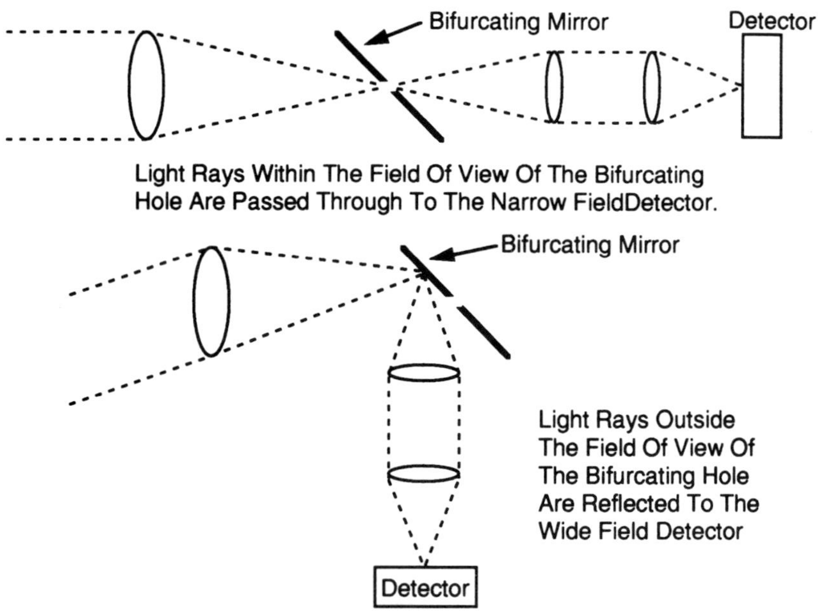

Figure 3.8 Bifurcating field splitter.

levels range from tens of milliwatts to more than one watt for semiconductor devices, and up to megawatts for solid state lasers. Transmitter powers are then ten orders of magnitude stronger than receiver sensitivities. Without isolation, the transmitters would self-jam the receivers, resulting in a nonperforming system design. Typically, a "good" laser communication system design would allow for 10- to 20-dB additional attenuation of the transmit source below the required signal level. Thus, the designer must achieve 100- to 120-dB isolation between the transmit and the receive signal levels.

The major contributors to isolation problems are backscatter off of optical surfaces, which reside in common transmit and receive optical elements, and signal coupling via radio frequency interference inside the transmitter and receiver electronics. The first can be mitigated using a variety of techniques. The second requires the use of shielded and isolated electronic circuitry and possibly separate pulse or signal frequencies on transmit and receive links. Mitigating this effect can be accomplished in a variety of ways as described below.

3.4.3.1 Spatial Isolation

As mentioned, signal backscatter off of common elements in the optical system is a major contributor to isolation issues. First and easiest is the use of separate transmit and receive apertures (i.e., spatial isolation). This keeps the outgoing transmit beam from backscattering off of common optical elements altogether. The spatial isolation technique is not used in practice except for very short links that use large beam divergences that can accommodate the large transmit-to-receive beam misalignments normally associated with separate transmit and receive aperture systems.

3.4.3.2 Spectral Isolation

Another technique involves using separate transmit and receive wavelengths. By using optical bandpass filtering, good optical design practices, and sufficient wavelength separation, the transmit and receive beams can be effectively isolated. This spectral separation of optical signals has been demonstrated to achieve greater than 120 dB of isolation and allows use of common optical elements in the system design. Typically, this means the same telescope aperture is used to transmit and receive the optical signals. This is necessary in systems where narrow beams and coalignment are required. Most long-range crosslink systems require these narrow beam and coaligned features. In system designs where two wavelengths are available, this is the preferred isolation approach.

3.4.3.3 Temporal Isolation

In system designs that use only a single wavelength, but still require the narrow beam and coaligned optical systems, a technique using temporal isolation can be employed. Usually this approach, which effectively blanks the receiver while the transmitter is outputting energy, can only be used in systems where the data rates are amenable to such timing constraints. The laser crosslink subsystem (LCS), which was a single wavelength design, utilized timing as one means of transmit to receive isolation. Another, or cascaded, technique used by this system was polarization.

3.4.3.4 Polarization Isolation

Laser outputs are typically linearly polarized in one of two states, horizontal or vertical. Through the use of optical techniques, the linearly polarized laser energy can be circularized. Using appropriate polarization-sensitive optical elements to circularize (right hand or left hand) allows for isolation of the transmit and receive energy if it is transmitted in different polarizations. Test have shown this technique is good to about six to eight orders of magnitude limited by physical constraints.

3.4.3.5 Combined Isolation

One final word on isolation techniques is offered to the system designer. Any combination of the above techniques is also possible. This will allow solutions to isolation issues not reachable by one technique alone. Isolation is not something that should be taken lightly. The entire system design is affected by the choice of transmit-to-receive isolation. Improper isolation can lead to self-jamming of the receive signal by the transmit source. This is a situation that needs to be avoided.

3.4.4 Beamsteering Technology

Several approaches are available to the system designer for steering the transmit or receive lines of sight. For coarse pointing, gimbals are used. These can take on many configurations, from single axis to triple axis. Gimbal movement is used to move the line of sight over large angles. For example, many laser communication systems must be able to operate over a field of regard that is very large. In some cases, hemispherical, or greater, angular coverage is required. Gimbals provide this large angular movement. Chapter 8 will discuss in greater detail the gimbal configurations available to the designer.

3.4.4.1 Coarse Beam Steering

Gimbal systems usually support one of two types of optics. For large angular coverage systems, most approaches use a gimbaled optical telescope with an optical relay through the gimbal axes. Figure 3.9 illustrates one such gimbaled telescope design. For some small fields of regard applications, it may become more cost and weight effective to use a fixed telescope and an external gimbaled flat mirror. Figure 3.10 illustrates a system that uses a gimbaled flat mirror.

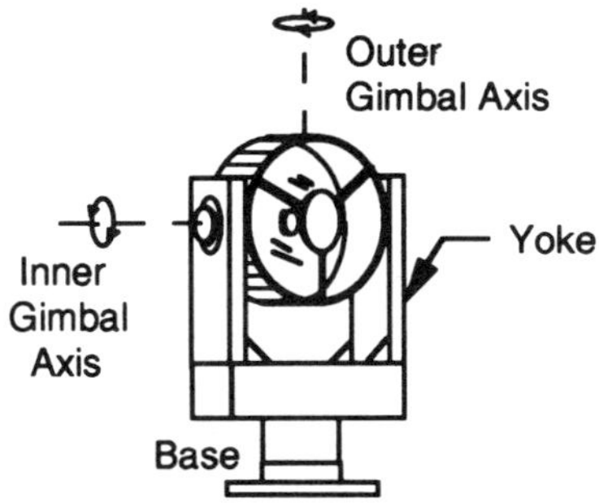

Figure 3.9 Gimbal telescope system.

The system designer must determine the trade impact of one type of coarse pointing approach over another based on the requirements of the systems and the optical effects of gimbaled flats over gimbaled telescopes. Many design features must be considered, including weight, field coverage, optical scattering, design implementation, and integration into the host, to name a few.

3.4.4.2 Vernier Beam Steering

For systems that require gimbal-only tracking (i.e., many wide-beamwidth systems), using the above beamsteering approach will suffice. But when narrowbeam systems are being considered for design solutions, vernier beam steering is required to maintain the tight tolerances on pointing and tracking. Many approaches to vernier, or fine, beamsteering exist.

When considering the fine beamsteering devices, the system designer must be concerned with several parameters that determine the controllability of the device as well as their optical performance. Usually, fine beam steering devices are used in closed-loop control to either steer or track the optical beams. Control parameters of

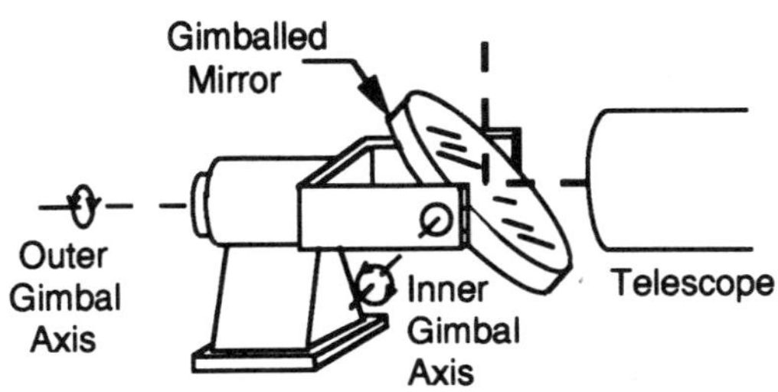

Figure 3.10 Gimbal flat system.

each device are important in determining its ability to perform as required in the control system. Parameters such as the natural frequency, the damping coefficient, and the open-loop bandwidth all must be analyzed to determine if the device can be used in the design being postulated. In addition to the control parameters, the optical performance in the form of transmission (or reflection) efficiency, wavefront quality, and angular coverage must be considered. Only devices that meet all the system designer's criteria can be considered for use.

Historically, single-galvanometer-driven mirror devices have performed well in vernier applications. These devices are limited by their single-axis nature. In a system design, two-dimensional coverage is almost always required. Two single-axis galvanometers are required to perform to the desired mission requirements. Using single-axis galvanometers in pairs complicates most optical system designs.

Recently, dual-axis actuator-driven mirror devices with required performance parameters have become available. This devices are preferable over single-axis devices because they reduce the amount of optical elements in the system design. Most reflective devices, such as the galvanometer mirrors and actuator mirrors, meet the optical performance requirements when proper optical coatings are deposited onto the mirror surfaces.

Piezo-electric devices, which have been widely used in adaptive optics and deformable mirror elements, have not been used in most laser communication system designs to date. Bandwidth limitations and angular coverage are the two factors limiting their use.

Nonreflective beam steering includes the use of nematic liquid crystals and acousto-optic beam deflectors. Both of these approaches yield performance degradation in the area of optical transmission and beam quality. The liquid crystals are very new in their consideration as beamsteerers. Many controllability issues need to be identified and investigated before these devices will becomes acceptable as beam steering devices. Acousto-optic beamsteerers have been available for a long time. The high voltages required to establish the acoustic field in the crystal have limited their use for space applications.

A final technique for steering beams lies in the use of distributed phased-array source technology. Using phased-array emitters, in an N by N configuration, it is theoretically possible to steer the output beam by varying the phase across the emitters. This is the same as an RF phased-array antenna, which steers the output beam over a defined area. The resolution of the steering and the amount of steering available is strictly governed by the laws of physics, the wavelength of the output, the spacing between emitters, and the total number of emitters. Packing densities are not available today that allow the construction of N by N laser diode arrays that meet all the criteria necessary for a laser communication system beamsteering device. The advice to any system designer today is to "steer clear" of phased-array beam steering technology and use the alternative techniques.

3.4.5 Acquisition

Spatial acquisition of the two companion terminals using very narrow laser beams has long been perceived as a very difficult task. This can be nothing but far from the truth. Terminal acquisition has been demonstrated in the laboratory, in the field from an aircraft to a ground station [12], and in both a space to ground experiment and in a space to space experiment [13]. Using a systematic approach to acquisition, the designer can trade from a variety of options for implementing acquisition. The pressing requirements associated with acquisition involve detection of the optical energy with a certain probability within a given amount of time.

3.4.5.1 Acquisition Uncertainty Budgets

Acquisition, while implemented systematically, is still a statistical process. The initial uncertainty solid angle is defined in terms of errors. Usually these errors, which are a combination of attitude and ephemeris uncertainties in the satellite navigation system, are expressed as azimuth and elevation uncertainties. These "x" and "y" axis errors are referenced to the satellite navigation system and transferred to the laser communication system line of sight via the satellite structure. This structure, while nominally stable on a "macroscale," is usually quite unstable on a "microscale." Bending of the structure due to thermal deformation, uncertainty in the mounting of the laser crosslink payloads onto the satellite structure, and other errors associated with payload line-of-sight uncertainty all contribute to this solid angle error. A representative acquisition open-loop pointing budget is identified in Figure 3.11.

The individual error contributors are usually uncorrelated and random. Slow varying errors, such as the thermal deformations are usually considered bias errors and the full value of the estimated deformation angle error is what is estimated. Other errors are more random in nature at a much higher frequency. Three sigma values are generated for these errors. In any event, an estimate of the three-sigma uncertainty is what is normally defined as the uncertainty area. As is seen, this is a statistical model. The azimuth and elevation errors are most commonly considered Gaussian and the resultant uncertainty probability takes on the Rician form when a bias offset is considered. This uncertainty probability is defined mathematically by Papoulis [14] in a radial sense as

$$P_{\text{area}} = \int_0^r \frac{z}{\sigma^2} e^{-(z^2+\eta^2)/2\sigma^2} I_o\left(\frac{z\eta}{\sigma^2}\right) dz \tag{3.11}$$

where I_o is the modified Bessel function of order zero and is given again by

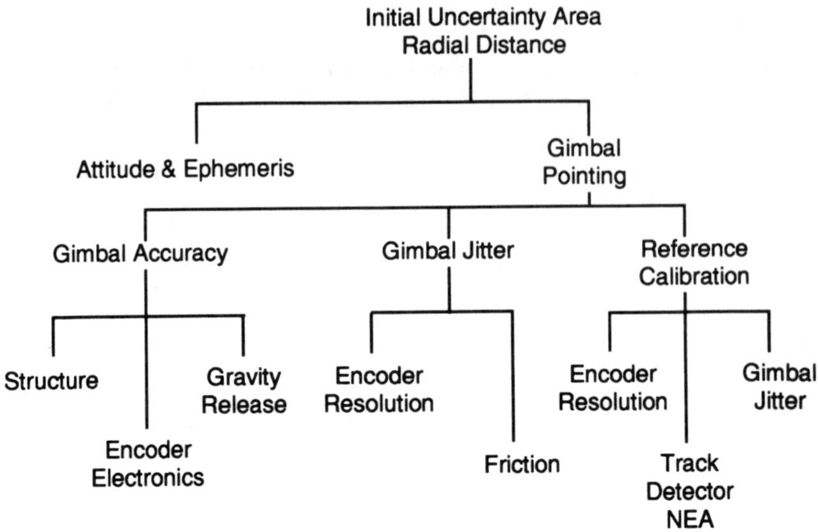

Figure 3.11 Representative acquisition open-loop pointing budget.

Papoulis as

$$I_o = \frac{1}{2\pi}\int_0^{2\pi} e^{x\cos\theta}\, d\theta = \sum_{n=0}^{\infty} \frac{x^{2n}}{2^{2n}(n!)^2} \tag{3.12}$$

The acquisition coverage probability is given by the solution of (3.11) for any radial distance, r, and mean and variance of the axial components of uncertainty. Equation (3.11) reduces to a Rayleigh distribution when the means of the axial components are identically zero in each axis.

3.4.5.2 Scanning / Staring Approaches

Acquisition scenarios all have one particular aspect in common. That is, to point a laser beam to cover an uncertainty solid angle and to detect that beam at the receiving end. A variety of approaches can be envisioned to accomplish this task. In each case, however, the initial uncertainty solid angle must be covered. Several techniques to accomplish this solid angle coverage are identified below.

Stare / Stare. In this approach to acquisition, the laser beam divergence is of sufficient size to illuminate the entire uncertainty area. At the same time, the receiver FOV is also large enough to view the entire uncertainty. In this scenario, acquisition occurs instantaneously with a probability equal to the area statistical coverage times the probability of detection. This is represented by

$$P_{\text{acq}} = P_{\text{det}} \cdot P_{\text{area}} \tag{3.13}$$

In only a few cases is this a practical approach to uncertainty coverage. In most cases, except those of very short range, sufficient laser power is not available to allow the transmit beam to be stared over the uncertainty area. Other techniques must be used.

Stare / Scan. This technique, one of the most common, involves staring the receiver viewfield and scanning a narrowbeam laser over the uncertainty area. By using a narrowbeam divergence laser, sufficient power is available to enable detectability of the signal. A high probability of detection is usually designed into the signal strength (beam divergence size) model. In this approach, acquisition probability is the same as (3.13); however, acquisition does not occur simultaneously. The time to acquire depends upon the ratio of uncertainty area to detectable beam size (usually an inscribed square at the 1/e beamwidth) times the dwell time of each "spot." An approximate relationship for acquisition time in this case is given as

$$T_{\text{acq}_{\text{stare/scan}}} \approx \left[\frac{\theta_{\text{unc}}^2}{\theta_{\text{beam}}^2}\right] \cdot T_{\text{dwell}} \cdot N_t \tag{3.14}$$

where θ_{unc} is the uncertainty area diameter, θ_{beam} is the 1/e beam divergence of the transmit laser source exiting the aperture, T_{dwell} is the amount of time the transmitter beam dwells in any location, and N_t is the number of total transmitter scan area repeats.

Normally the acquisition designer adds overlap to the beams to provide some additional margin of safety against large high-frequency jitter fluctuations on the order of the scan rate. With the overlaps of 10 to 15% as a general rule of thumb, the ratio of uncertainty to beam divergence is modified by the overlap factor as

$$\xi_t = (1 - \kappa)^2 \tag{3.15}$$

where κ is the overlap factor. Substituting back into (3.14) yields an acquisition time of

$$T_{\text{acq}_{\text{stare/scan}}} \approx \left[\frac{\theta_{\text{unc}}^2}{\theta_{\text{beam}}^2 \cdot \xi_t}\right] \cdot T_{\text{dwell}} \cdot N_t \tag{3.16}$$

Scan / Scan. Scanning of both the transmitter beam and the receiver FOV is another technique for covering the uncertainty area. This technique is used sparingly since the additional requirement to scan the receiver field increases the acquisition time by the ratio of uncertainty area to FOV. Here the FOV is the inscribed square in the detector viewfield. Once again, overlap is a consideration so the acquisition time is defined as

$$T_{\mathrm{acq}_{\mathrm{scan/scan}}} \approx \left[\frac{\theta_{\mathrm{unc}}^2}{\theta_{\mathrm{beam}}^2 \cdot \xi_t}\right] T_{\mathrm{dwell}} \cdot N_t \left[\frac{\theta_{\mathrm{unc}}^2}{\theta_{\mathrm{fov}}^2 \cdot \xi_r}\right] \cdot R_{\mathrm{dwell}} N_r \tag{3.17}$$

where θ_{unc}, θ_{beam}, T_{dwell}, N_t, and ξ_t are as before. The value θ_{fov} is the angular receiver viewfield, R_{dwell} is the receiver scan dwell time, N_r is the number of receiver scan repeats, and ξ_r the receiver FOV scan overlap.

Notice the dependence on both the beam divergence and FOV size, and how a minor change in either increases acquisition time significantly. Acquisition probability again takes on the form of (3.13).

Scan / Stare. A very seldom used technique involves scanning the detector viewfield and staring the transmitter. Once again, sufficient power must exist to allow signal detectability. Most often, wide beam divergences do not provide sufficient optical energy at the receiver. Additionally, most receivers are capable of covering large areas in low-background conditions. However, for completeness the equation for acquisition time is given for this case and is represented including the overlap factor as

$$T_{\mathrm{acq}_{\mathrm{scan/stare}}} \approx \left[\frac{\theta_{\mathrm{unc}}^2}{\theta_{\mathrm{fov}}^2 \cdot \xi_r}\right] \cdot R_{\mathrm{dwell}} \cdot N_r \tag{3.18}$$

where the variables are as before.

3.4.5.3 Who's the Boss?

Master-Slave. Acquisition processes can evolve in a variety of ways. In some cases, one terminal will start the process at a predetermined time and scan its beam. At the same predetermined time, the receiver may begin staring. In this scenario, the receiver waits to detect emissions from the transmitter before returning the signal. This technique is referred to as a "master-slave" approach. Typically, one terminal is designated the initiator, or master. The other terminal is coined the replier, or slave. Acquisition under these conditions usually takes longer since the protocol has to be established. This technique is recommended, however, in scan/scan cases, since it allows for a systematic approach to acquisition. Not identifying one terminal as the initiator and the other the replier in this scenario could result in an oscillatory condition where acquisition never occurs.

Master-Master. In stare/stare, scan/stare, or stare/scan cases, another process, referred to as a "master-master" approach, works very well. In this technique, each terminal is operating independently of the other until a detection is received. Once a detection is identified, at either end, the receiving terminal turns its attention towards the direction of the detected light and continues. Eventually the companion will be illuminated and turn its attention also towards the direction of the detected signal. This continues until, in the two scanning cases, the system has reduced the scan to a stare.

3.4.5.4 Scan Techniques

Scanning can take on several forms. The most common in use in laser communication systems are the classical raster and spiral scans, and subsets of these manifested as subfield scans.

Continuous Spiral Scan. The spiral scan is the most efficient technique for covering the uncertainty area defined by the radial uncertainty. A spiral scan is depicted in Figure 3.12. The width of the spiral is determined by the beam divergence, the desired overlap, and the uncertainty area. The classical equations for a spiral scan are given as

$$X_s = V_r t \cdot \cos(V_\theta t) \tag{3.19}$$

$$Y_s = V_r t \cdot \sin(V_\theta t) \tag{3.20}$$

where $V_r = R_{\max}/T_{\max}$, the maximum radial distance traveled over the maximum time, and $V_\theta = 2\pi N/T_{\max}$, N = number of rings, and is given considering overlaps

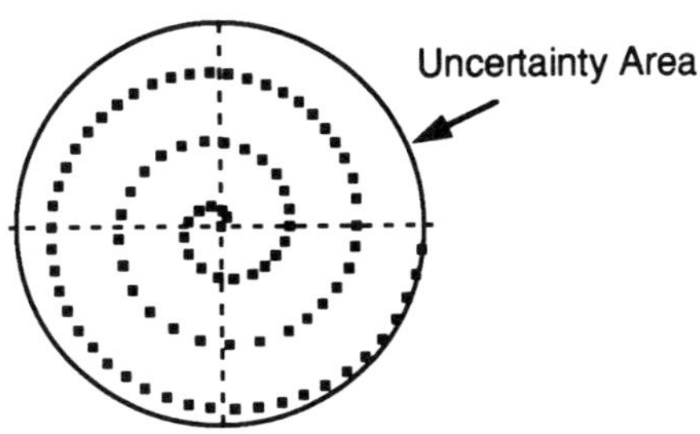

Figure 3.12 Spiral scan pattern.

as

$$N = \frac{R_{\max} - \dfrac{\theta_{1/e}}{2}}{(1 - F_o)\theta_{1/e}} \tag{3.21}$$

where $\theta_{1/e}$ is the $1/e$ beam divergence of the source or is the detector viewfield, which ever is being scanned, and F_o is the overlap factor.

Equations (3.19) and (3.20) describe a spiral that has a constant angular velocity, V_θ. In many cases, the scan is sized to allow illumination of a particular "spot" for a given duration. This requires a constant linear velocity. In this case, V_θ and V_r are modified as follows.

$$V_r = \sqrt{\frac{R_{\max}\left(\dfrac{\theta_{1/e}}{t}\right)}{2\pi \cdot N \cdot T_{\text{dwell}}}} \tag{3.22}$$

$$V_\theta = \sqrt{\frac{2\pi N \cdot \theta_{1/e}}{R_{\max} \cdot t \cdot T_{\text{dwell}}}} \tag{3.23}$$

Spiral coverage time is calculated from the rate of the scan and is given by

$$T_{\text{scan}-\text{cav}} = \frac{R}{V_r} \tag{3.24}$$

for the constant angular velocity case and by

$$T_{\text{scan}-\text{clv}} = \frac{2\pi N T_{\text{dwell}} \cdot R}{\theta_{1/e}} = \frac{2\pi}{(1 - F_o)}\left[\left(\frac{R}{\theta_{1/e}}\right)^2 + \frac{1}{2}\left(\frac{R}{\theta_{1/e}}\right)\right] \cdot T_{\text{dwell}} \tag{3.25}$$

for the linear velocity case.

Typical times to cover the uncertainty area are shown in Figure 3.13 for divergences (or viewfields) that are a ratio of the total uncertainty area.

Raster Scan. Raster scanning the laser beam is another approach to covering the uncertainty areas with the laser beam or the receiver viewfield. The raster approach is less efficient in that is overscans the circular uncertainty area at the edges. This overscan increases the coverage time. The raster involves scanning one axis over the full range and incrementing the other axis at the end of the field. The first axis is

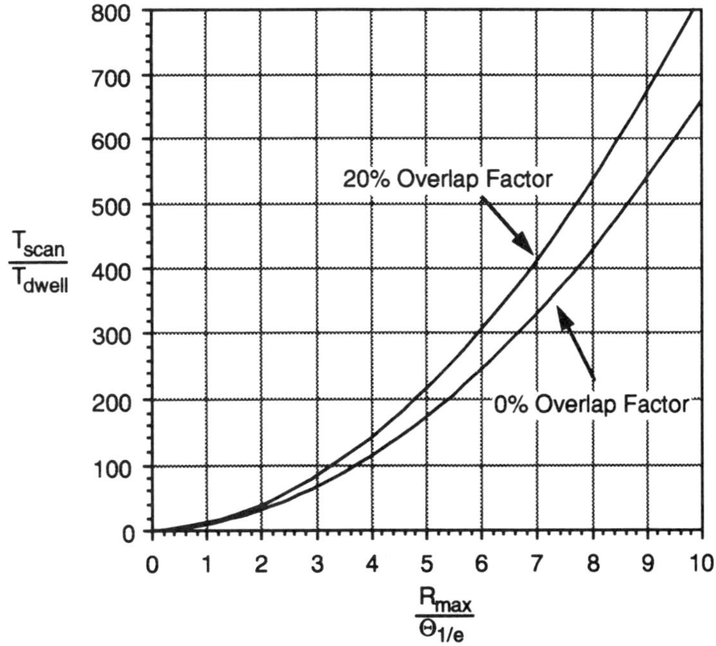

Figure 3.13 Spiral scan coverage times.

turned around and returned in the direction from where it initiated. This type of scan is illustrated in Figure 3.14.

The raster scan can be described as a function that consists of N_r horizontal scans and $N_r - 1$ vertical steps. The raster scan is a discontinuous function and therefore must be represented by a series of equations rather than just a single continuous function. Raster scans are typically implemented to dwell on a certain

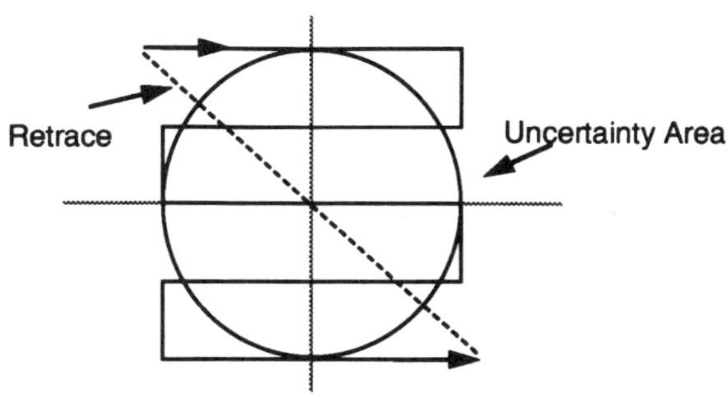

Figure 3.14 Raster scan pattern.

spot for a period of time. This dwell time equates to a linear velocity that can best be determined by

$$V_{\text{raster}} = \frac{N_r \cdot \theta_{\text{unc}} + (N_r - 1) \cdot \dfrac{\theta_{\text{unc}}}{N_r - 1}}{T_{\max}} = \frac{(N_r + 1) \cdot \theta_{\text{unc}}}{T_{\max}} \tag{3.26}$$

The motion of the raster scan, which can be described by a triangle waveform in the X-axis and a step function in the Y-axis, is given for the condition of five raster lines starting from the initial positions, X_o and Y_o, as

$$\begin{aligned}
&X_r = X_o + V_{\text{raster}} t && 0 \le t \le t_a \\
&Y_r = Y_o && \\
&X_r = X_o + V_{\text{raster}} t_a && t_a \le t \le \frac{N_r}{(N_r - 1)} t_a \\
&Y_r = Y_o - V_{\text{raster}}[t - t_a] && \\
&X_r = X_o - V_{\text{raster}}\{t - [(2N_r - 1)/(N_r - 1)]t_a\} && \frac{N_r}{(N_r - 1)} t_a \le t \le \frac{(2N_r - 1)}{(N_r - 1)} t_a \\
&Y_r = Y_o - V_{\text{raster}}[t_a/(N_r - 1)] && \\
&X_r = X_o + V_{\text{raster}} t_a && \frac{(2N_r - 1)}{(N_r - 1)} t_a \le t \le \frac{(2N_r)}{(N_r - 1)} t_a \\
&Y_r = Y_o - V_{\text{raster}}\left[t - \frac{(2N_r - 1)}{(N_r - 1)} t_a\right] && \\
&X_r = X_o - V_{\text{raster}}\{t - [(2N_r)/(N_r - 1)]t_a\} && \frac{(2N_r)}{(N_r - 1)} t_a \le t \le \frac{(3N_r - 1)}{(N_r - 1)} t_a \\
&Y_r = Y_o - V_{\text{raster}}[2t_a/(N_r - 1)] &&
\end{aligned} \tag{3.27}$$

where t_a is given by $T_{\max}/(N_r - 1)$.

The raster scan equations above will continue until the following holds true:

$$\text{Number of equation} = 2N_r - 1 \tag{3.28}$$

Figure 3.15 illustrates the X-axis and Y-axis signals needed to implement a raster scan. Note the triangle waveform in the X-axis and the step function in the Y-axis.

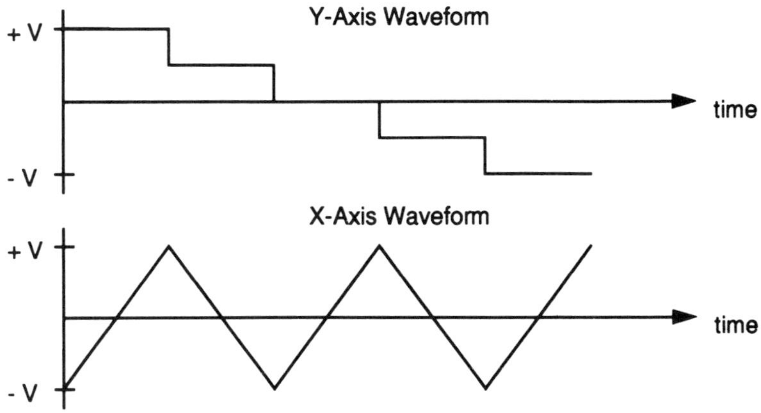

Figure 3.15 Raster scan X-axis and Y-axis drive signals.

Now, using (3.26) the raster scan coverage time can easily be computed for any size square raster area as

$$T_{\text{raster}} = \frac{[N_r + 1]}{V_{\text{raster}}} \cdot \theta_{\text{unc}} \tag{3.29}$$

One additional note on raster scanning is the use of multiple rasters. Just as in the raster scans associated with television traces, the raster scan must return to its original position for the start of the next raster frame. This time must be added to the total raster time if multiple frames are used. Again, assuming a square raster scan area and the same raster velocity yields the retrace time as

$$T_{\text{retrace}} = \sqrt{2}\, t_a = \sqrt{2}\, \frac{T_{\max}}{N_r - 1} = \sqrt{2} \left[\frac{N_r + 1}{N_r - 1} \right] \cdot \frac{\theta_{\text{unc}}}{V_{\text{raster}}} \tag{3.30}$$

Step Spiral Scan. A scan technique that is a subset of the continuous spiral scan is the step spiral scan. The step spiral scan effectively turns sharp corners as it scans the beam outward from the center. A step spiral scan is depicted in Figure 3.16. The width of the spiral steps is, like before, determined by the beam divergence, the desired overlap, and the uncertainty area. As with the continuous spiral, the step spiral requires a constant linear velocity to facilitate a constant illumination time as the spot step spirals outward.

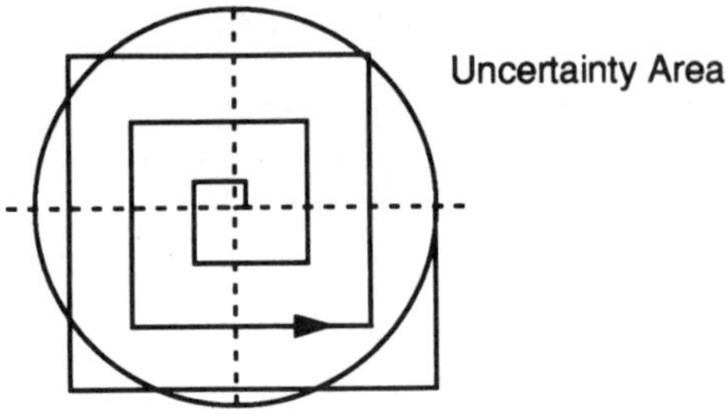

Figure 3.16 Step spiral scan pattern.

Segmented Scans. In some acquisition scenarios, the scans are broken into segments or sections of the total uncertainty area. This is most common in raster approaches. The center of the uncertainty area, where the highest probability of spatial detection exists, is covered first to try to reduce acquisition time. An example of this type of segmented scan coverage is shown in Figure 3.17.

Multiple Scans. Trades are frequently made to determine the optimum beam size for coverage and acquisition time. Recall that the pressing requirements associated with acquisition involve detection of the optical energy with a certain probability within a given amount of time. Usually scans can occur quite quickly. Because of this, multiple scans can be used to increase the detection statistics. By scanning the uncertainty area more than once, the probability of detection is increased by the M-out-of-N statistics. Additionally, if the false alarm rate needs to be reduced, multiple detections can be required to identify the signal. Equation (3.31) illustrates

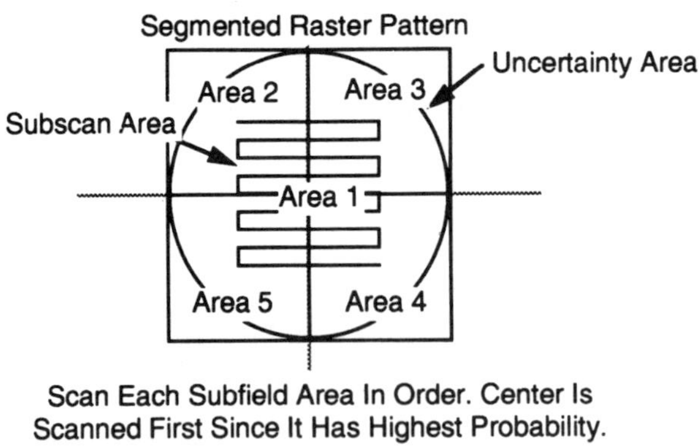

Figure 3.17 Segmented scan pattern.

how the detection statistics can be improved by multiple scans.

$$P_{\text{det}} = \sum_{m=1}^{n} p_d^m (1 - p_d)^{n-m} \tag{3.31}$$

where

$$\binom{n}{m} = \frac{n!}{m!(n-m)!}$$

and p_d is the individual scan detection probability.

Sometimes the designer finds it useful to scan high probability of spatial acquisition areas multiple times. This not only increases the detection probability but increases the chances of overall detection by covering these high probability areas more rapidly in the initial phases of the scan. The hope here is that the average acquisition time will also be reduced.

3.4.5.5 Field of Regard

Covering the uncertainty area is just one aspect of acquisition. The other is pointing towards the line of sight. In most laser communication system designs, gimbals are used to provide the coarse pointing along the estimated line of sight. Acquisition times must include gimbal slew times to move from a parked or random location to the line of sight. In the worst case, this could mean hemispherical travel of a gimbal. Gimbal rates of 10 deg/sec are not uncommon for moving telescopes that weigh only a few pounds. More massive payloads may require slightly slower slew times. Nevertheless, the slew times over a hemispherical (180°) angle must be factored into any reportable acquisition time.

3.4.5.6 Field of View

To facilitate rapid acquisition, large-format detectors are most commonly used over large field angles. In most scenarios, separate detectors are used for acquisition. The large field angle formats, while beneficial to acquisition, become burdensome to the communication and tracking systems. These systems are normally required to operate under higher background conditions or need large physical spot sizes to facilitate efficient coherent mixing.

Several types of large-format acquisition detectors have been used in laser communication systems. All fall into the category of segmented detectors; however some are simple quadrant and others large pixel format devices.

Simple Quadrant Detectors. The QAPD and *quadrant p-intrinsic* (QPIN) detector are the two most common simple quadrant devices used for acquisition processing. The uncertainty area is generally optically mapped onto the detector. Quadrant information is used by the acquisition logic to allow the beamsteerers, internal to the optical system, to boresight the received energy onto the tracking detector. Most quadrant detectors have a finite linear range, which is defined by the optical spot size on the detector. For acquisition, the detector is typically operating "bang-bang." Bang-bang implies the quadrant signal has saturated and the only information obtained is that which indicates the quadrant the signal was detected in. No linear information is obtained from the quadrant acquisition detector. The transfer function of the this type of device is shown in Figure 3.18.

Large-Pixel-Format Detectors. The CCD, the CID, and detector arrays larger than the quadrant format are often considered for use as acquisition devices. Detector arrays, especially in the $N \times N$ type formats, have suffered from large dead regions between the elements. These large dead zones, which are larger than most spot sizes, swallow the signal when it is incident in the zone. The result is a loss of signal, which is not a desirable effect during acquisition. For this reason, until dead zones can be reduced, detector arrays have not been seriously considered as acquisition detectors.

Both CCD and CID devices have been used in various design concepts. The pixel elements allow for large viewfields to be mapped onto the array. The background energy is reduced since each individual pixel element is a separate detector with a very small angular subtense. The linear range of these devices extends all the way to the edge of the field and is not limited by spot size. These devices are ideal for acquisition, but have been used sparingly in laser crosslink designs because of known limitations.

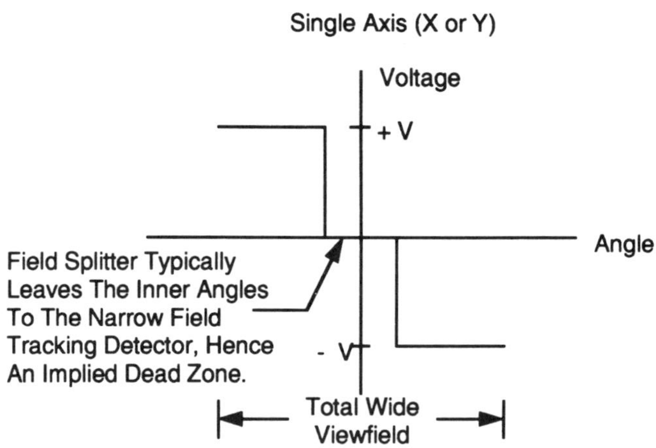

Figure 3.18 Bang-bang detector transfer function.

Pixel devices are typically low bandwidth. The devices are integrators and because of the sheer number of pixels that must be interrogated, the detection bandwidth is quite low. Additionally, the excess readout electronics associated with these devices have been a drawback in many space applications. On a positive note, the European Space Agency is developing, through contracting agencies, a 14 × 14 pixel CCD device [10] that from all accounts is a reasonable acquisition device.

3.4.5.7 Acquisition Signal Types

The temporal signal used during acquisition can take on many forms. Most often, except when separate acquisition sources may be required, the acquisition waveforms are derivatives of the communication and tracking waveforms. This follows from laser design principles. Pulsed lasers are designed for just that. Attempting to operate a pulsed laser in a *continuous wave* (CW) or quasi-CW mode usually results in a difficult if not impossible design task. The thermal design for such a device is usually complex. It is for these reasons that typically, if a pulsed laser source is used for communication, most often a pulsed acquisition source is used. The same analogy holds true for quasi-CW lasers. To this end, the discussion evolves toward the types of signals typically used for acquisition.

Pulsed Waveforms. The simplest waveform for acquisition is the laser pulse. Pulsing the laser at a specific pulse repetition frequency allows the designer to paint the scan area with a series of discrete-interval pulses. The scan fields are usually matched to the pulse rate to deliver at least a single pulse to each scan "spot." As the uncertainty is decreased and the scan is reduced, the dwell times are increased and each spot of the scan pattern now is illuminated by a higher pulse rate. In this manner, the bandwidth of the receive signal is increased and the scan pattern is further reduced, leading to the steady-state tracking condition.

Another advantage of pulse acquisition is the high peak power that is available to the designer. With high peak powers, larger beam divergences can be used by the designer in covering the uncertainty area in a smaller amount of time. Of course, the bandwidth of the signal information must be traded against the additional dwell times to achieve the overall acquisition time performance.

In some cases, separate wide beam, high peak power, low duty cycle pulsed lasers are used to initiate acquisition. Typically, cases involving large initial uncertainty areas may use this technique. These "call-up" lasers are used to initially reduce the uncertainty "bucket," after which more conventional techniques then take over the acquisition process.

Tone Waveforms. Other waveforms used for acquisition are tones. These can take on the form of pure sinusoids or squarewave single-frequency tones. Using tone frequencies that exceed the tracking bandwidth allows the receiver to acquire and perform high-bandwidth tracking without requiring multiple pulses. The receiver locks onto the frequency information in the tone and begins tracking immediately in a high-bandwidth fashion. Tone acquisition is very sensitive from a receiver point of view. It is used in cases where CW or quasi-CW sources are available and low-rate linear (or squarewave) modulation is achievable. These cases exist with single-mode diode lasers and externally modulated CW lasers.

One drawback with tone waveform acquisition is that the scans are usually slowed down to allow several cycles of the tones to be incident on the receiver. Slowing the scan down increases the time required to scan the uncertainty areas, which can increase acquisition time. Experience has shown, however, that tone acquisition performs as well or better than pulsed-waveform acquisition. This is due mainly to the requirement that a pulsed system has to scan multiple times to achieve the high bandwidth necessary to track accurately.

dc Waveforms. A final waveform worth mentioning is simply dc. Some receivers are of the integrating signal type (namely, CCD and CID detectors). With these types of receivers, a simple dc waveform is detectable by the system. The advantages of this type of system involve a simple laser design. However, the integration time and algorithms involved in determining if the signal is indeed the laser or just a dc-interfering background source can be acquisition time prohibitive. Techniques for discriminating the dc source from background point sources can become quite involved. In most cases where a pulse or tone source is available, they are preferred to the CW source. The extra dimension that the temporal information provides is invaluable in discriminating the acquisition signal from the background radiation.

3.4.5.8 Acquisition Receiver Types

Techniques for detecting the acquisition signal are numerous. Depending upon the type of signal transmitted, several techniques can be implemented. In some cases, the detection method depends upon both the transmitter and detector types used in the system. The most important criteria usually involves the transmitted waveform type and the dwell times available.

Recall that acquisition requires that the signal be detected within a certain time with a probability greater than some value. In an acquisition system, the probability of detecting the signal, the number of scans (or looks) at the signal, and the amount of time spent servicing false detections all contribute to the requirement. Ideally, false detections should be kept to a minimum. If the number of false detections can be made greater than the scan time of the total acquisition requirement, this would be ideal. Usually, the statistics will not allow such small false alarm rates. False alarms are typically considered part of the acquisition process and are accounted for in the addition of time to service them.

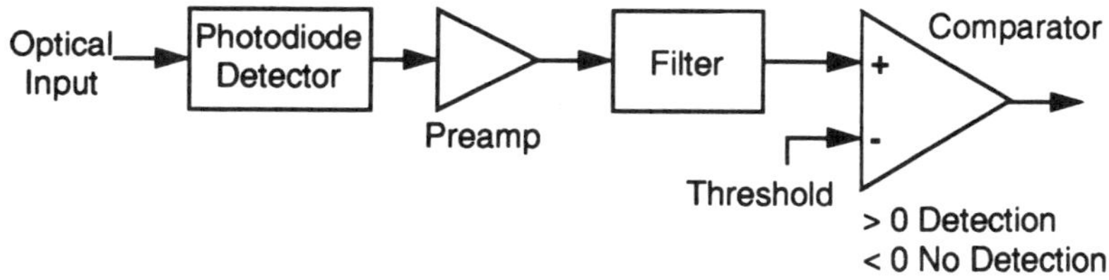

Figure 3.19 Pulsed receiver.

Pulsed Receivers. One of the simplest acquisition receivers involves direct detection of a single pulse of energy. This simple receiver is depicted in Figure 3.19. The required signal for this type of detection system follows from classical detection theory [15] and Gaussian detection statistics. The two classical Gaussian distributions illustrate noise only and signal plus noise. The equations derived from Figure 3.20 are given for probability of detection and false alarm probability. Note that the term under the integral is just the Gaussian expression.

$$P_{\text{det}} = \frac{1}{\sqrt{2\pi}\,\sigma_{sn}} \int_{\mu_{\text{th}}}^{\infty} e^{-[(x-\mu_s)^2/2\sigma_{\text{sn}}^2]}\, dx \tag{3.32}$$

where σ_{sn}^2 is the signal plus noise variance, μ_{s} is the required signal level necessary to achieve the P_{det}, and μ_{th} is the threshold setting of the receiver.

$$P_{\text{fa}} = \frac{1}{\sqrt{2\pi}\,\sigma_n} \int_{-\infty}^{\mu_{\text{th}}} e^{-[x^2/2\sigma_n^2]}\, dx \tag{3.33}$$

where σ_n^2 is the noise only variance and μ_{th} is as before.

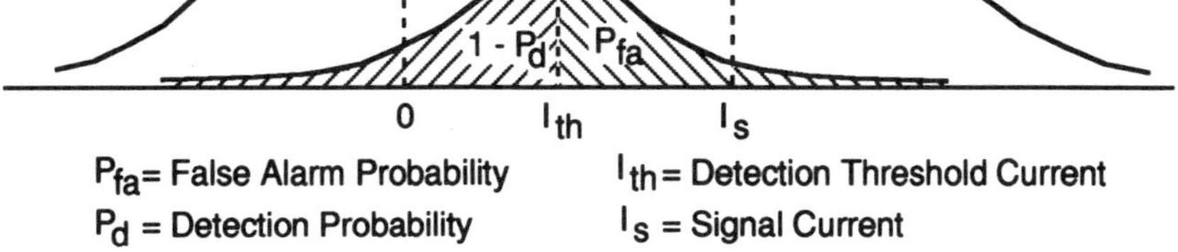

Figure 3.20 Classical detection signal distributions.

The pulsewidth of the signal determines the required bandwidth of the receiver. The number of serviceable false detections (i.e., the false alarm rate) along with the receiver bandwidth determines the probability of false alarm required. From the solutions to (3.32) and (3.33), the values of K_1 and K_2 are determined and are given as

$$K_1 = \frac{\mu_{\text{th}}}{\sigma_n} \tag{3.34}$$

$$K_2 = \frac{\mu_s - \mu_{\text{th}}}{\sigma_{\text{sn}}} \tag{3.35}$$

Substituting μ_{th} from (3.34) into (3.35) yields the expression for required signal, μ_s, as

$$\mu_s = K_1 \sigma_n + K_2 \sigma_{\text{sn}} \tag{3.36}$$

In a pulsed optical detection system using direct detection, noise contributions evolve from a variety of sources. One of these sources, which cannot be overlooked, results from the dc content of the optical signal incident upon the detector. This noise source is known as signal shot noise and is a function of the received average signal intensity on the detector surface. Equation (3.36) can be rewritten to include this term and results in

$$\mu_s = K_1 \sigma_n + K_2 \sqrt{\left(\sigma_n^2 + \Gamma \mu_s\right)} \tag{3.37}$$

where Γ consists of receiver performance parameters. Solving for μ_s then becomes a solution to the quadratic equation and is given by

$$\mu_s = \frac{(2K_1 \sigma_n + K_2 \Gamma)}{2} \pm \sqrt{\left[\frac{(2K_1 \sigma_n + K_2 \Gamma)}{2}\right]^2 + \sigma_n^2\left(K_2^2 - K_1^2\right)} \tag{3.38}$$

M-out-of-N Detection. Using multiple detections to reduce the false alarm probability or to increase the detection probability is a common technique to achieve the required performance with less received signal. M-out-of-N receiver statistics use the pulsed detection described above and process the combination of signal detections. A typical technique to increase the false alarm probability (or in effect increase the tolerable level, allowing the system to operate with a higher single-pulse false alarm probability) is to achieve the detection while requiring M-out-of-N pulses to be detected. Typically M is greater than one but less than N. In this way, the probability that two (or more if desired) pulses are false is exceedingly small in the window of N pulses. Equation (3.31) applies here too and is rewritten as (3.39) for both detection and false alarm probabilities. The binomial distribution set up by the M-out-of-N trials illustrates this point. Note, however, that achieving the false alarm reduction does not come free. The price paid is an increase in required signal power, manifested by an increased detection probability requirement on a pulse-to-pulse basis.

$$P_d = \sum_{m=k}^{n} \binom{n}{m} p_d^m (1 - p_d)^{n-m} \tag{3.39}$$

$$P_{\text{fa}} = \sum_{m=k}^{n} \binom{n}{m} p_{\text{fa}}^m (1 - p_{\text{fa}})^{n-m}$$

where p_{fa} is the individual pulse window false alarm probability, P_{fa} is the overall false alarm probability, p_{det} the individual pulse detection probability, and P_{det} the overall pulse detection probability.

Similarly, the individual pulse detection probability can be reduced by requiring one out of N pulses to be detected. Here the system can miss $N - M$ pulses and still determine the pulse has been detected. The price one pays here, as is evident by (3.39), is an increase in the false alarm probability, which requires additional signal power to overcome. Both conditions require additional signal power; however, the increase is minimal compared with trying to achieve the desired results in a more brute-force manner. The limiting factor on N is the number of pulses available in the acquisition time window of interest, and k the number of pulses detected.

Tone Receivers. Several techniques can be used to detect the signal when transmitting a tone during the acquisition process. Two of the most frequently used techniques are illustrated here. The first approach involves the use of a *phase locked loop* (PLL) to lock onto the received signal tone. In this approach, the PLL locks onto the received tone within a specified lock time determined by the frequency offset of the received signal tone to the PLL frequency set point and the bandwidth of the PLL.

The designer is usually concerned with acquiring or detecting signals within a specified amount of time. From Gardner [16], the loop lock time of a PLL can be approximated by

$$T_{\text{lock}} = \frac{4.2(\Delta f)^2}{B_L^3} \tag{3.40}$$

where Δf is the frequency offset between the PLL and the received signal and B_L is the PLL bandwidth. As can be seen, the lock time has a strong dependence upon loop bandwidth and the frequency stability of the transmitting source and PLL reference. Equation (3.40) can be rewritten solving for the loop bandwidth and then B_L can be used to determine the *signal to noise ratio* (SNR) of the PLL using

$$\mathrm{SNR} = \frac{[I_{\mathrm{st}}(1 - 1/N_e)]^2}{4N_o B_L} \tag{3.41}$$

where of course I_{st} is the peak received signal current, N_e the extinction ratio of the transmit source, and N_o the noise spectral density of the receiver including signal shot noise.

Reference [16] also yields a rule of thumb for the loop SNR to assure good loop performance in the presence of noise. Having loop SNR $\geq$ 6–10 dB usually provides this measure of assurance.

Another approach to acquiring a tone signal involves using linear envelope detection. Figure 3.21 illustrates tone acquisition using linear envelope detection. The input signal to the receiver, usually in the form of a squarewave, may be approximated, after ac coupling, by the first term in its Fourier series as

$$I_{\mathrm{tone}} = I_{\mathrm{st}} \cdot \frac{2}{\pi}(1 - 1/N_e)\cos(\omega_T t) \tag{3.42}$$

where the term $2/\pi$ represents the power ratio in the first harmonic of the received squarewave and N_e the extinction ratio of the transmit source. Figure 3.22 represents the classical detection diagram for conditions where signal and noise are present and where only noise alone is present. From [17], envelope detection results in a Rayleigh probability density function when noise alone is present in the detection process and results in a Rician probability density function when both signal and noise are present. These distributions are given as

$$P_{R_n} = \frac{x}{\sigma_n^2} e^{-(x^2/2\sigma_n^2)} \tag{3.43}$$

$$P_{R_{s+n}} = \frac{x}{\sigma_{\mathrm{sn}}^2} e^{-(I_{\mathrm{st}}^2 + x^2)/2\sigma_{\mathrm{sn}}^2} I_o\left(\frac{I_{\mathrm{st}}\, x}{\sigma_{\mathrm{sn}}}\right) \tag{3.44}$$

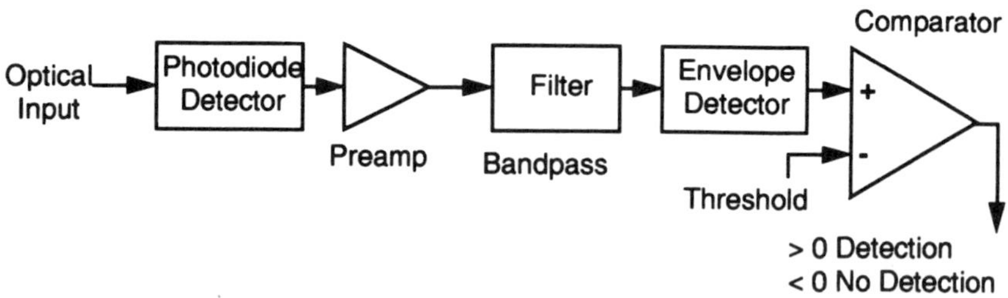

Figure 3.21 Tone modulation with linear envelope detection.

where σ_{sn} is the signal plus noise variance and I_o the modified Bessel function of order zero.

From classical theory, it follows that the probability of detection and probability of false alarm are given as

$$P_{\text{fa}} = e^{-(I_{\text{th}}^2/2\sigma_n^2)} \tag{3.45}$$

$$P_d = \int_{I_{\text{th}}}^{\infty} \frac{x}{\sigma_{\text{sn}}^2} e^{-(I_{\text{st}}^2+x^2)/2\sigma_{\text{sn}}^2} I_o\left(\frac{I_{\text{st}}\,x}{\sigma_{\text{sn}}}\right) dx \tag{3.46}$$

where I_{th} is the threshold setting in the detection process and is shown in Figure 3.22, P_{fa} is thc false alarm probability and P_d the probability of detection.

3.4.5.9 Reducing the Acquisition Uncertainty Area

Satellite navigation systems can be categorized from a laser communication's perspective simply as highly accurate and relaxed. The highly accurate systems, which are deployed on attitude-critical satellites such as sensor satellites, involve the use of star trackers to yield precise (arc second) accuracy of the satellite attitude. The relaxed systems are deployed on nonattitude-critical satellites, such as RF communication satellites. These types of systems employ horizon sensors that achieve the desired accuracy (tenths of degrees) for these satellites.

Satellites With Highly Accurate Attitude Determination. The major contributor to acquisition uncertainty is the transfer of attitude from the satellite's navigation unit to the crosslink line of sight. The bending motion of the satellite due to thermal distortions is the major culprit. In this case, several options exist to the satellite designer. One involves calibration, on-orbit, of the bending to reduce these transfer

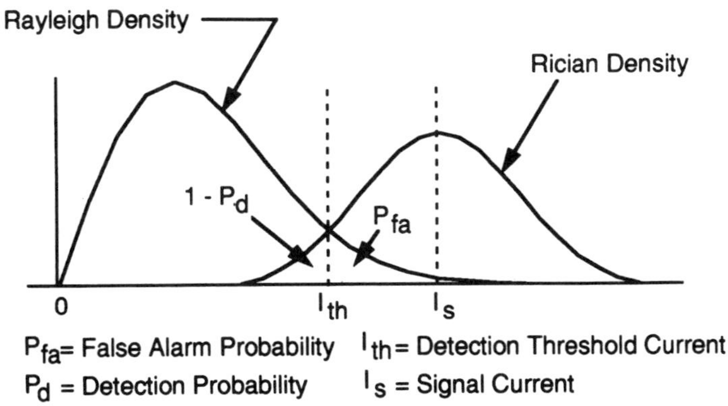

Figure 3.22 Classical detection distributions for linear envelope receivers.

errors. This can take a long time and the motions need to be determined for all solar angles and diurnal variations of each satellite's individual orbits. In the worst case, it may take a year to determine all the variability in this value. More practically, however, the crosslink will perform its acquisition using a worst case situation and the calibration will occur while the links are tracking. In this way, decreased acquisition times will exist for subsequent link establishments, which are required in order to alter constellation connectivity or to circumvent structural or solar outages.

Another technique to accurately transfer the attitude information involves an active optical alignment. A form of autocollimation that measures angular movement in all three axes from the navigation system to the crosslink line of sight has been envisioned for some satellite systems. The added weight of this type of system versus an increased field-of-view acquisition system must be traded at the satellite level to determine the best choice for the system design. A trade such as this has not been made to date.

Satellites With Relaxed Attitude Determination. In other cases, however, the satellite's attitude reference system is not accurate enough to achieve the desired uncertainty. It has been postulated in these cases that the laser crosslink systems integrate into their optical systems a simple star tracker. By gimballing over to a known star's location and locating its position, and then repeating this in a nearly orthogonal axis to the first, the inertial knowledge of the line of sight is determined. This information is then used to acquire the far satellite.

Integrating star trackers into laser communication payloads is not new. By employing a simple chopper wheel in the optical receive path, a quadrant avalanche photodiode can be made to track a known stellar object. Here, the difficulty in system design revolves around the limited viewfields and narrow wavelength bands typical of laser crosslink receivers. The magnitude of the stellar objects must be considered with these constraints in mind.

3.4.6 Pointing and Tracking

Being able to track the received signal line of sight through disturbances that are both externally generated by the satellite and coupled into the laser communication system and internally generated by the laser communication system itself is tantamount to being able to field an effective crosslink system. Without the ability to return a beam along the line of sight towards the companion terminal, communication cannot take place. To that end much attention has been given to understanding the disturbance environment of the satellite and understanding the internally generated noise in the laser communication system's tracking and pointing subassemblies.

A typical laser communication pointing and tracking system is depicted in Figure 3.23. This system illustrates a nested tracking system with a gimbal and fine

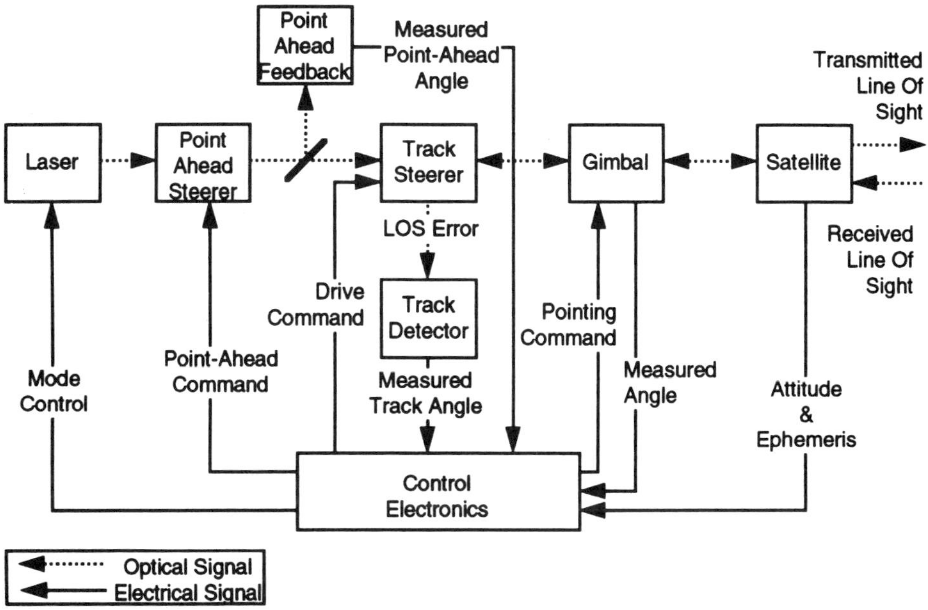

Figure 3.23 Representative pointing and tracking block diagram.

tracking loop and the additional forward correction offered by a point-ahead loop. Noise contributors to the closed loop pointing system are identified in Figure 3.24. Low-bandwidth disturbances (usually handled as biases) are normally added linearly, while higher frequency disturbances are root-sum squared to achieve an estimate of the pointing uncertainty. The total pointing error is the contributions of these bias and the random terms.

3.4.6.1 Disturbances

Internally Generated Disturbances and Residual Errors. A contributor of noise which is internally generated includes the receiver *noise equivalent angle* (NEA). This is a measure of the angular jitter in the tracking system that results from the detection process. The NEA is a function of the received spot image size, the received spot spatial intensity profile, and the received SNR in the tracking bandwidth. Typically, the NEA can be reduced by increasing the received signal level. A practical limit is reached when the spatial resolution of the optical system is reached. The NEA, or σ_{rms}, is described by

$$\sigma_{\mathrm{rms}} = \frac{1}{\mathrm{SF}\sqrt{\mathrm{SNR}}} \tag{3.47}$$

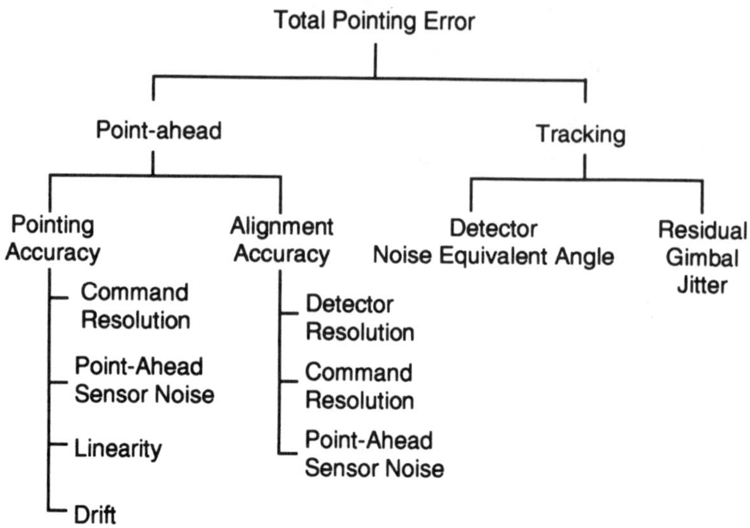

Figure 3.24 Closed-loop pointing budget.

where the *slope factor* (SF) is defined as the transfer function through the origin and converts the angular offset to a linear voltage. Typically SF has units of 1/radian. The SNR is defined in the tracking loop bandwidth. SNR is dimensionless and can be written as

$$\mathrm{SNR} = \frac{[P_r R_d]^2}{N_o B} \tag{3.48}$$

where N_o is the receiver noise in A^2/Hz, B is the bandwidth of the tracking loop, P_r is the required optical signal power in watts, and R_d is the detector responsivity in A/W.

Another contributor of internally generated disturbances includes the residual jitter associated with the gimbal system, which cannot be tracked out by the fast steering mirrors. The gain of the high-bandwidth tracking systems may be unable to accommodate the large angular jitter of the gimbal position and rate loops. In this case, the system must be able to "ride out" these residual disturbances. An example of residual gimbal jitter in a typical laser communication system is depicted in Figure 3.25. The disturbance spectrum also indicates the rejection offered by a typical high-bandwidth tracking system and the amount of residual jitter present.

In systems where internal pointing elements are used to compensate for point-ahead angles, this pointing system must be examined for its own NEA.

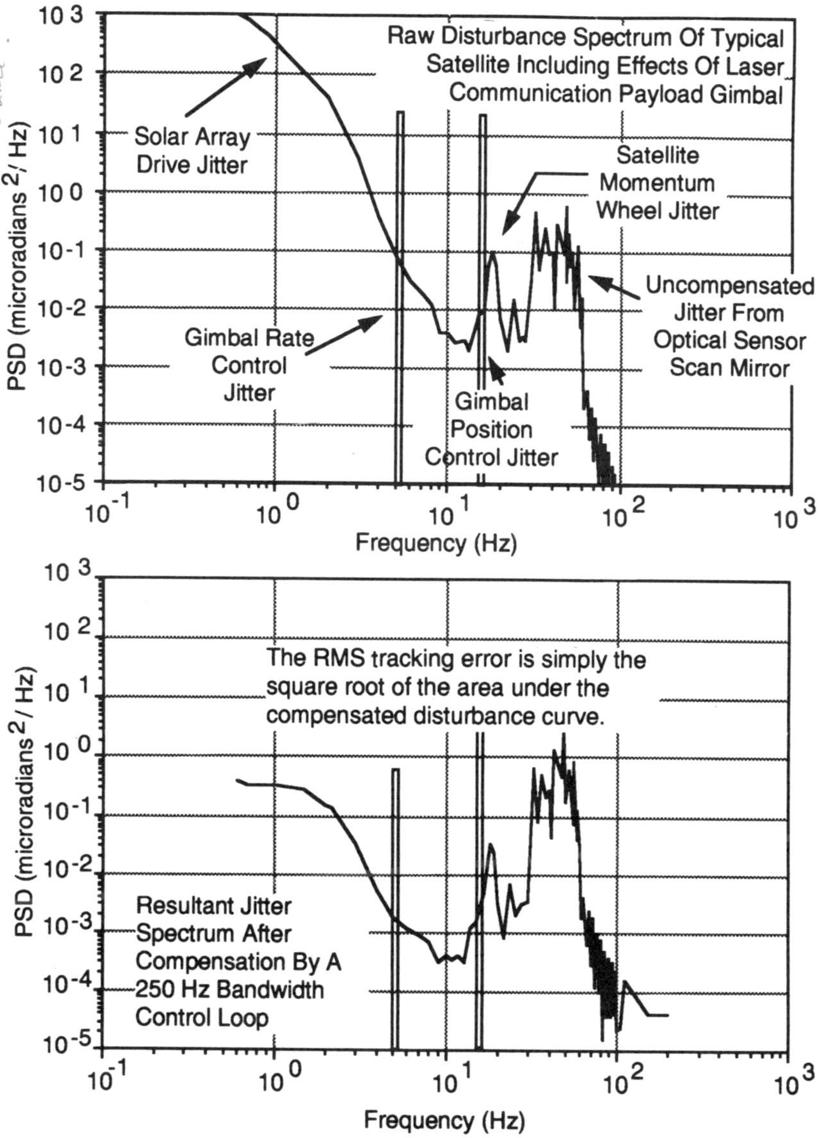

Figure 3.25 Disturbance spectrum of a typical 3-axis stabilized sensor satellite.

Point-ahead is used to compensate for the time of travel of the light over the long crosslink ranges. The point-ahead angle is calculated from the tangential angular velocity differential of the two satellites. As a simple example, consider two satellites in orbit, as shown in Figure 3.26. Each is traveling at an angular velocity, V, that is 360 deg per 24 hours or 72 μrad per second. For satellites at an altitude, h, above Earth's surface and traversing circular orbits, the range between satellites when spaced at orbital angle β is given as

$$\text{Range} = \sqrt{2 \cdot (R_{\text{earth}} + h)^2 \cdot (1 - \cos(\beta))} \tag{3.49}$$

where R_{earth} is the radius of Earth. The motion along the line of sight between the two satellites is in opposite directions and therefore the differential angular velocity is given by

$$V_{\text{diff}} = V + V\cos(\beta) \tag{3.50}$$

In many cases the cosine term is omitted to yield a more conservative estimate of the point-ahead angle. From (3.49) and (3.50) and using the speed of light in a

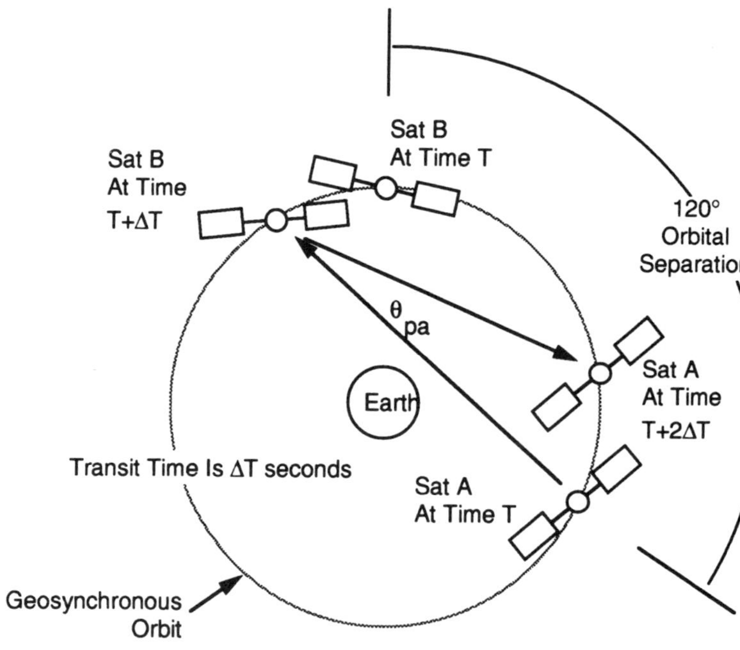

Figure 3.26 Simple point-ahead geometry.

vacuum, c, the point-ahead angle θ can be calculated as

$$\theta_{pa} = \frac{2V}{c} \tag{3.51}$$

The pointing system is usually commanded by the laser communication system's pointing and tracking controller. Feedback of the angular position of the pointing elements has some residual NEA just like the tracking system. These errors are typically small and very low frequency since the pointing system typically is only compensating for point-ahead errors and alignment offsets.

A final error in the pointing system results from the alignment offsets of the transmit and receive paths. In many laser communication system designs, the tracking detector boresight is considered the system boresight. All other alignments are referenced to it. During manufacture and build-up of the optical system, the transmit and receive boresights are mechanically aligned as precisely as possible. In a separate-aperture short-range system this may be only to 50 μrad (which is typically a tenth of the beam divergence). In this case, the wide beam can accommodate a large alignment error. In a single-aperture system, typical of long-range narrowbeam crosslinks, alignment errors must be much smaller. A narrowbeam system cannot tolerate large alignment errors.

In most cases, narrowbeam divergence systems are aligned precisely in manufacturing to less than one tenth of the beam divergence. On-orbit, an active "off-line" alignment technique, referred to as static alignment, is exercised to minimize static errors that could have crept into the system via launch vibration or microcreep of metallic elements. Additional active link techniques can also be employed to minimize drift contributions due to thermal variations and aging. The residual errors in these on-orbit measurements also add to the ability to point the beam accurately. Each of these contributors has been identified in Figure 3.24.

Externally Generated Disturbances and Residual Errors. The satellite is generally considered a noisy environment. On-orbit vibration is coupled through the satellite structure into the laser communication system. Many of the major contributors of vibration and their relative frequencies are given in Table 3.5. Satellite solar panels contribute very low frequency disturbances, which are a function of the solar array drive motors and the damping of the panels. Momentum wheels can be quite noisy in many satellites. Typically, momentum wheels are spun at rates in the hundreds of hertz to kilohertz region. While the amplitude of these disturbances are small, most laser communication tracking systems have very low rejection at bandwidths out in the kilohertz region. This could result in residual jitter being added to the pointing system. Additional satellite contributors include other payloads (especially gimbaled packages or antennas, which are not momentum compensated), scanning mirrors on optical sensors, and of course reaction jet or thruster firings.

Table 3.5
Satellite Vibration Sources

Source	*Frequency Range*
Solar panel drive	0.1–1.0 Hz
Momentum wheels	20–1,000 Hz
Gimbal jitter (residual)	5–15 Hz
Sensor scan mirrors	20–40 Hz
Thrusters	Step response

Each of these external disturbances must be analyzed and their particular contributions into the laser communications payload must be understood. Unfortunately for the designer, this is not an easy task. The satellite structure must be modeled and the coupling of the disturbances must be analyzed from the source, through the structure (which sometimes provides gain), to the laser communication optical payload mounting feet. Most of the time in any new satellite design the structure is not defined well enough to perform this in the early stages. It is during this time that the crosslink designer needs to begin formulating the pointing budgets required. Thus, some educated engineering experience must come into account to accurately estimate the bounds on these disturbances. With bounds, the estimates can be made in both tracking system bandwidths required and the residual tracking error that exists.

3.4.6.2 Tracking Techniques

The discussion now turns to understanding the type of tracking systems that can be used in a laser communication terminal designs. Recall how acquisition system design was more beneficial if the acquisition source was the same as the tracking and/or communication laser source? The same is true for the tracking system. A tracking system laser source that is the same or very close to the communication source simplifies the system design. Several approaches to tracking have been used in laser communication system designs. These will be discussed below.

Tracking systems can be separated into two distinct categories. The first involves those systems which derive the track information from the communication signal. The second set concerns those systems which use a separate laser "beacon" to track. No right or wrong answer as to which category is better is made here. Usually the designer must pick a system that is compatible with the overall design issues and constraints. For example, if a pulsed laser is used, tracking on the pulses is an acceptable solution. However, if a quasi-CW laser is used for communication, using the pulses may not be the most sensitive technique. Each technique must be explored and the best approach for the design at hand utilized. Factors to consider must also include compatibility with existing standards or designs and growth paths to higher data rates.

dc Tracking. The first technique that can be used to derive track information from the communication signal is dc tracking. This term is used to describe tracking the laser source by integrating the received amplitude-modulated signal over a large number of cycles or pulses. Usually an integrating type of detector, such as a CCD, would be used to track the beam. The integration time of the detector would be optimized to the track bandwidth, if possible. The drawback with this type of tracking is the susceptibility to optical background, especially point sources in the FOV. It is not recommended that dc tracking be used because unique discrimination is not possible without very narrow linewidth filtering of the signal.

Pulse Tracking. A second technique for tracking the communication signal is the use of pulse tracking. This type of tracking is typically performed when the communication source is also a pulsed waveform, but can be used also as an independent beacon channel. In a pulse tracking system, the receiver threshold detects each pulse and uses this information to generate a high-bandwidth tracking error signal from the track quadrants. Pulse tracking normally has a high-bandwidth receiver front end to effectively detect very temporally short pulses. The bandwidth of the system is dependent upon the communication system pulsewidth and pulse rate.

From (3.47), it follows that the rms tracking error for pulse tracking is given by

$$\sigma_{\text{rms}} = \frac{1}{\text{SF}\sqrt{\text{SNR}}} \times \sqrt{\frac{f_c}{f_R}} \tag{3.52}$$

where f_c is the tracking control-loop bandwidth and f_R the received pulse rate. The noise bandwidth of the SNR is that required to detect each individual pulse.

Square-Law Tracking. This term identifies the category of tracking systems that derives a track signal by squaring the communication waveform to generate a tracking signal. An example of this type of tracking system, shown in Figure 3.27, evolves from a biphase-modulated waveform whose frequency content has no distinct component with which to track on. By squaring the incident signal waveform, a harmonic is generated at twice the signal bandwidth. This harmonic can then be phase-locked and used to generate the quadrant track errors. One drawback with this type of system is that the track signal is two times the communication bandwidth and, more importantly, the tracking system is very dependent upon the data rate. Any increase in data rate effectively requires an increase in the tracking system noise bandwidth, making the tracking system less sensitive as data rate is increased. This technique can be used most effectively when a single quasi-CW modulated source is used for communication.

The NEA for square-law tracking is also calculated from (3.47), but the SNR value is compensated by the additional loss factors associated with the second

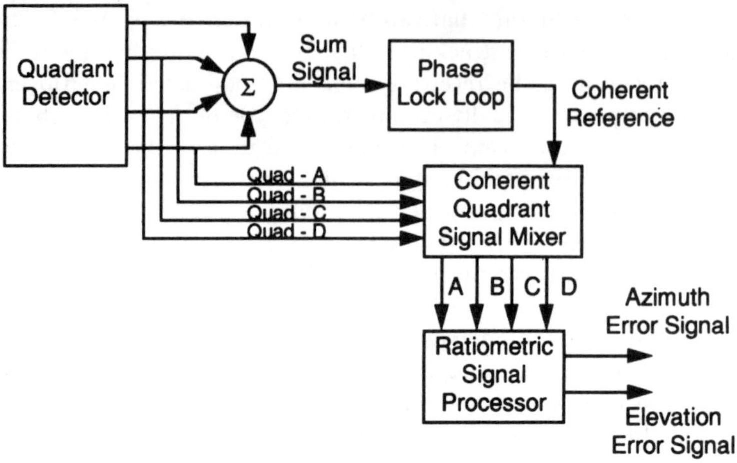

Figure 3.27 Square-law tracking system.

harmonic generation [18]. This results in

$$\mathrm{SNR} = \frac{4 \cdot \left[I_{\mathrm{pk}}(1 - 1/N_e)k_1\right]^2}{B_n \cdot \left[I_{\mathrm{pk}}(1 + 1/N_e)k_1 N_o + 4N_o^2 W\right]} \tag{3.53}$$

where I_{pk} is the peak received signal current, k_1 is the first harmonic filtering factor for a square wave, B_n is the track loop bandwidth in hertz and W the detected signal bandwidth, also in hertz. The value N_o is the noise spectral density of the receiver in $\mathrm{A/Hz}^{1/2}$ and N_e is the transmitter extinction ratio.

Tone Tracking. Tone tracking involves transmitting a separate tone beacon via an additional laser source or modulating the tone onto the communication waveform. Figure 3.28 illustrates the waveforms for a typical tone tracker with a quasi-CW communication signal. This type of modulated tone is usually at a frequency that does not interfere with the message content of the communication waveform. It could involve a separate detector if wavelength separation is available at the receiver. However, filtering the signals to separate the low-frequency tone-track signal from the high-bandwidth communication signal has been performed even when both signals are incident upon the same optical detector. Tone tracking is very efficient and very sensitive. The bandwidth of the tracking receiver can be no wider than the tone frequency, which is typically a few kilohertz (or at least ten times the tracking bandwidth required).

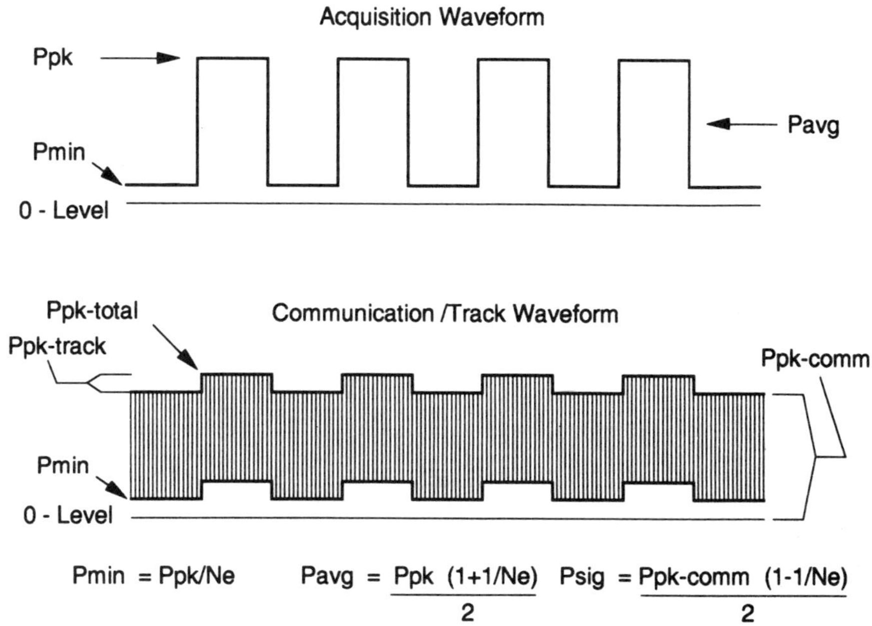

Figure 3.28 Typical tone-tracking waveforms.

Once again (3.42) is used to calculate the rms track error and is given as

$$\sigma_{\text{rms}} = \frac{1}{\text{SF}\sqrt{\text{SNR}}}\sqrt{\frac{f_c}{f_t}} \tag{3.54}$$

where f_c is the track control-loop bandwidth and f_t is the track-tone frequency. The SNR is calculated in the bandwidth of the tone frequency.

Coherent Tracking. Spatial tracking can also be achieved using coherent waveform techniques. Coherent techniques offer advantages not available with direct-detection signal tracking. Background radiation and detector noise can limit the receiver sensitivity. Coherent techniques use the high front-end local-oscillator gain to compensate for the downstream noises. In addition, Swanson and Chan [19] have reported that the local-oscillator spatial intensity distribution can be tailored to yield the desired angular slope factor. They have also shown that a uniform local-oscillator spatial intensity pattern yields approximately −1.7-dB performance differential from the radial derivative local oscillator intensity pattern pictured in

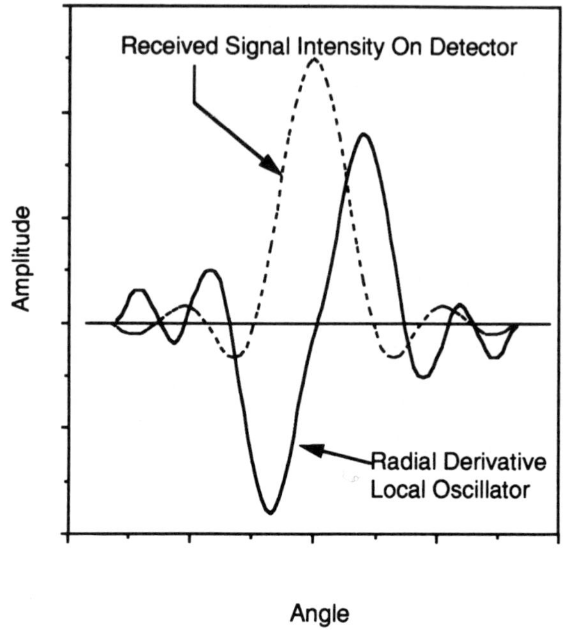

Figure 3.29 Differential radial local-oscillator intensity profile. (After [19].)

Figure 3.29 [19]. Using a matched local oscillator (i.e., signal and oscillator intensities ideally matched) yields a -4.4-dB performance differential.

In general, heterodyne tracking can outperform direct-detection tracking. However, the theory must be tempered by realism. The additional complexities associated with coherent optical detection must also be considered.

3.4.6.3 Nonconventional Tracking Techniques

Some nonconventional approaches to tracking deserve mention here. The first involves the use of gimbals only to perform tracking. This type of system would normally be used where wide beam divergences are available, as is typically the case with short-range systems. A second nonconventional approach is one which uses only a point-ahead beamsteerer and gimbals to perform the necessary jitter compensation. This has been referred to as "feed-forward" tracking.

Gimbals-Only Tracking. Using the gimbals as the only tracking mechanism in an optical communication system has been postulated in designs where large beam divergences are available. Normally, the gimbal jitter is large enough that narrow beams cannot be supported in a gimbals-only configuration. When using gimbals only, the same types of tracking techniques identified above can be used. Gimbal control bandwidths are usually much lower than beamsteerer system control bandwidths, so the frequency content of the track signals can also be lower. The designer must be aware of the higher frequency jitter terms if their amplitudes are significant. Usually this is not the case, however, and the system beam divergence is driven primarily the gimbal jitter and transmit to receive alignment accuracy.

Feed-Forward Tracking. A technique that uses only one beamsteerer in the transmit path has been postulated for use in laser communication system designs. Figure 3.30 is a block diagram of such a tracking system. In this type of design, the linear range on the receive detector must be large enough to accommodate the range of travel of the beamsteerer in the transmit path. Tracking is performed at a designated point that is not on boresight and this offset is dialed into the pointing beamsteerer to compensate for the angular errors.

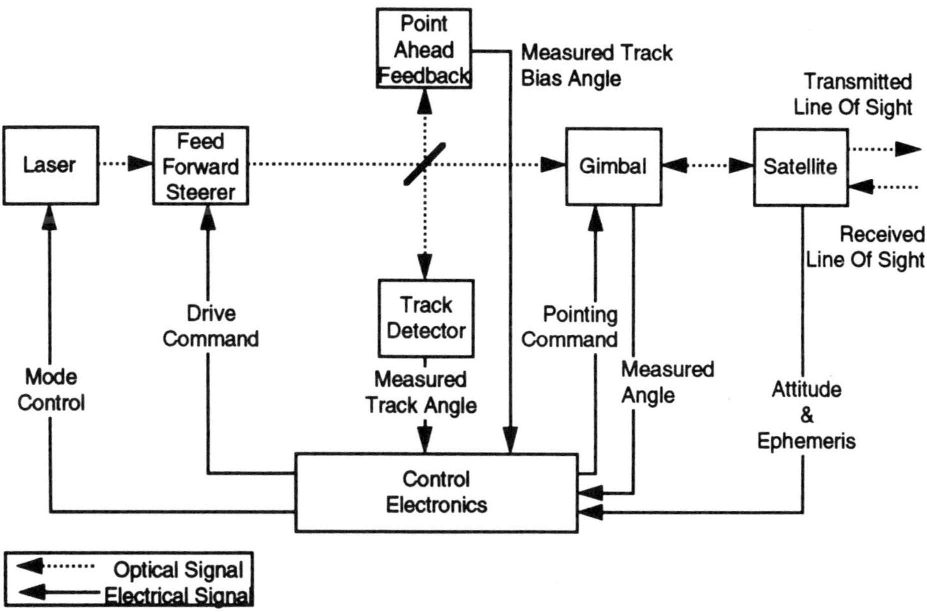

Figure 3.30 Feed-forward tracking system implementation.

This technique has the potential to reduce the number of active elements in the optical train. One drawback to this type of system is the large linear range required by the tracking system. To accomplish this, the tracking system must have a very large received spot size, which is usually many times diffraction-limited. By increasing the size of the spot, the angular gain of the tracking system is reduced, making it less sensitive. Sensitivity must be traded and compensated for by increasing the received energy levels. To do this when keeping the aperture size constant, the transmit power must be increased. In many cases, the requirement for more power negates the system benefits of reduced optical elements.

3.4.7 Communication Techniques

Not surprisingly, the modulation formats for optical communication are very similar to conventional RF techniques. In fact, many were adapted directly from the RF techniques. In addition to modulation techniques, a brief overview of coding types applicable to these modem is presented. Coding gain on the communication link is usually beneficial in that it reduces the required signal energy at the receiver for a given data rate. It should be noted that this is not an exhaustive list of modulation and coding techniques. Other modulation techniques and coding types may also provide beneficial gains to the communication system. As a designer, you are encouraged not to be limited solely by these approaches.

The discussion begins by assembling the modulation techniques into several general categories. First and foremost, modulation waveforms are separated into two major groupings, namely direct and coherent detection waveforms. Subsets of these are also identified. For direct-detection system designs, digital baseband modulation is the most often used and most straightforward. Analog modulation of the waveforms has also been explored for some applications. These all will be discussed below.

Coherent-detection techniques fall into two primary categories. Homodyne and heterodyne modulation techniques for coherent detection systems will be also identified below.

Finally, coding techniques will be discussed and their benefit to the overall link performance will be analyzed. While most coding requires additional bandwidth to accomplish, the signal gain realized has a positive effect on the system design. Care must be taken to not increase the complexity of the laser source or receiver to realize this gain. The designer must trade this complexity with the overall system design drivers like weight and power. Many simple coding techniques do exist that provide good gain.

3.4.7.1 Direct-Detection Digital Baseband Signaling

The simplest form of modulation used in a direct-detection laser communication linking is binary baseband signaling. Here, one state is identified as a binary "one"

Table 3.6
Binary Modulation Techniques

NRZ	Nonreturn to zero
RZ	Return to zero (biphase)
PPM	Pulse position modulation
PBM	Pulse binary modulation
PDBM	Pulse delay binary modulation
PPBM	Pulse polarization binary modulation
PIM	Pulse interval modulation
PQM	Pulse quaternary modulation

and another as a binary "zero." Many different approaches to binary baseband modulation exist. Table 3.6 identifies some of the more common techniques. Depending upon the type of laser source present (usually limited to pulse and quasi-CW sources), different binary modulation techniques are chosen.

Quasi-CW Source Modulation Formats. For the quasi-CW sources, the simplest and most straightforward techniques are the amplitude variant *nonreturn to zero* (NRZ) and split phase (or Manchester) waveforms. In NRZ, as illustrated in Figure 3.31, a binary one is identified by a high signal level over the bit period. A binary zero is identified by no signal present over the bit period. The probability of bit error for this type of signaling is given from [17] as

$$P_E = \frac{1}{2}\text{erfc}\left(\sqrt{\frac{z}{2}}\right) \tag{3.55}$$

where erfc(x) is the complementary error function given by 1 − erf(x) as

$$\text{erfc}(x) = 1 - \frac{2}{\sqrt{\pi}}\int_0^x e^{-t^2}\,dt \tag{3.56}$$

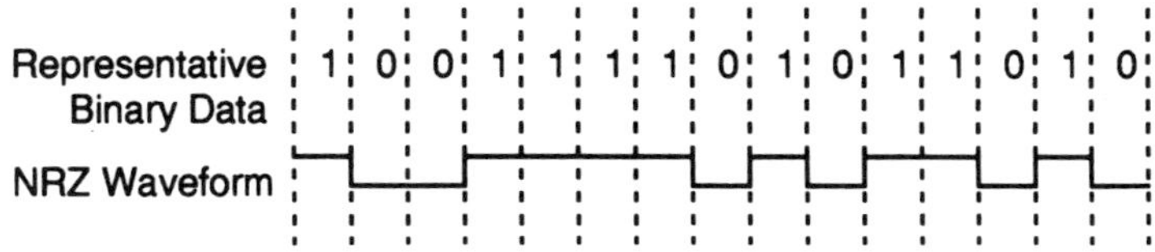

Figure 3.31 Representative NRZ waveform.

and z is the SNR defined as

$$z = \frac{\left[P_{\text{pk}} R_d \cdot (1 - 1/N_e)\right]^2}{4 N_o B} \tag{3.57}$$

where P_{pk} is the peak optical power intensity, R_d is the detector responsivity in A/W, N_e is the extinction ratio of the source, N_o is the spectral density of the noise in A^2/Hz, and B is the bandwidth in hertz.

The current European laser communication experiment is being designed using NRZ modulation [20].

Similarly, the split-phase waveforms identify a binary one by having the signal high in the first half of the bit period and low (usually zero) during the second half. A binary zero is just the opposite, where it is zero in the first half of the bit period and high in the second half. This waveform is illustrated in Figure 3.32. Again, the bit error probability is given from [17] as

$$P_E = \frac{1}{2}\operatorname{erfc}(\sqrt{z}) \tag{3.58}$$

A unique superset of waveforms involves the use of what is known as M-ary signaling, sometimes also referred to as *pulse-position modulation* (PPM). In M-ary modulation, each signal's high level is used to depict two or more bits of binary information. This is numerically illustrated by noting

$$k = 2^m \tag{3.59}$$

where k is the number of bits per pulse and m is the number of receiver slots.

Using this technique, more information can be transmitted using the same number of laser "on" states. M-ary signaling is illustrated in Figure 3.33 for two, four, and eight bits per pulse. Using M-ary signaling increases the information bandwidth, but also requires higher pulse power to offset the increased noise bandwidth. The classic bit error probability equation for M-ary signaling is given

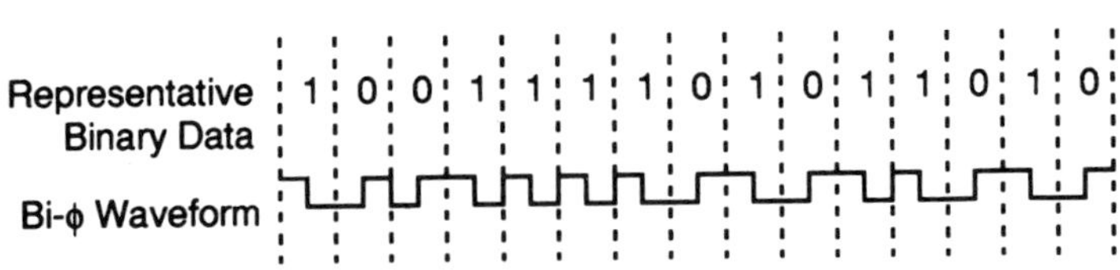

Figure 3.32 Representative bi-phase (Manchester) waveform.

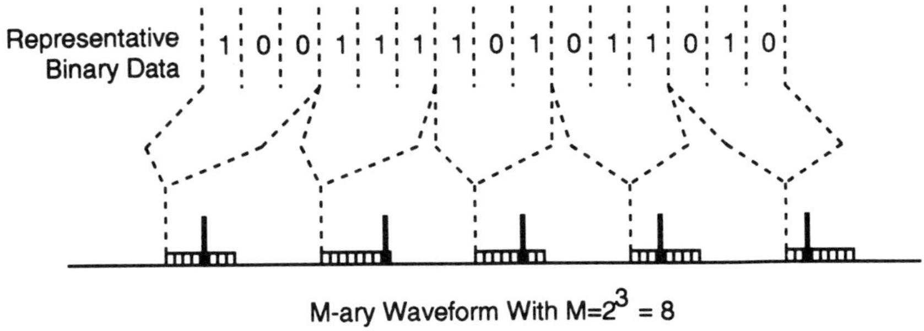

Figure 3.33 Representative pulse-position modulation waveform. Pulse resides in one of eight slots conveying 3 bits of information.

from [21] as

$$P_{\text{be}_{\text{PPM}}} = \frac{M/2}{M-1} P_{\text{we}} \tag{3.60}$$

$$P_{\text{be}_{\text{PPM}}} = \frac{M/2}{M-1}\left\{1 - \left\{\int_{-\infty}^{\infty} \frac{1}{\sqrt{2\pi}\,\sigma} e^{-(x-\mu_1)^2/2\sigma^2}\left[\int_{-\infty}^{x} \frac{1}{\sqrt{2\pi}\,\sigma} e^{-(x-\mu_0)^2/2\sigma^2}\right]^{M-1}\right\}\right\} dx$$

where μ_0 is the mean noise, typically zero, and μ_1 is the mean signal plus noise, typically just signal when mean noise is zero.

M-ary signaling is inherently more efficient than either of the two binary schemes presented earlier, but does put some constraints upon the laser source technology. In quasi-CW systems, the sources are usually nearly peak power limited. Driving the source power to higher and higher levels to accommodate more and more bits per pulse is not recommended. In fact, a practical limit may be 4-ary, or *quaternary PPM* (QPPM), for most diode sources available today.

Pulse-Laser Modulation Formats. Digital information can also be transmitted via very narrow pulse laser beams. Analogous to the *return to zero* (RZ) and NRZ formats above, two types of modulation know as *pulse binary modulation* (PBM) and *pulse delay binary modulation* (PDBM) can be identified. In PBM, the pulse is either present or not in the receiver window. A present pulse is identified as a binary one and the absence of a pulse identifies a binary zero. The probability of bit error for this type of modulation is identical to that given in (3.55). This modulation format is depicted in Figure 3.34.

PDBM employs a laser pulse in the first half-window of the receiver as a binary one and a laser pulse in the second half-window as a binary zero. The classical equation for probability of bit error is exactly (3.58). While RZ and PDBM,

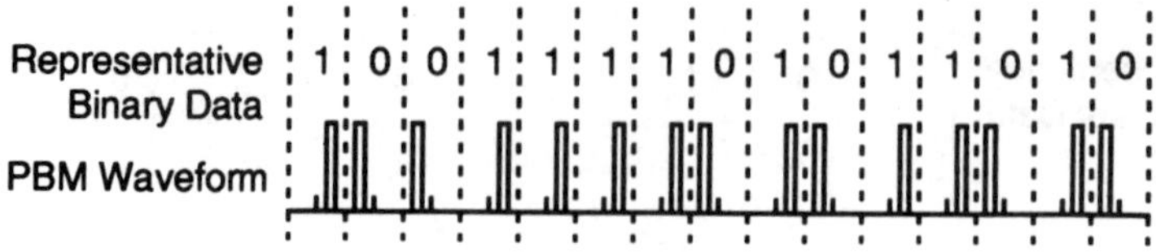

Figure 3.34 Representative pulse binary modulation waveform.

and NRZ and PBM, have the same classic shape to the bit-error-rate curves, it is obvious that the narrow pulsewidth associated with the pulse-laser modulation formats results in significantly larger receiver noise bandwidth and, thus, requires higher peak signal powers than the quasi-CW waveforms.

An additional subset of these modulation formats can result from the use of an external polarization modulator in the optical system. Known as *pulse-polarization binary modulation* (PPBM), this format can transmit binary information by varying the polarization of a single pulse. Then, by utilizing two receivers sensitive to orthogonal polarizations, a temporally fixed pulse can be made to have two states (or one bit of information). Figure 3.35 illustrates the PPBM format and depicts a typical PPBM receiver configuration. In most instances, however, this receiver configuration is less sensitive than the PDBM format but is more sensitive than the PBM format.

As with the quasi-CW laser source, M-ary modulation formats can be envisioned for pulse-laser waveforms. Pulse sources, which typically increase their peak

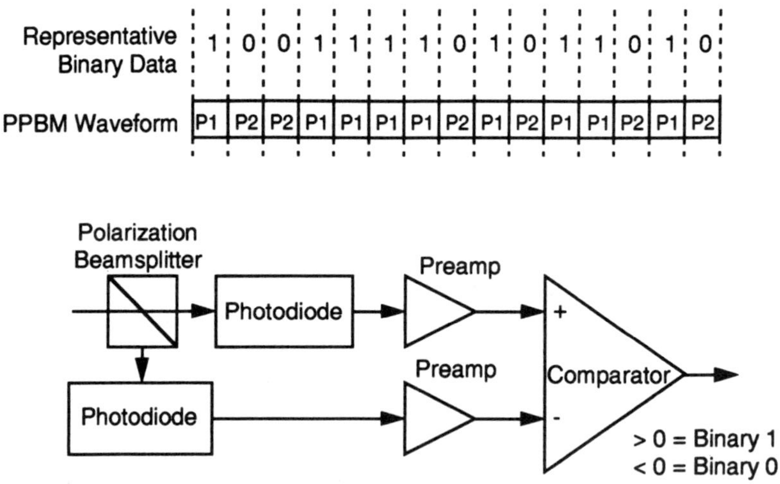

Figure 3.35 Representative pulse-polarization binary modulation waveform.

power as the pulsewidth is decreased, are ideal sources for M-ary signaling. Pulse sources are not limited to 4-ary signaling like the quasi-CW laser sources. Systems using 32-ary [22] and 256-ary [1] pulsed modulation have been produced. Two types of PPM are possible. The first uses the classical signaling synchronized to a fixed reference time frame. The bit error probability for this PPM system follows as (3.60). A similar PPM system, known as *pulse interval modulation* (PIM), uses the known time of the last detected pulse to establish the time windows for the next pulse. In this PIM system design, the bit error probability is given by

$$P_{\text{be}_{\text{PIM}}} = \frac{M}{M-1} P_{\text{we}} \tag{3.61}$$

where the probability of word error, P_{we}, is exactly that given in (3.60).

Notice that if a pulse is undetected or detected in error, then the next pulse in sequence is also not demodulatable. In this type of modulation signaling system, two consecutive M-ary words are in error. While this may seem to be a drawback, it does simplify the timing synchronization required between the receiver and transmitter. The LCS [22] uses this type of modulation.

As a final note, use of polarization discrimination at the receiver could add an additional bit to any binary or M-ary modulation format. A scheme using this was employed during the Airborne Flight Test System demonstration [1] to increase the data rate in a binary signaling system by a factor of two. This effectively became a 4-ary format and was known then as pulse quaternary modulation (PQM). PQM has, to the authors' knowledge never been used or postulated in any system designs since.

3.4.7.2 Direct-Detection Analog Signaling

In addition to digital signaling, several analog modulation formats can also be postulated for use in a laser communication system. While digital signaling is clearly the most efficient way to transmit binary data, some systems still use analog techniques to transmit information. Analog techniques are important to laser communication systems that link to RF bent-pipe information sources. Recall that many of today's satellites use downconverting and upconverting to relay the information from one source to the destination. No demodulation of the signal takes place onboard the satellite. On these satellites it may be more efficient to transmit the analog intermediate-frequency waveform directly without demodulation and remodulation. This is where analog signaling plays an important role in space laser crosslinks.

Analog signaling is not adaptable to pulsed lasers, and as such will be discussed for CW-type lasers with external modulators and for directly current-driven laser diode sources. The linearity of the modulator output is directly proportional to the efficiency of these analog modulation formats. Care must be taken to use modulators and modulatable sources, which have good linear modulation characteristics.

Amplitude Modulation. Classical amplitude modulation (AM) can be used to modulate the laser. The carrier optical signal is intensity modulated by the information content. The received SNR in an amplitude modulated signal is given by [17] as

$$\mathrm{SNR} = \frac{P_t}{N_o W} = \frac{\left[P_{\mathrm{pk}} R_d (1 - 1/N_e)\right]^2}{4 N_o W} \tag{3.62}$$

where W is the bandwidth of the message signal in hertz and the other parameters are as previously given.

At the receiver, the photodiode acts like an envelope detector, stripping the amplitude information off of the optical carrier and passing the information waveform directly to the electronics for further processing. Classical demodulation techniques are then used to detect this AM signal.

Phase Modulation. Similarly, *phase modulation* (PM) is typically achieved by externally phase modulating the optical waveform using an *electro-optic* (EO) modulator. Recently, PM has also been performed on-chip in laser-diode power amplifier arrangements. The phase of the carrier is continuously varied allowing the information to be transmitted. The SNR of PM is given by [17] as

$$\mathrm{SNR} = k_p \langle m_n^2 \rangle \frac{P_t}{N_o W} \tag{3.63}$$

where k_{p} is the phase deviation of the message signal and $\langle {m_n}^2 \rangle$ is the mean square power in the message signal. For sinusoidal message signals $\langle {m_n}^2 \rangle$ is one-half.

Frequency Modulation. Continuous linear-frequency modulation of a laser source has been very difficult to achieve over large frequency deviation ranges. For narrow bandwidth signals, some form of frequency modulation could be considered; however, other techniques usually are more efficient. Until an effective and efficient *frequency modulation* (FM) technique is developed, FM will usually not be in the trade space.

3.4.7.3 Coherent-Modulation Signaling

In a coherent-detection system the incident optical phase front is summed (or mixed) with a local oscillator at the receiver. In an optical communication system,

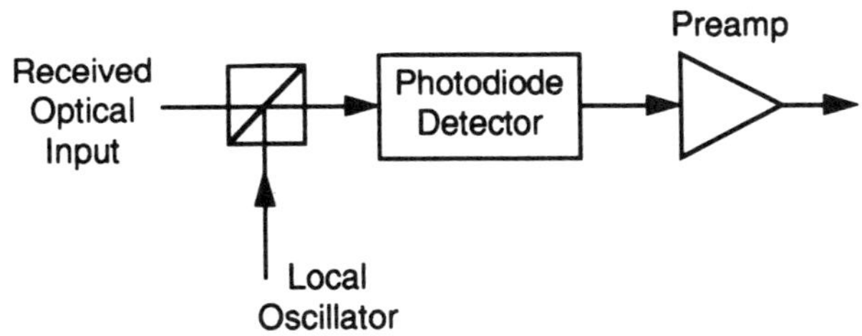

Figure 3.36 Coherent-detection receiver.

the mixing takes place at the detector. Figure 3.36 illustrates a technique used to mix the local-oscillator signal with the received signal. A simple optical beam beamsplitter is used to coalign the two signals and optical elements are used to image the signal onto the detector. The detector in a coherent system is typically a PIN photodiode since front-end receiver gain is provided by the amplitude of the local oscillator waveform.

The mixing process is described by (3.6) and (3.7) for homodyne and heterodyne detection. Several classical modulation schemes are used with coherent detection. These are discussed below.

Heterodyne Signaling. Several forms of heterodyne signaling are identified and their bit error probabilities are given. In *amplitude shift keying* (ASK), the amplitude of the signal is varied as a function of time. Each bit of information is represented by either the presence or absence of the optical waveform. Typically, ASK information is imparted on the optical signal using either directly current-driven laser diodes or external modulation using either acousto-optical or electro-optical modulators. When using heterodyne techniques to demodulate the waveform, the classic bit error probability for ASK results from

$$P_E = \frac{1}{2}\operatorname{erfc}\left(\frac{1}{2}\sqrt{\frac{z}{2}}\right) \tag{3.64}$$

where z is the SNR given previously and again erfc(x) is the complementary error function.

In binary *frequency shift keyed* (FSK) systems, two distinct frequencies are transmitted and the signal is received by two receivers tuned to two distinct transmitted frequencies. Modulation is accomplished by chirping the direct-drive current of laser diode sources, resulting in two distinct frequencies being transmitted . The bit error probability for FSK can be expressed as

$$P_E = \frac{1}{2}\operatorname{erfc}\left(\frac{\sqrt{z}}{2}\right) \tag{3.65}$$

In binary *phase shift keyed* (PSK) systems, two distinct signals, identified by their phase difference, are transmitted. Modulation is accomplished either by an external electro-optical phase modulator or by a technique reported by Mecherle [23] whereby a signal preconditioning occurs, allowing FSK-type modulation techniques to yield a direct-phase-modulated signal. The receiver identifies the binary state by detecting the phase relationship of the signal to a phase reference. The probability of bit error for PSK is given by

$$P_E = \frac{1}{2}\mathrm{erfc}\sqrt{\frac{z}{2}} \tag{3.66}$$

Similar to PSK is *differential* PSK (DPSK). In DPSK, the information is transmitted when the signal changes phase states. The bit error probability for DPSK is given as

$$P_E = \frac{1}{2}e^{-(z^2/2)} \tag{3.67}$$

Homodyne Signaling. For homodyne detection of ASK and PSK waveforms, the required bandwidth can be reduced to approximately half that of a comparable heterodyne system. This results in a 3-dB performance improvement in SNR for homodyne over heterodyne.

The bit error probability for homodyne ASK can be expressed as

$$P_E = \frac{1}{2}\mathrm{erfc}\left(\frac{\sqrt{z}}{2}\right) \tag{3.68}$$

which, when comparing to (3.65), is equivalent to heterodyne FSK. Similarly, the bit error probability for homodyne PSK can be expressed as

$$P_E = \frac{1}{2}\mathrm{erfc}(\sqrt{z}) \tag{3.69}$$

Coherent M-ary Signaling. M-ary signaling can also be used in coherent systems. Increases in SNR sensitivity for 4-ary, 8-ary, and 16-ary FSK and PSK can be estimated as 3, 5, and 6 dB, respectively [24]. However, fluctuations in transmitter frequency and phase can reduce the overall gains of M-ary signal sets over their binary root signal sets.

3.4.7.4 Coding Techniques as a Link Aid

The system link designer must not ignore coding when analyzing the links for any communication system. This includes laser communications. Coding usually involves adding extra bits of information that, when processed at the receiver, aid in correcting errors that were made in transmission. Thus, data encoding enables a reduction in the link error probability, which can be traded for a reduction in the required signal energy to close the link. Many coding types and rates exist. A few types will be mentioned here and their advantages examined. Coding is just one type of technique used to offset channel effects. Others, like interleaving, can also be used effectively in a laser communication link environment, especially in a channel whose nature is bursty.

The most frequent use of coding on a communications link is to reduce the required signal power at the receiver. The measure of this reduced signal requirement is commonly referred to as coding gain. Code gain will be defined for eight reasonably powerful, nearly rate one-half codes and is defined as

$$\Gamma_{\text{code}} = 10 \log_{10} \left[\frac{P_{\text{req}}(\text{uncoded})}{P_{\text{req}}(\text{coded})} \right] \tag{3.70}$$

The eight common codes explored are compared at two commonly used reasonable bit error probabilities of 10^{-6} and 10^{-8}. These codes have been identified for their ease of implementation into hardware and, in fact, many are available in chip form. The eight codes are given in Table 3.7.

Golay Codes. The Golay code (23, 12, 3) is a nearly perfect code and the bit error probability can be expressed in terms of the symbol error probability as

$$P_{\text{be}} = (6/11) \sum_{\varepsilon=4}^{23} \binom{23}{\varepsilon} p_s^{\varepsilon} q_s^{23-\varepsilon} \tag{3.71}$$

where p_s is the code symbol error rate and $q_s = (1 - p_s)$. Table 3.8 identifies the symbol error probabilities for the two bit error probabilities of interest and shows the coding gain achieved for a direct-detection laser link with Manchester encoding.

Table 3.7
Coding Types for Laser Crosslinks

Golay	Viterbi-soft decision
Extended Golay	Viterbi-hard decision
Reed Solomon	Concatenated (RS-Viterbi-soft)
Concatenated (RS-Golay)	Concatenated (RS-Viterbi-hard)

Table 3.8
Coding Gains for Representative Rate One-Half Codes

	$P_{be} = 10^{-6}$		$P_{be} = 10^{-8}$	
Code	p_s	Γ_{code} (dB)	p_s	Γ_{code} (dB)
Golay	.004389	1.45	.001388	1.69
Extended Golay	.0052	1.36	.00165	1.57
Viterbi (hard decision)	.014	2.17	.0059	2.40
Viterbi (soft decision)	.040	3.33	.023	3.61
Reed Solomon	.012	1.64	.01122	2.91
RS/Viterbi (hard decision)	.0739	3.99	.0683	4.61
RS/Viterbi (soft decision)	.115	4.89	.109	5.71
RS/Golay	.029	2.26	.019	2.74

The extended Golay code (24, 12, 3) detects four errors and has a decoded bit error rate of

$$P_{\text{be}} = (4/24)(5/6)\binom{23}{4}p_s^4 q_s^{20} + (8/24)\sum_{\varepsilon=5}^{24}\binom{24}{\varepsilon}p_s^{\varepsilon} q_s^{24-\varepsilon} \tag{3.72}$$

Once again, the coding gain and symbol error probabilities are given in Table 3.8.

Viterbi Decoding of Convolutional Codes. The rate one-half, constraint length seven convolutional code has become an industry standard and is decoded using *Viterbi algorithms* (VA). This device has been implemented on a chip [25]. The decoded error rate to symbol error rate relationship is rather complex, but can be expressed over the region of interest by the following approximation.

$$p_s \approx (.757/2)\text{erfc}\left[\sqrt{(-\ln 2 \cdot P_{\text{be}}) \cdot 10^{\beta/10}}\right] \tag{3.73}$$

where erfc(x) is the complementary error function, $\beta = 0$ for soft-decision decoding, and $\beta = 0.2$ for hard-decision decoding. Table 3.8 gives the symbol error rates and coding gains for convolutional codes with both soft and hard-decision Viterbi decoding.

Reed Solomon Codes. The *Reed Solomon* (RS) family of codes are block codes and as such are M-ary rather than binary-based codes. That is to say they act upon blocks or segments of data as opposed to each data element. The decoded probability of bit error is given by

$$P_{\text{be}} = \frac{n+1}{2n^2}\left[(2t+1)\sum_{i=t+1}^{2t+1}\binom{n}{i}p_s^i q_s^{n-i} + \sum_{j=2t+2}^{n} j\binom{n}{j}p_s^j q_s^{n-j}\right] \tag{3.74}$$

Table 3.9
Rate One-Half Reed Solomon Codes

n	k	t	$p_s\ (P_{be} = 10^{-6})$	$p_s\ (P_{be} = 10^{-8})$
32	16	8	.007513	.004477
64	32	16	.01057	.00801
128	64	32	.01198	.01037
256	128	64	.01208	.01122
512	256	128	.01152	.01110
1024	512	256	.01074	.01054

where n is the code length and $2t + 1$ the number of correctable bits. A number of RS(n, k, t) codes, where the code rate is given by k/n, are given in Table 3.9. As can be extracted from the table, the best RS code is the RS(256, 128, 64). The code gain for this code is shown in Table 3.8.

Concatenated Codes. Concatenated codes are the family of codes whereby a coded link is further encoded. In other words, the codes are nested with inner and outer codes. An efficient concatenated code is that recommended by [26], namely the RS(255, 223) outer code with a Viterbi VA(2, 1) inner code. The code gains are given in Table 3.8 for both hard and soft-decision inner codes.

Another possible concatenated code is the RS(40, 36, 2) outer code with a Golay (19, 8, 3) inner code. This concatenated code's gain is also shown in Table 3.8 for comparison. Concatenated codes are the most powerful codes given here, but are also the most complex to implement. Failure on the inner code could have deleterious effects upon the outer code, causing long link dropouts until resynchronization can occur.

Interleaving. One additional technique that is sometimes implemented to overcome channel effects, namely fading (or bursty) channels, is interleaving of the data bits. The interleaving process involves systematically scrambling the transmit bits in time such that fades of duration T_{fade} will not corrupt adjacent bits in the original stream. In this way, the channel fade errors can be corrected by a codec that might normally be susceptible to burst error effects. Interleaving is a powerful tool against fading channels. Unfortunately, it brings with it some ill effects. The two most noticeable effects of interleaving are increased data buffering capability, which manifests itself as additional memory, and data latency, which can be at least as long as the expected fade times.

Today, many codecs exist in single-integrated-chip form. As integrated-circuit technology moves toward high densities and speeds ($\geq$ 1 Gbps), codecs (in integrated packages) will become common place. It is suggested that the system designers consider coding in all their link designs. These codecs are easily adaptable to the optical channel.

3.4.7.5 Channelization Approaches

Employing channelization in an optical communication system is achieved in ways already used in conventional RF signaling. Channels can be configured and separated via several means. Techniques, such as wavelength division multiplexing, time division multiplexing and code division multiplexing can and have been used with optical systems. The use of polarization can also be used, just as in RF technology, to increase system capacity twofold by allowing an additional state to exist. The advantage that channelization brings is an increased data capacity without using higher speed electronics. However, a penalty is incurred in increased receiver complexity.

Wavelength Division Multiplexing. The most common form of channelization that exists in the optical arena involves using distinct wavelengths to send channelized information across the link. The receiver splits the composite signal into the transmitted wavelengths and routes each to a separate detector. Each channel is handled independently. Successful demonstration of wavelength control of laser diodes has allowed center-to-center spacing of $\leq 30\text{Å}$ [7]. This permits the use of a large number of channels in a small wavelength band.

Time Division Multiplexing. Another technique used for channelization involves temporal slots for each channel's information. While this requires additional per-link bandwidth, it is an efficient means of allowing multiple low-rate users access to a single link. Time division multiplexing can occur at many levels, from packets to individual bit intermixing. The one constraint on time division multiplexing is the data latency that could occur for very large numbers of channels.

Code Division Multiplexing. Recently code division multiplexing has been examined in a multiple access scenario [4]. Here, multiple signals are allowed to collide with each other in time. The receiver utilizes a correlation function, much like a convolutional decoder decision tree, to extract the most likely signals for each channel. This technique appears reasonably useful for many users into a single receiver system at moderate to low data rates.

3.5 THE OPTICAL LINK EQUATION

The link equation for laser communication is very analogous to the link equation for any RF communication link. Starting with the transmit source power, the designer identifies all sources of link degradation (losses) and improvements (gains) and determines the received signal level. Based on the background and receiver noise and the type of signal modulation which is to be detected, a required signal is generated. The ratio of received signal to required signal is the system link margin.

Identifying these gains and losses requires intimate knowledge of the system design, including both the internal constraints and design choices and knowledge of the external factors, including range, data rate, and required signal criteria.

The link equation can be simply written as

$$P_r = P_t G_t L_t L_R G_r L_r \tag{3.75}$$

where

P_r is the receive signal power;
P_t is the transmitted signal power;
G_t is the effective transmit antenna gain;
L_t is the efficiency loss associated with the transmitter;
L_R is the free space range loss;
G_r is the receive antenna gain;
L_r is the efficiency loss associated with the receiver.

Detailed understanding of each entry into the link equation is given in the remainder of this chapter. Both active and dynamic link considerations are identified. The detailed required signal calculations are left until Chapters 5, 6, and 7.

3.5.1 Transmit Source Power

The amount of transmit power generated by the source is determined and entered into the link equation. Additional information about the bandwidth of the transmitted signal, pulsewidth, extinction ratio (measure of on to off power level), and wavelength are all gathered from the transmit source. This additional information is used in calculating the antenna gains and in determining the required signal. Typically, the power is stated as peak power in watts, although as long as the designer is consistent, the power may be booked as average or even referenced to milliwatts (as is the case for RF links).

Transmit power is usually a straightforward entry into the link analysis table. The designer must be careful, however, to enter the power out of the laser device. This device could consist of a hermetically sealed laser diode, for example. Here, the facet power should not be used, but rather the power out of the hermetic package. Similarly, if the laser source is a power-summed beam, the power should be the output of the combined sources, taking into account the optical losses required to combine the beams. In using the highest common denominator for the source, the power will always be consistently booked in the link equation.

On occasion, the source is a purchased item. In this case, if the power is booked at what the deliverable item is, it can readily be specified to the vendor and identified in the link directly. Some flexibility should exist for the link designer to allow for specific bookkeeping at the natural interfaces. In this way, the link is more easily managed during the design and, more importantly, easily specified and flowed down as a highly visible requirement.

3.5.2 Transmitter Optical Losses

The transmission of the laser energy through the optical system results in two important phenomenon that need to be characterized. First, the optical transmission at the source wavelength needs to be estimated using good design practices, superior optical coatings, and via an estimate of the number and type of optical surfaces being traversed. Each optical surface contributes a multiplicative loss factor to the overall transmission budget. This loss factor degrades the transmit power by the derived factor.

In addition to optical transmission, the quality of the optical energy is determined by estimating the wavefront quality of the beam exiting the telescope. Aberrations due to imperfections in the manufacturing process and mechanical stresses on the optics result in a nonperfect optical wavefront. Typically, the wavefront error is determined for each optical element or group of elements and the aggregate is *root sum squared* (rss) to determine an rms wavefront error value. The wavefront error is used in determining the gain of the transmitting antenna in the far field.

3.5.3 Effective Transmit Gain

The gain consists of three distinct parts. The first is the spatial distribution of energy in the far field, which is based on the aperture size (feed beam size), the near-field-energy profile, and the wavelength of the laser crosslink system. The second part involves the off-axis loss factor due to pointing errors in the optical system. The third is simply a geometric reduction of the far-field gain due to the wavefront errors determined above. Each of these will be discussed in further detail below.

3.5.3.1 Transmit Gain

The classical gain equation of a radiating source is derived from the ratio of isotropic (radiating into 4π sr) to directional energy. Figure 3.37 illustrates the geometry that defines this gain scenario. From this scenario it can be easily seen that the ratio of radiated solid angles yields the gain. This is numerically represented as

$$G_t = \frac{4\pi}{\Omega} \tag{3.76}$$

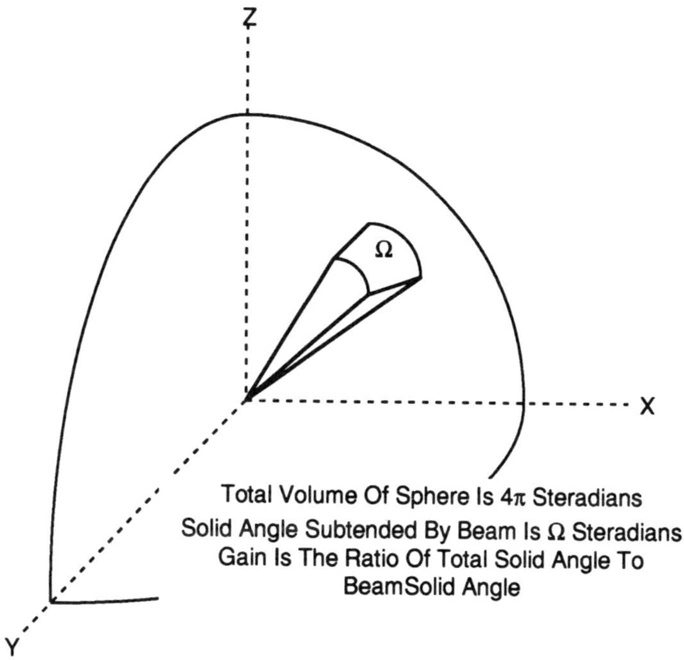

Figure 3.37 Far-field transmitter gain geometry.

where Ω is the solid angle subtended by the source whose full angle divergence angle is θ_{div}. For small θ_{div}, Ω can be determined using a small angle approximation as

$$\Omega \approx \frac{\pi}{4}(\theta_{\text{div}})^2 \tag{3.77}$$

Substituting (3.77) into (3.76) yields

$$G_t = \frac{16}{\theta_{\text{div}}^2} \tag{3.78}$$

From scalar diffraction theory, the angular divergence of a plane wave spatially filtered by an aperture of diameter, a, is proportional to the wavelength, λ, and inversely proportional to the aperture size. This relationship can be described by the numerical expression

$$\theta_{\text{div}} = \frac{4\lambda}{\pi a} \tag{3.79}$$

Now, if (3.79) is substituted into (3.78) the gain equation can be rewritten as

$$G_t = \frac{4\pi A}{\lambda} \tag{3.80}$$

where A is the aperture area and is expressed as $A = \pi a^2/4$.

The above gain equation applies to laser sources whose near-field intensity can be described by a uniform plane wave. However, laser communication transmitters have near-field beam profiles which are Gaussian, or nearly Gaussian, in intensity. Knowing this, the gain for a Gaussian amplitude plane wave with waist size, ω, and wavelength, λ, impinging onto a circular aperture a with obscuration b takes on the form of the classical equation from Klein and Degnan [27], namely

$$G_t(\alpha, \beta, \gamma, X) = (4\pi A/\lambda^2) g_t(\alpha, \beta, \gamma, X) \tag{3.81}$$

where $\alpha = a/\omega$ (ω is the Gaussian feed beam waist diameter at the $1/e^2$ point), $\gamma = b / a$, $X = (2\pi/\lambda) \cdot a \cdot \sin\theta_1$, and $\beta = 0$ for Gaussian plane waves in the far field. By examining (3.81) closely, one can see the first term is the classic gain equation for a uniform plane wave. The second term is a gain efficiency, which accounts for obscuration, truncation, off-axis intensity, and near field and defocusing effects. This efficiency factor is also given by [27] as

$$g_t(\alpha, 0, \gamma, X) = 2\alpha^2 \left| \int_{\gamma^2}^{1} J_0(X\sqrt{u}) e^{-(\alpha^2 u)}\, du \right|^2 \tag{3.82}$$

where J_0 is the Bessel function of order zero.

On-Axis Gain. When analyzing far-field on-axis gain, the value for X in (3.82) becomes zero. Again, from [27], the on-axis gain is found to be

$$G_t = \frac{4\pi A}{\lambda^2} \left[\frac{2}{\alpha^2} \left[e^{-\alpha^2} - e^{-\gamma^2 \alpha^2} \right]^2 \right] \tag{3.83}$$

Suboptimum performance results when the aperture to feed beam diameter (at the $1/e^2$ intensity) does not follow the relationship established by Klein and Degnan, namely

$$\alpha \approx 1.12 - 1.30\gamma^2 + 2.12\gamma^4 \tag{3.84}$$

which is valid for $\gamma \leq 0.4$ to within 1% [27].

Equation (3.83) defines the effective on-axis far-field gain for circular apertures perturbed by obscuration and truncation effects. In the limiting case, where no obscuration exists (i.e., $\gamma = 0$), (3.83) reduces to

$$G_t = \frac{4\pi A}{\lambda^2}\left[\frac{2}{\alpha^2}\left[e^{-\alpha^2} - 1\right]^2\right] \tag{3.85}$$

And in the case where truncation effects are minimized (i.e., $a \gg \omega$), the gain equation takes on the form of

$$G_t = \frac{4\pi A}{\lambda^2}\left[\frac{2}{\alpha^2}\right] = 2\left[\frac{\pi\omega}{\lambda}\right]^2 \tag{3.86}$$

Now, substituting (3.79) with ω defining the limiting aperture yields the familiar expression for the gain of an untruncated and unobscured Gaussian beam in the far field

$$G_t = \frac{32}{\theta_{\text{div}}^2} \tag{3.87}$$

Equation (3.87) is useful for approximating the gain for feed beams, which are only a fraction of the limiting aperture. This can be the case when an acquisition beam of larger divergence than the communication beam is used in the same physical antenna aperture. In many laser communication system designs, this is true.

Off-Axis Gain. Since laser communication beam divergences are very small angularly, it is important to understand the nature of the gain profile off-axis. Returning to (3.82), the gain off-axis can be characterized. Figure 3.38 illustrates the far-field gain profile for a circular aperture and optimized Gaussian feed beam diameter. For most laser communication system applications, the designer is only concerned with off-axis gain down to the $1/e^2$ intensity level. Equation (3.82) can be approximated off-axis by a Gaussian profile out to this level. Thus, the off-axis gain can be approximated by

$$G_t(\theta_{\text{off}}) \approx \frac{4\pi A}{\lambda^2} e^{-8(\theta_{\text{off}}/\theta_{\text{div}})^2} \tag{3.88}$$

where θ_{off} is the off-axis angle and θ_{div} is the $1/e^2$ beam diameter.

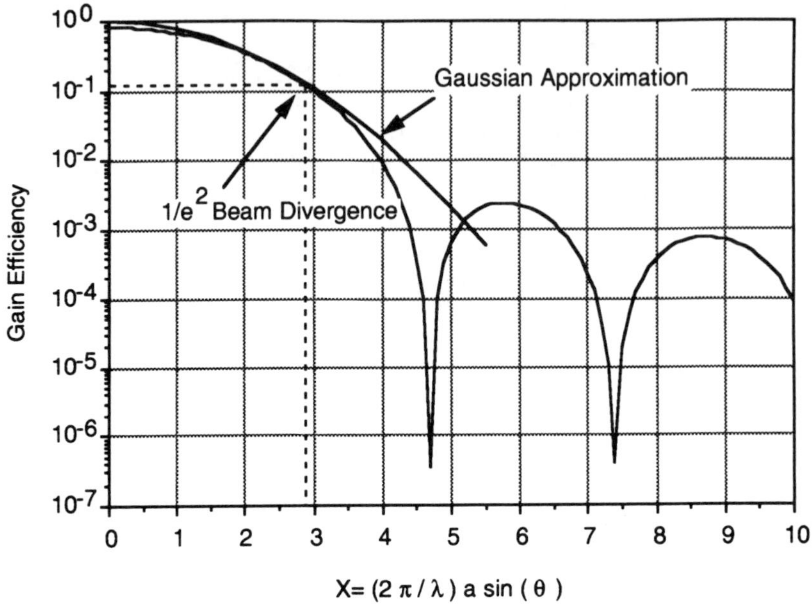

Figure 3.38 Far-field gain efficiency profile with Gaussian approximation [27].

Equation (3.88), normalized by $4\pi A/\lambda^2$, is also plotted on Figure 3.38 as a comparison of the actual and approximated gain equations.

3.5.3.2 Pointing Loss

The X and Y-axis pointing errors in a communication system are typically assumed to be statistically independent random variables with Gaussian statistics. The joint density function describing the composite X and Y pointing error is given by

$$f_{xy} = \frac{1}{2\pi\sigma_x\sigma_y} e^{-(x-\mu_x)^2/2\sigma^2} e^{-(y-\mu_y)^2/2\sigma^2} \tag{3.89}$$

where μ_x and μ_y are the mean of the X and Y axes and σ_x^2 and σ_y^2 are the variances. If σ_x equals σ_y then substituting $r^2 = x^2 + y^2$, $x = r\cos\theta$, $y = r\sin\theta$ and $\theta = \tan^{-1}(y/x)$ into (3.89) yields the polar form for the pointing error probability which is

$$f_{R\theta}(r, \theta) = \frac{r}{2\pi\sigma^2} e^{-(r^2+\mu_x^2+\mu_y^2+2\mu_x r\cos\theta+2\mu_y r\sin\theta)^2/2\sigma^2} \tag{3.90}$$

For a given communication system with pointing error, the received signal power can be expressed as a function of the pointing error r (converted to angular as θ_p), and can be determined by the expected value of the bit error probability. This is accomplished by integrating over all possible pointing errors and is described by

$$P_{\text{be}} = \int_0^{\infty} P_E | r \cdot f_R(r)\, dr \tag{3.91}$$

$f_{\text{R}}(r)$ is expressed by integrating over all θ and is given as

$$f_R(r) = \frac{r}{\sigma^2} e^{-(r^2+\mu_x^2+\mu_y^2)/2\sigma^2} I_0 \left[\frac{r \cdot \sqrt{\mu_x^2 + \mu_y^2}}{\sigma^2} \right] \tag{3.92}$$

where I_0 is the modified Bessel function of order zero. This expression is recognized as the classical Rician density function. For the special case where $\mu_x = \mu_y = 0$ (i.e., only random pointing errors, no bias component), (3.92) reduces to

$$f_R(r) = \frac{r}{\sigma^2} e^{-r^2/2\sigma^2} \tag{3.93}$$

which is the familiar Rayleigh density function.

Now, substituting (3.92) into (3.91) yields

$$P_{\text{be}} = \int_0^{\infty} P_E | r \cdot \frac{r}{\sigma^2} e^{-(r^2+\mu^2)/2\sigma^2} I_0 \left[\frac{r\mu}{\sigma^2} \right] dr \tag{3.94}$$

With solutions to (3.94) for various communication bit error probabilities, the expected pointing loss can be determined. A rigorous solution to (3.94) allows the designer the ability to optimize the transmit gain based on the pointing and bit error probability statistics. Chen and Gardner [21] have analyzed this case for the M-ary PPM channel and zero mean (no bias component) pointing statistics. Their results indicate that pointing errors can have a significant impact on the link performance and, thus, bit error probability.

While (3.94) is an exact solution, quite frequently a solution which bounds the design is more easily determined and facilitates the system design. Instead of solving for P_{be} directly, the bit error contributions from pointing and system noise are determined independently. To do this, the bit error probability is divided into two components. One contribution is from system noise and is the classical $\text{P}_{\text{benoise}}$

found by solving equations for each modulation format chosen. The second contribution, P_{beptg} is determined ideally from

$$P_{\text{beptg}} = \int_0^\infty f_R(r) L_p(r)\, dr \tag{3.95}$$

The contributions of each error source to the total is found from

$$P_{\text{no error}} = (1 - P_{\text{be}}) = (1 - P_{\text{benoise}})(1 - P_{\text{beptg}}) \tag{3.96}$$

and then solving for P_{be} gives

$$P_{\text{be}} = P_{\text{benoise}} + P_{\text{beptg}} - P_{\text{benoise}} \cdot P_{\text{beptg}} \tag{3.97}$$

Setting $P_{\text{benoise}} = P_{\text{beptg}}$ and ignoring the multiplicative term to give an upper bound yields

$$P_{\text{benoise}} = P_{\text{beptg}} = \frac{P_{\text{be}}}{2} \tag{3.98}$$

Therefore each component of the bit error probability is handled independently and is considered to contribute equally to the total error. From this assumption, (3.95) is then expanded to yield

$$P_{\text{beptg}} = \int_0^\infty e^{-8}(\theta/\theta_{dn})^2 \cdot \frac{\theta}{\sigma} \cdot e^{-(\theta^2+\mu^2)/2\sigma^2} I_o\left(\frac{\theta\mu}{\sigma}\right) d\theta \tag{3.99}$$

Then by defining the loss factor α_0 as

$$\alpha_0 = e^{-8(\theta/\theta_{\text{div}})^2} \tag{3.100}$$

(3.99) can be expressed as

$$P_{\text{beptg}} = \left(\frac{\theta}{\sigma}\right)^2 e^{-0.5[(\mu/\sigma)^2 - 0.125(\theta/\sigma)^2]} \int_0^1 \alpha_0 I_o\left(\frac{\mu}{\sigma}\sqrt{\frac{-\ln \alpha_0}{8}}\left(\frac{\theta}{\sigma}\right)\right) d\alpha_0 \tag{3.101}$$

3.5.3.3 Wavefront Loss

The wavefront loss associated with aberrations of the optical signal by the optical elements in the transmit path is booked as the rms wavefront error. This error is modeled at the link level as a reduction in the on-axis intensity due to high-frequency "ripples" in the far-field intensity pattern. Because the receiver could collect energy in the valley of one of these ripples (or on a peak), an rms quantification of these error contributors is typically used to determine the loss due to wavefront error. Since each optical element in the path contributes to this wavefront quality measure, and since the number of elements is usually large, the contributions are root sum squared to yield a total rms wavefront error value. From this value an approximation to the on-axis intensity loss is used. Equation (3.102) is used for this purpose. It should be noted that most laser communication systems operate with an rms transmit wavefront error of approximately $\lambda/10$. Also, note that the approximation facilitated by (3.102) becomes inaccurate as rms wavefront errors become large ($> \lambda/4$ rms).

$$L_{\mathrm{wf}} = e^{-(k\sigma)^2} \tag{3.102}$$

where $k = 2\pi/\lambda$ and σ is the rms wavefront error represented as λ/x, $x =$ wavefront error.

3.5.4 Range Loss

Link range loss results from the diverging wavefront of the optical energy as it traverses the link distance. A convenient way of representing the this loss is via a unitless expression for range loss. Calculation of the classical range loss is given by

$$L_r = \left[\frac{\lambda}{4\pi R}\right]^2 \tag{3.103}$$

3.5.5 Receive Antenna Gain

The effective collecting aperture of the receiver constitutes the receive antenna gain. The receive antenna gain is calculated from the collecting area of the antenna and the wavelength of the incident optical energy and is expressed (again, conveniently) by a unitless parameter as

$$G_r = \left[\frac{\pi D}{\lambda}\right]^2 \tag{3.104}$$

3.5.6 Receiver Optical Losses

The transmission of the received optical energy through the optical system to the detector comprises the receiver optical losses. As before, the optical transmission at the wavelength of operation is required to be estimated using good design practices, superior optical coatings, and the number and type of optical surfaces being traversed. Recall, each optical surface contributes a multiplicative loss factor to the overall transmission budget. Again, this loss factor degrades the transmit power by the derived factor.

3.5.7 Receiver Pointing Loss

Similar to the loss associated with the transmitter pointing error, a loss term must be considered for the mispointing at the receiver. For direct-detection links, this is normally not a concern since the receive FOV and spot size are oversized and not diffraction-limited. The receiver in these cases is just a "photon" bucket.

However, for coherent links, where the received spot size is nearly diffraction-limited and the received electric field is summed with the local oscillator field to perform mixing of the signal at the detector, pointing errors are of concern. Any angle-of-arrival fluctuations of the receive electric field results in reduced mixing efficiency at the receiver. Thus, tracking of the received signal to very tight accuracies becomes important.

3.5.8 Received Signal Level

The received signal level P_r is then calculated from the aggregate of all gains and losses. It is identically (3.75) and is repeated here as a more rigorous expression as

$$P_r = P_t \cdot \frac{4\pi A}{\lambda^2}\left[\frac{2}{\alpha^2}\left[e^{-\alpha^2} - e^{-\gamma^2\alpha^2}\right]^2\right] \cdot e^{-(k\sigma)^2} \cdot e^{-8(\theta_p/\theta_{\text{div}})^2} \cdot \left[\frac{\lambda}{4\pi R}\right]^2 \cdot \left[\frac{\pi D}{\lambda}\right]^2 \cdot L_{\text{pr}} \cdot L_r \tag{3.105}$$

3.5.9 Required Signal Levels

The required signal level, P_{req}, is calculated for a variety of different conditions in Chapters 5, 6, and 7. Each condition requires special considerations and will be discussed in detail in these chapters. The resultant required signal is very much dependent upon the link details discussed above. For example, background radiation is collected by the receive aperture and is transferred to the detector via the same optical system. Thus, the optical collection area, detector view field, and optical transmissions all play a role in determining background noise levels.

3.5.10 Link Margin

Link margin is calculated by comparing the received signal level, P_r, with the calculated required signal level, P_{req}. The ratio of received to required signal is the link margin. This is typically expressed in terms of dB and is calculated from

$$\text{Link margin} = 10\log_{10}\left(\frac{P_r}{P_{\text{req}}}\right) \tag{3.106}$$

Most space-based communication systems have requirements for excess margin in the design of the link. For satellite crosslinks, link margins of greater than 3 dB at launch are typical. This excess margin accounts for the unexpected on-orbit link degradation that may occur due to larger than normal contamination or radiation levels or degradation in other areas of the system.

3.6 CONCLUSION

This chapter has presented a disciplined approach to laser communication system design. The generic functions of a laser communication system were presented. Design and technology options and implementation techniques were identified and compared. Finally, the details of system level link analysis were discussed. With the basic skills outlined in this chapter it is hoped the system designer now has the proper understanding and tools required to design laser communication links.

References

[1] Ross, M., Freedman, P., Abernathy, J., Matassov, G., Wolf, J., and Barry, J. D. "Space Optical Communications With the Nd:YAG Laser," *IRE Proc.*, Vol. 66, No. 3, March 1978.

[2] Van Hove, P., and Chan, V.W.S. "Spatial Acquisition Algorithms and Systems for Optical ISL," *IEEE Trans. on Communications*, 1983, pp. 1208–1214.

[3] Guranatz, I, Gardner, S., Viterbi, A., and Zbik, E. "Multiple-Access Communication Using Coded Pulse Interval Modulation," *IEEE Proc. of The National Telecommunications Conf.*, *Vol.* 1, 1980.

[4] Nicholson, D. and Bondurant, R., "CDMA Waveforms in Laser Crosslinks," *Proc. From the Military Communications Conf.*, MILCOM '92, Oct. 1992.

[5] Dreisewerd, D., Fritz, W., Begley, D., Schwedt, S., and Elliott, G. "Reliability of AlGaAs/GaAs Laser Diodes Grown by Metalorganic Chemical Vapor Deposition," *Proceedings Of SPIE – Free Space Laser Communication Technologies*, Vol. 885, Jan. 1988.

[6] "200 mW CW Single Mode Laser Diode – SDL-5430 Series," *Spectra Diode Labs Product Sheet.*

[7] Jackson, J. E., Dreisewerd, D. E., Lambert, S. G., Morris, T. R. and Casey, W. L. "Evolution and Future of the Multiple Diode Laser," *SPIE Proc. – Free Space Laser Communication Technologies V*, Vol. 1866, Jan. 1993.

[8] Livas, J. C., Alexander, S. B., Kintzer, E. S., Swanson, E. A., and Paul, T. J. "GBPS Class Optical Communications Systems for Free-Space Applications," *SPIE Proceedings – Free Space Laser Communication Technologies V*, Vol. 1866, Jan. 1993.

[9] "1 Watt CW Single Mode MOPA Laser Diode – SDL-5760 Series," *Spectra Diode Labs Product Sheet.*

[10] Mathur, R. P., Beard, C. I., and Purll, D. J. "Analysis of SILEX Tracking Sensor Performance," *SPIE Proc. – Free Space Laser Communication Technologies* II, Vol. 1218, Jan. 1990.

[11] Green, S. I. "Tracking Atmospheric Optical Communication System," *SPIE Proc. – Free Space Laser Communication Technologies II*, Vol. 1218, Jan. 1990.

[12] Katzman, M., Editor. *Laser Satellite Communications*, Englewood Cliffs, NJ: Prentice Hall, Inc., 1987.

[13] Davis, D. J., Deadrick, R. B., and Stahlman, J. R. "ALGS: Design and Testing of an Earth to Satellite Optical Transceiver," *SPIE Proc. – Free Space Laser Communication Technologies V*, Vol. 1866, Jan. 1993.

[14] Papoulis, A. *Probability, Random Variables and Stochastic Processes*, New York, NY: McGraw Hill, USA, 1965.

[15] Van Trees, H. L. *Detection, Estimation and Modulation Theory – Part I*, New York, NY: John Wiley and Sons, 1968.

[16] Gardner, F. M. *Phase-Lock Techniques*, New York, NY: John Wiley and Sons, 1966.

[17] Zeimer, R. H., and Tranter, W. H. *Principles of Communications–Systems, Modulation and Noise*, Boston, MA: Houghton Mifflin, 1976.

[18] Spilker, James J., Jr. *Digital Communications by Satellite*, Englewood Cliffs, NJ: Prentice Hall, 1977.

[19] Swanson, E. A., and Chan, V.W.S. "Heterodyne Spatial Tracking System for Optical Space Communication," *IEEE Trans. on Communications*, Vol. Com-34, No. 2, Feb. 1986.

[20] Oppenhauser, G., and Wittig, M. E. "European SILEX Project: Concept, Performance, Status and Planning," *Proc. of SPIE—Free Space Laser Communication Technologies II*, Vol. 1218, Jan. 1990.

[21] Chen, C.-C., and Gardner, C. S. "Impact of Random Pointing and Tracking Errors on the Design of Coherent and Incoherent Optical Intersatellite Communication Links," *IEEE Trans. on Communications*, Vol. 37, No. 3, March 1989.

[22] Deadrick, R., and Deckelman, W., "Laser Crosslink Subsystem-an Overview," *Proc. From the Military Communications Conference*, MILCOM '92, Oct. 1992.

[23] Mecherle, G. S. "Direct Phase Modulation of Laser Diodes," *SPIE Proc.—Free Space Laser Communication Technologies IV*, Vol. 1635, Jan. 1992.

[24] Okoshi, T., Emura, K., Kikuchi, K., Kersten, R. Th., "Computation of Bit-Error Rate of Various Heterodyne and Coherent-Type Optical Communication Schemes," *Journal of Optical Communications*, 2 (1981) 3.

[25] "Q1401, k = 7, Rate 1/2 Single Chip Viterbi Decoder—Technical Data Sheet," Qualcomm, Sept. 1987.

[26] Consultative Committee for Space Data Systems, "Recommendations for Space Data Standards: Telemetry Channel Coding Blue Book," May 1984.

[27] Klein, B. J., and Degnan, J. J. "Optical Antenna Gain. 1: Transmitting Antennas," *Applied Optics*, Vol. 13, No. 9, Sept. 1974.

Chapter 4

Optical Design

4.1 OPTICAL CROSSLINK SYSTEM OVERVIEW

Whenever optical signals pass through a laser communications system, or through space, optical laws are obeyed. We will discuss the behavior of these signals and their effect on system design. There are numerous internal elements in a system, many of which perform unique functions. Figure 4.1 is a simplified block diagram of a laser communications system, which will be used to discuss the top level optical functions.

Referring to Figure 4.1 and starting at the transmitter end, the input signal from the satellite is electronic and contains information impressed upon it. The transmitter converts the electronic signal into an optical signal. The transmitter source in our illustration is a diode laser, which is modulated directly by turning the diode drive electronic signal on and off. The modulation of the diode current results in the diode output light following in similar fashion. The optical signal is routed through the imaging optics to the pointing element (in this case, a telescope) where it is expanded into the desired transmit beam and sent across space to the cooperative receiver terminal. Once received by the receiver telescope, the optical signal is routed through the imaging optics to the detector where it is converted to an electrical signal for processing by the receiving satellite and sent to the ultimate user of the data. To be received, the receiver terminal must be pointed directly at the transmitter and vice versa. To create this coalignment condition, both terminals simultaneously go through a process of pointing, acquisition, and tracking of each other. The beam pointed out of the laser communications transmitter is exceedingly narrow so that the pointing, acquisition, and tracking of the cooperating terminals are not trivial and require considerable effort to accomplish. A description of the major blocks of the system will be desirable. The pointing element types are discussed first.

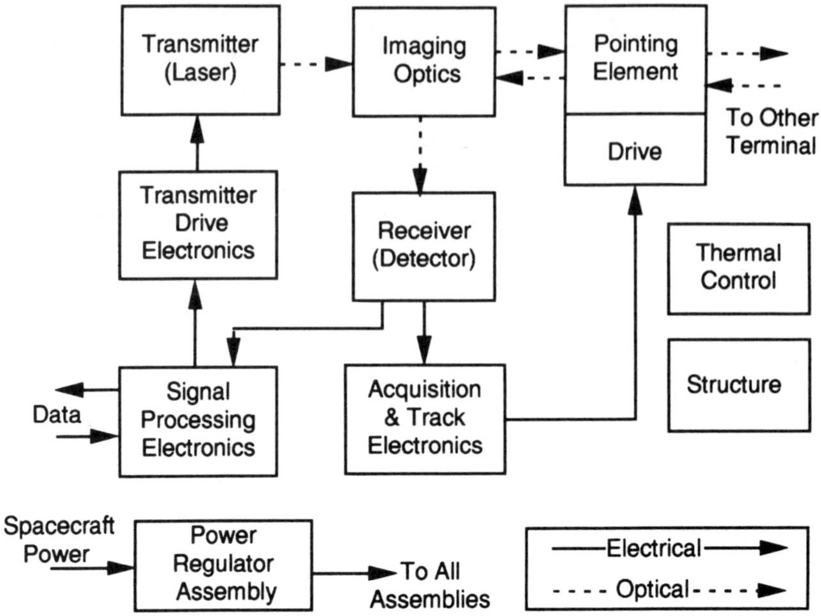

Figure 4.1 Laser communications block diagram.

4.2 POINTING ELEMENTS

The pointing element of a system (see Figure 4.1) performs the function of redirecting the line of sight such that the system has access to a large solid angle (field of regard) for initial pointing prior to acquisition. In addition, the pointing element can be redirected over small angles in a systematic fashion to aid in the acquisition process. Lastly, the pointing element is commanded to point directly at the cooperating terminal while it is being tracked. There are three major types of pointing elements.

4.2.1 Gimbaled Flat

The gimbaled flat, or gimbaled mirror as it is sometimes known, is illustrated in Figure 4.2. The gimbaled flat is positioned in front of the receiving telescope because the flat mirror has no capability to gather and focus energy like the telescope but, rather, can only redirect the line of sight.

The gimbaled flat offers the advantage that the telescope and imaging optics, detector, laser, and so forth are all fixed; that is, they need not move to alter the line of sight. There are shortcomings to this approach. The combined weight of the flat

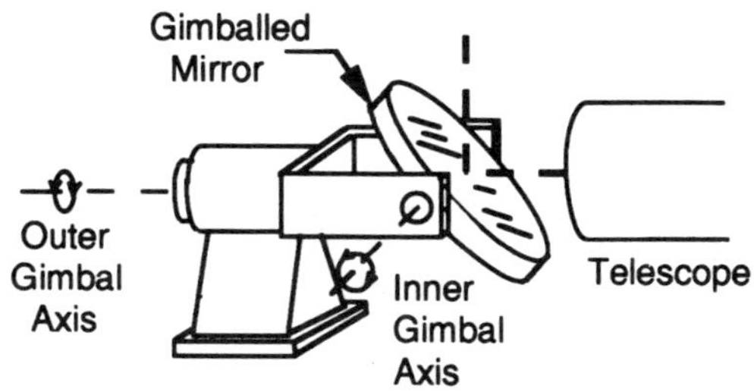

Figure 4.2 Gimbaled flat.

mirror and the telescope is generally greater than that of a gimbaled telescope. The length of the flat inherently must be larger than the aperture, usually by a factor of 1.4 (for 45°) or more depending on the angle through which it must reflect the line of sight. Further, the gimbaled flat has a limited field of regard, considerably less than a hemisphere. In addition, there are shortcomings to this approach for space application. The exposure of the flat mirror and its reflective coating to micrometeoroids, ultraviolet rays, energetic radiation, atomic oxygen, or illumination from laser weapons can cause degradation in the coatings. For a gimbaled telescope or a gimbaled package, the collecting/transmitting optics are less exposed to the external space environment. Also, the impingement of solar energy on the mirror can cause thermally induced mirror bending, resulting in distortion of the reflected transmitter signal. This distortion lowers the received intensity at the far terminal. This is not a good situation since signal power is usually considered at a premium and must be conserved in order to "close the link."

4.2.2 Gimbaled Telescope

The gimbaled telescope is the form of pointing element most used for the longer range crosslinks. Again, as with the gimbaled flat, the rest of the system is fixed; however, with a gimbaled telescope less mass must be moved to redirect the line of sight. The field of regard can be hemispherical or greater. Both the *laser crosslink subsystem* (LCS) [1] and its conceptual derivatives [2–4] use this design pointing element. A gimbaled telescope is shown in Figure 4.3.

The LCS telescope is of Cassegrain construction, which means that it is a two-mirror telescope with a concave primary mirror and a convex smaller secondary mirror. There is a third (tertiary) fold (flat) mirror to direct the beam through the inner gimbal and down though relay optics until the path exits through the azimuth (outer) axis. The passage of the optical signal through the gimbals is called a Coudé path and is illustrated in Figure 4.4. The transmit and receive paths are common.

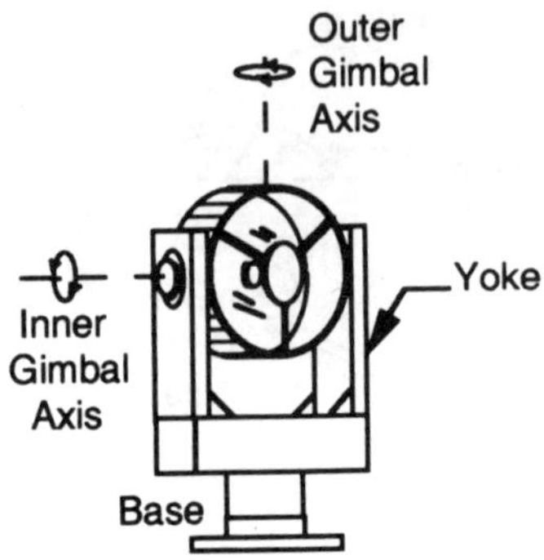

Figure 4.3 Gimbaled telescope.

For illustration, a second-generation gimbaled telescope is also a Coudé design. Figure 4.5 shows the gimbaled telescope on a pedestal as part of an optical assembly that includes all optical elements, the laser sources, and the detectors. The telescope is a 5.5-inch aperture off-axis Gregorian. In a Gregorian design, the energy comes to a focus before it hits the secondary. The optical characteristics of the Gregorian, the LCS Cassegrain telescope, and other telescopes will be discussed later in this chapter when our attention focuses on the telescopes. Gimbaled telescopes can be designed to be lighter than the gimbaled flats since none of the optical elements exceed the collection aperture in size.

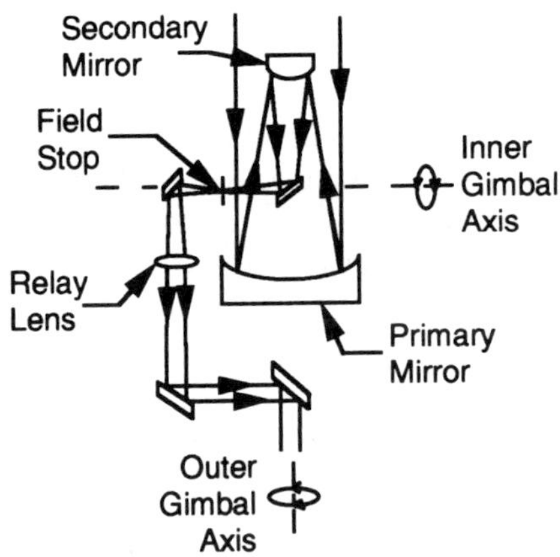

Figure 4.4 Gimbaled telescope Coudé path.

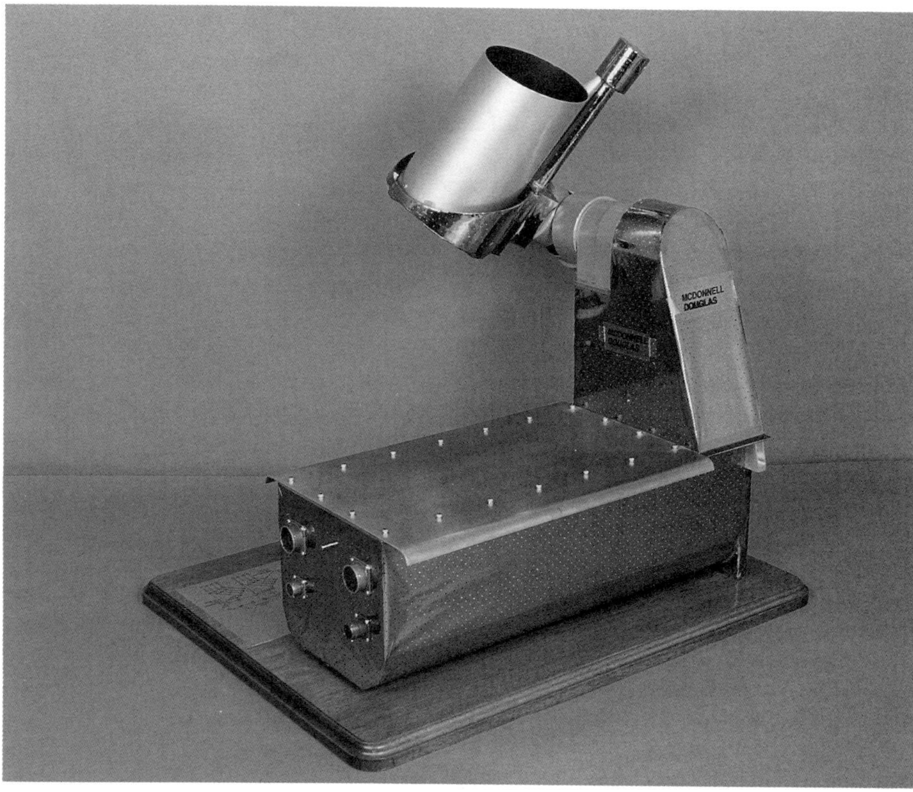

Figure 4.5 Second-generation crosslink system. (Courtesy of McDonnell Douglas Aerospace.)

4.2.3 Gimbaled Package

Gimballing the entire package or a major portion of it is another approach to the design of the terminal. Because it is undesirable from a dynamics viewpoint to swing a large mass that can affect the stability of the carrying satellite, this approach appears to be most suitable for smaller aperture (3 or 4 inches in diameter), shorter range systems [5]. The gimbals to support the mass of a larger aperture telescope are also heavy unless of unusual design. Several lightweight, unusual designs have surfaced recently. Both conventional and unusual gimbal designs are discussed in Chapter 8. A gimbaled package design is illustrated in Figure 4.6. Note the separate transmit and receive apertures, which serve to isolate the transmit power from the receiver. The electronic functions residing on-gimbal must be supplied with power and signals. The cabling for this system can be

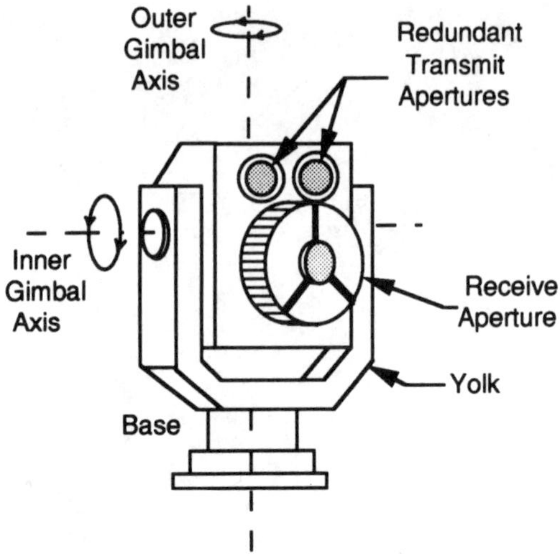

Figure 4.6 Gimbaled package.

challenging. Many conductors are required. As previously stated, this design, if heavy, can impart a significant torque on the carrying vehicle when it accelerates in either of the two axes shown. The field of regard for this design is nearly hemispherical being limited only by the clearance present for the package to swing through the inner gimbal.

A variation of this design mounts the transmitter off the gimbal and routes the transmitter signal to the transmit aperture on the gimbal via a fiber-optic link. The communication receiver can also be mounted separately, coupled to the receive aperture by a fiber. In both these cases, the fiber losses must be considered. The tracking detector would normally stay on the gimbaled package since a fiber would not normally carry the spatial tracking information unless an approach similar to the Massachusetts Institute of Technology Lincoln Lab nutating fiber approach is used [6].

4.3 LASER BEAMS

In order to design a laser communication system, it is necessary to understand the optical behavior of laser beams. Laser light is monochromatic; that is, it is one exact color. Of course, this is an ideal characteristic and lasers actually do produce a range of colors, albeit a very narrow range. A measure of the pureness of the laser light is the linewidth, expressed as a frequency. For coherent systems, the linewidth must be extremely narrow while direct-detection systems don't have such a strin-

Figure 4.7 Perfect wavefront.

gent requirement. The linewidth need be only narrow enough to pass through the background rejection narrowband filter in front of the detector.

4.3.1 Transmitted Wavefront Error

A laser, if it is built to operate in a single mode (TEM_{00}), will produce a beam that is nearly "perfect." This means that the light emanates from the laser as from a point source, resulting in the spherical wavefront illustrated in Figure 4.7. The further away from the laser the light travels, the flatter the wavefront becomes, which is obvious from an intuitive viewpoint. The beam illustrated in Figure 4.7 is perfect or very nearly perfect. The beam of Figure 4.8 is not perfect but rather has unevenness in the spherical envelope of the wavefront. Note that the rays at any point are perpendicular to the wavefront. This unevenness is quantitatively referred to as wavefront error and is expressed in fractions of a wave (e.g., one-tenth of a wave ($\lambda/10$ wave), referenced to the color of light of the laser). This one-tenth wave error means that there is an error in the wavefront corresponding to a distance equal to one-tenth of a wavelength. For 830-nm laser energy, this corresponds to an error of 83 nm (83×10^{-9} m), a small dimension indeed. The wavefront error is usually expressed as *root mean square* (rms) value, which is an rms value for the entire wavefront and accounts for all its errors. A beam quality such as just

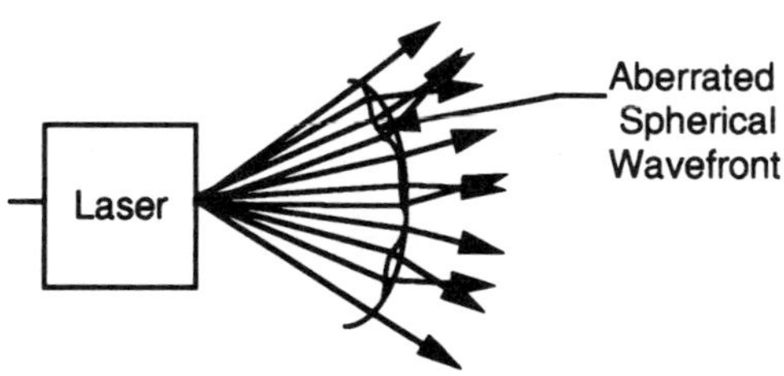

Figure 4.8 Wavefront with errors.

discussed (one-tenth wave rms) or better is referred to as diffraction-limited. Actually it is not truly diffraction-limited but is referred to this way. System wavefront errors are usually required to be no worse than one-tenth wave rms so that the transmitted beam emanating from the telescope will be as intense as possible at the receiver many thousands of kilometers away. If the wavefront error were great (that is, the wavefront quality were poor), the received power would be insufficient and it would be necessary to increase the transmit power to make up for the power lost in the beam as it travels to the other terminal.

4.3.2 Transmitter Gaussian Amplitude Distribution

The amplitude (power) profile across the aperture of a laser transmitter is frequently Gaussian, which means it follows a Gaussian curve. The beam emanating from the aperture is much brighter at the center and falls off rapidly, in accordance with the Gaussian curve illustrated in Figure 4.9. When transmitting a laser beam from an aperture, the beam shape must be configured to optimize the energy in the far field. The beam size is tailored to the aperture such that as the beam falls off in

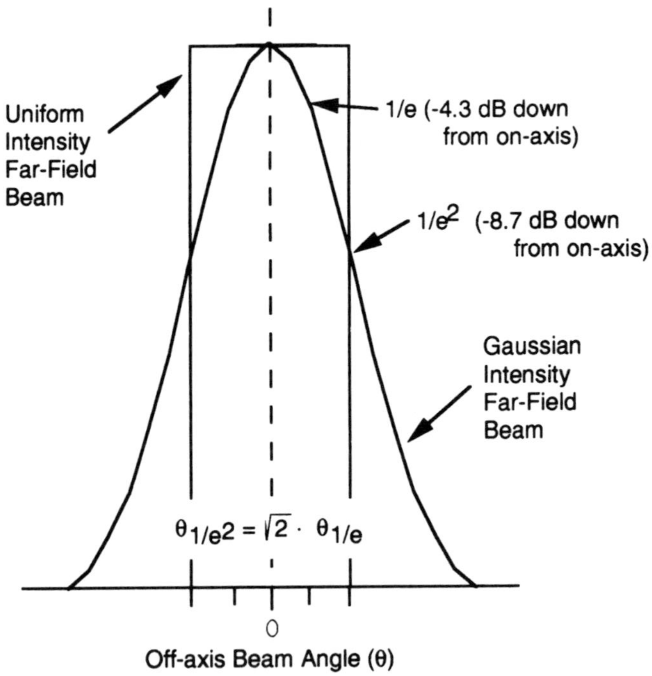

Figure 4.9 Gaussian amplitude curve.

amplitude with distance from the center, no more than a few percent of the beam power is wasted beyond the edge of the primary mirror. Two points of reference here are the points at which the beam amplitude falls off to either 37% or 13% of its amplitude at the center. These points are called the $1/e$ and $1/e^2$ points respectively and will be referred to throughout this book. These points are illustrated in Figure 4.9. To achieve this diffraction-limited output from the transmitter, the laser must operate in the lowest cavity mode, the TEM_{00}, as it is called. This TEM_{00} mode emits a diffraction-limited beam, and so as long as the image quality is preserved and significant wavefront error is not present, the output beam should be diffraction limited. By using the highest quality optics in the transmit path, this is achieved. The received optical system diffraction-limited levels are a function of the type of system being designed. For coherent systems, diffraction-limited performance is required of the receive optics. In a direct-detection system, where receive image quality is not of major concern, the optical quality can be less than diffraction limited and still provide the necessary imaging onto the detector. This is a drawback for coherent systems because additional weight is required to achieve diffraction limited receiver performance.

4.3.3 Strehl Ratio

Another measure of transmitter wavefront quality is the Strehl ratio. This is simply the ratio of the irradiance (W/m^2) of the center of the beam in question to the irradiance of a perfect beam at the center.

$$\text{Strehl ratio} = e^{-(ka)^2} \tag{4.1}$$

where a is the rms wavefront error and k is given by $2\pi/\lambda$, λ the wavelength in meters.

This calculation is good for wavefront error equal to or less than $\lambda/5$ rms and a telescope obscuration ratio of 0.3 or less. The wavelength cancels in the above equation so that it need not be included. The Strehl ratio measure can be represented by a decibel loss by simply converting the Strehl number to decibels. For example, a Strehl ratio of 0.8 corresponds to a decibel loss of $10 \log_{10} 0.8$ or -0.97 dB.

4.3.4 Receive Spot Diameter

As the wave travels through space, the radius of curvature approaches infinity and the wave is flat, or at least nearly flat as affected by the errors in the wavefront. As it hits the primary mirror of the receiver at the opposite terminal, it is flat and as a consequence, the amplitude distribution is not Gaussian but is the same at every point across the aperture. As a result, this plane wave, when it is imaged by the telescope, will produce a spot to the first minimum, with an angular diameter, α, of

$$\alpha = 2.44\frac{\lambda}{D} \qquad \text{radians} \tag{4.2}$$

where D is the receive aperture diameter and λ the wavelength.

We can find the spot size produced by the optical system by multiplying α by the focal length (in meters) of the image forming optics. This works accurately if the convergence angle is less than 10 or 15 deg., which it frequently is.

4.3.5 Transmitter Beam Divergence

The beam divergence of an unobscured aperture transmitted laser beam which has a Gaussian near-field intensity distribution can be approximated by the following equation when the truncation of the beam is much greater than the $1/e^2$ beam diameter.

$$\theta = \frac{4\,\lambda}{\pi\omega} \qquad \text{radians} \tag{4.3}$$

where ω is the transmit beam diameter and λ the wavelength.

For a centrally obscured aperture telescope, the beam divergence decreases slightly with increasing obscuration. Figure 10 in [7] gives beam divergence as a function of obscuration.

4.3.6 Transmitter Gain

The ability of a not exceedingly powerful laser to produce a useful signal at the very large ranges required of space communication links lies at the heart of laser communications. In the previous chapter, a major input to the link margin analysis is the telescope and receiver gain. Gain is expressed in *decibels* (dB). As was pointed out in the link margin explanation of the previous chapter, an advantage of expressing the link parameters in dB is that they can be added since they are log functions. We may remember from earlier math that multiplication and division can be performed by adding or subtracting logarithms . So it is with link margins in laser communications. The transmit and receive telescopes can be characterized by converting their characteristics to a quantitative gain value, which can be used directly in the link equation. The gain of an unobscured (no secondary blockage) telescope was defined in Chapter 3.

As an example comparison of laser transmitters to RF systems, the gain of a 5.5-in aperture telescope, which produces a 9.5-μrad beam, is calculated. From

Chapter 3, the gain is then equal to 2.2×10^{11}. Converting to decibels by taking the log and multiplying by 10, we get a gain of 113.4 dB. This gain is compared with the gain from two RF antennas of 18-in diameter in Figure 4.10. As can be seen, this is a prime example of why low-power laser beams transmitted from small apertures can communicate across large distances even without the extra margin gained by using coherent detection.

4.3.7 Receiver Gain

If the aperture and the secondary obscuration are known, the gain of a receiver can be calculated. An obscuration is anything in the path of the collector or primary mirror that shadows or obscures the received signal in a Cassegrain telescope. The ratio of the diameter of the secondary to the diameter of the primary in a Cassegrain telescope is called the obscuration ratio.

The receiver gain of an unobscured 5.5-in receiver can be calculated again using the equation developed in Chapter 3. The secondary obscuration is zero for a telescope without a secondary and no other obstruction. The result is a gain of 2.7×10^{11} or 114.2 dB. Note that receiver gain differs from the 113.4-dB transmitter gain previously calculated.

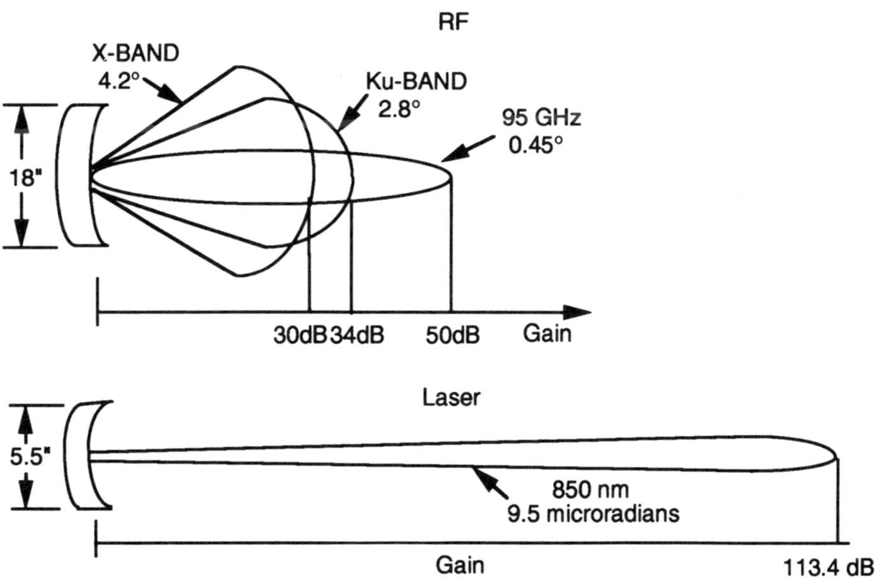

Figure 4.10 Antenna beam divergence comparison.

4.4 TELESCOPES

The telescope in a laser communication system is probably the most important optical component. It performs several major functions. First, it is the "launching platform" for the transmitted laser beam and as such it must properly form and point the beam and preserve its transmitted high image quality. Second, the telescope acts as the receive aperture, collecting the received signal from the companion terminal and relaying it in a manageable size to the imaging optics where it is subsequently relayed to the detector. In many laser communications systems a single telescope performs both the transmit and receive functions, although this is not essential. Smaller systems frequently use separate transmit and receive telescopes. An advantage to laser communications system optical design is the fact that lasers are used so that only a small range of wavelengths or even a single wavelength of light is used, greatly easing the optical designer's burden.

Telescopes used in laser communications systems derive primarily from astronomical telescopes [8–10]. Figure 4.11 illustrates a number of the more common configurations. Their characteristics will be discussed.

4.4.1 Telescope Types

4.4.1.1 Off-Axis Newtonian

Figure 4.11(a) uses a single mirror (paraboloid) with power (curvature). The off-axis characteristic yields a telescope that is unobscured, a desirable trait since the transmitted or collected signal power is not blocked. The deflecting mirror is flat and redirects the signal away from the telescope so that it can be conveniently handled. This type of telescope is not as compact as other designs. Since the off-axis primary mirror is, in essence, a section of the larger paraboloid indicated by the dashed line, the optical characteristics of the full mirror prevail. As a consequence, the focal length of the primary cannot be shortened without degrading the telescope image optical quality since very short mirror focal lengths are exceedingly hard to manufacture. The angular extent of the quality image is also limited to a fraction of a degree since only a single mirror forms the image. This type of telescope is also difficult to align since there is no symmetry.

4.4.1.2 Cassegrain

The Cassegrain type shown in Figure 4.11(b) was invented by a French physician of the same name many years ago. The classic configuration uses paraboloidal concave primary and hyperboloidal convex secondary mirrors. This configuration is very compact due to the folding employed. The length can be as short as one-third the focal length. This telescope type has been used in a number of laser communications programs including the first-generation crosslink, which used a 7.5-in aperture Cassegrain telescope with a flat window in front of the secondary to block

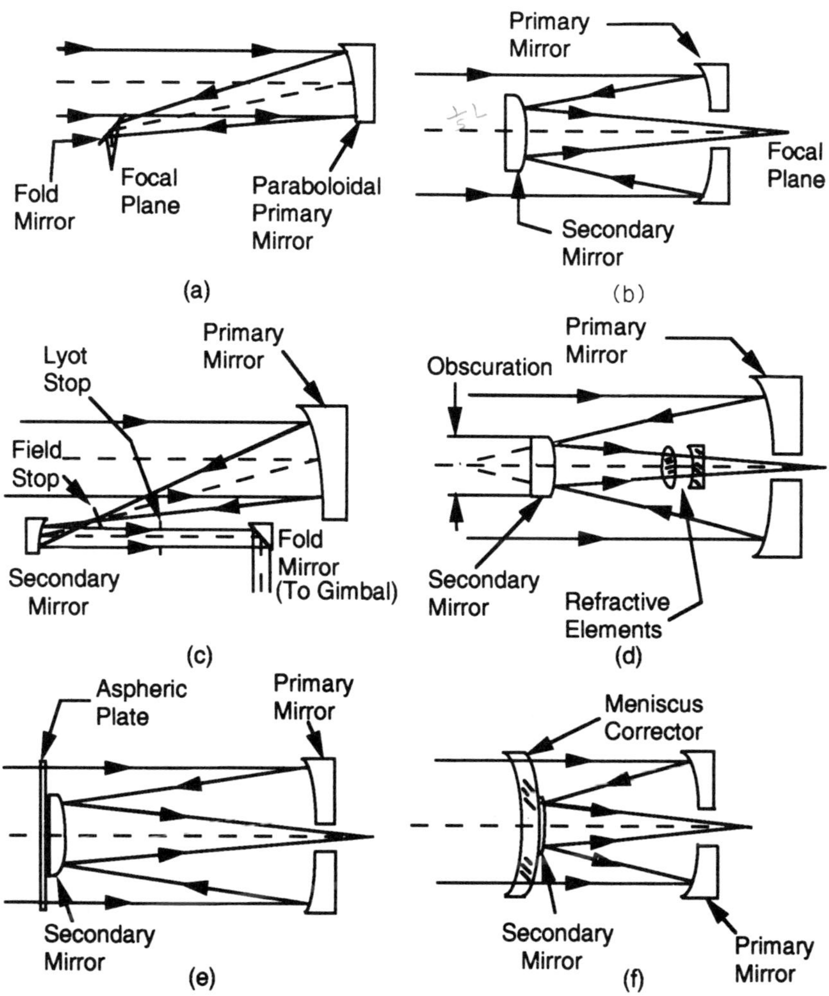

Figure 4.11 Telescope designs.

sunlight. The centration and spacing of the secondary relative to the primary mirror is very critical. Tolerances on the order of a fraction of a thousandths of an inch are common. Since two mirrors are used, the field of the Cassegrain is larger than that of a single-mirror system and a quality field approaching a half degree can be obtained. A wavefront quality of $\lambda/16$ rms or better has been specified and produced in the past. The secondary diameter is frequently on the order of 0.2 to 0.25 times the primary diameter.

4.4.1.3 Off-Axis Gregorian

Figure 4.11(c) illustrates this unobscured telescope type. Note that the rays from the primary mirror come to a focus before they diverge and reflect from the concave secondary. This focal position is a good place to put a field stop; in other words, a small, circular disk with a hole in the middle that defines the field. In this manner, off-axis stray light can readily be blocked, a task not easy with the conventional Cassegrain. The addition of a Lyot stop also blocks extraneous light. A Lyot stop is a stop that has the effect of rejecting light which may be reflected from portions of the tube, baffles, or edges of the primary mirror. A variant of this telescope, which is used in the second-generation crosslink, is the off-axis Gregorian. It combines the off-axis characteristic of being unobscured with the Gregorian stray light rejection capability to produce a good crosslink telescope. It is not as short for its aperture as is the Cassegrain; nevertheless, for a 5.5 in aperture, its length is acceptable.

4.4.1.4 Cassegrain With Refractive Elements

The field of view (FOV) can be improved to a degree or more (even up to a couple of degrees) by the addition of refractive elements (lenses) to a Cassegrain as shown in Figure 4.11(d). This type of telescope is called a catadioptric, which simply means it combines reflective (catoptric) elements with refractive (dioptric) elements. Historically, astronomers have always looked for larger high-quality fields so that larger star fields could be photographed. As a consequence, optical designers came up with the addition of lenses to existing astronomical telescopes to improve field extent. Eventually, new telescopes were designed with lenses that possessed large fields. The use of two (doublet) or three (triplet) refractive elements is common.

4.4.1.5 Schmidt-Cassegrain

A catadioptric telescope of unusual design is the combination of a Schmidt with a Cassegrain design. The Schmidt design uses an aspheric plate of the shape shown in Figure 4.11(e) to correct spherical aberration. When combined with a Cassegrain, a large flat-field telescope design results. A field of greater than a degree is attainable with diffraction-limited performance. This telescope type is available to amateur astronomers from a number of companies.

4.4.1.6 Maksutov-Cassegrain

Another type of comparatively wide-field catadioptric telescope is illustrated in Figure 4.11(f). The Maksutov, also called the Bouwers-Maksutov for the two gentlemen who are both credited for separately inventing it, possesses very wide field capability which, in its pure form, uses a spherical primary. The field size of the Maksutov is in tens of degrees but the penalty is a spherical focal plane. Variants provide a flat field at a somewhat less ambitious field size and in combination with a Cassegrain two-mirror configuration, a field similar to the Schmidt-Cassegrain of a degree or more results. For larger apertures, the meniscus corrector is much heavier for the same size telescope than a corresponding Schmidt plate. An interesting property of this telescope is that with the proper parameters assigned during the design process, the secondary can be placed on the spherical corrector plate by the simple expedient of reflectively coating the center of the plate as illustrated in Figure 4.11(f). In other words, the radius of the secondary is the same as the back side of the meniscus corrector. This telescope type is also available to amateurs from one company in a 3.5-in and a 7-in aperture model. These telescopes are of extremely high quality ($\lambda/40$ rms) and two have been used in laboratory test setups by one of the authors. Table 4.1 lists some of the characteristics of the above six types of telescopes used for laser communications.

4.4.2 Telescope Optical Characterization

Typical may not be a good word when used to describe a telescope used in a laser communication system; however, it does seem as though certain characteristics are typical. Cassegrain telescopes frequently produce a final focus that is just back of the primary, obviously a good location. An image inside the telescope would be awkward to deal with. The Cassegrain f/number, or f/#, is the ratio of the focal length to the diameter of the entrance aperture focus and ranges typically from f/8 to f/15. The primary f/number may range from f/4 down to f/1.5. These "f" numbers are the same as those encountered with photographic camera f/#. The faster (lower number), primaries are more difficult to produce and don't possess as good off-axis characteristics as the slower ones, but do result in a shorter telescope. The Cassegrain telescopes the writers are familiar with from application in systems are of nominal 7.5-in aperture with a primary f/# of f/1.7 and a nominal

Table 4.1
Telescope Characteristics

Type	*Figure Number*	*Central Obscuration*	*Field of View*	*Relative Length*	*Internal Field Stop*	*Relative Weight*	*Tube*
Off-Axis Newtonian	4.11(a)	No	Up to 1/4°	Long	No	Light	Open
Cassegrain	4.11(b)	Yes	About 12°	Short	No	Medium	Open
Off-Axis Gregorian	4.11(c)	No	About 1/2°	Long	Yes	Medium	Open
Cassegrain with Lenses	4.11(d)	Yes	> 1°	Short	No	Medium	Open
Schmidt-Casegrain	4.11(e)	Yes	> 1°	Short to medium	No	Medium to heavy	Closed
Maksutov Cassegrain	4.11(f)	Yes	> 1°	Medium	No	Heavy	Closed

Cassegrain f/# of about f/10. The telescopes for the long-range crosslinks to date have been gimbaled and the image passes through the gimbals, which as indicated previously is a Coudé path. The Coudé path was illustrated in Figure 4.4. Note that in this version the image is picked off for routing through the gimbals by a flat tertiary mirror. There is also a field stop in the path at the inner axis shown in Figure 4.4. The image is relayed down through the outer axis where it enters the imaging optics. There are fold mirrors in the path that must be of the highest quality and remain flat to very high tolerances when exposed to the space environment. The glasses used in the telescope must be radiation resistant (i.e., must not turn brown and become somewhat opaque to the laser signal when exposed to the energetic radiation of orbit).

Figure 4.12 illustrates an afocal telescope design of a type that could be used in a laser communication system. It is afocal since there is no focus, the exiting bundle of rays from a single point in the far field are all parallel (i.e., collimated). The diameter of the exiting bundle shown is d while the diameter of the entrance aperture is D. The ratio of D/d is the afocal magnification, or M. Interestingly enough, knowing M gives us insight into how the telescope will behave. If we steer a transmit beam of diameter d by an angle θ, then the beam exiting from the telescope will be steered by an angle that is θ/M. This relationship is used to rapidly steer the transmit beam through extremely fine angles by steering the beam where it is narrow in the imaging optics. Conversely, an error in the narrow portion of the beam is demagnified by M so that it has far less effect on the far-field beam pointing.

4.4.3 Off-Axis Stray Light Rejection

Crosslinks used by the military must be jam resistant. This ability to resist jamming is frequently referred to as *anti-jam* (AJ). Jamming is interfering with the operation

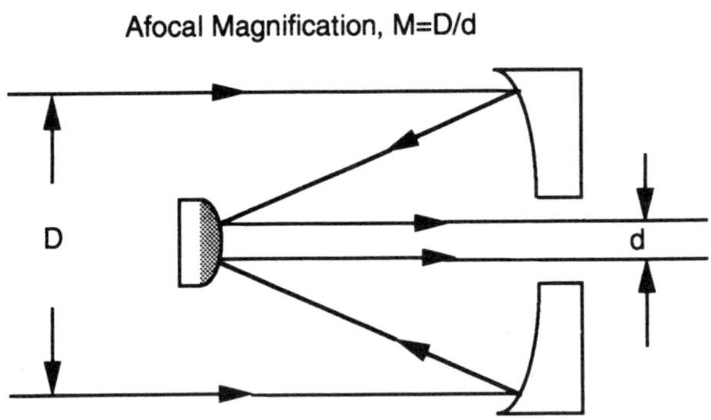

Figure 4.12 Afocal telescope.

of a link by transmitting energy of the same or similar properties at the receiver to be jammed. If the signal is strong enough and of the right wavelength and modulation, the received signal will be interfered with and the communication interrupted. The jamming can take place by interfering either with the acquisition, tracking, or the communication processes. Most frequently the jamming would be directed at the communication channel. The quantitative ability of a laser communication receiver to resist the transmission of a jamming signal to the detector is termed off-axis rejection. The most useful value for this property is the *point source rejection* (PSR). This is the ratio of the energy reaching the detector of the receiver to the energy entering the telescope aperture at the angle of interest. This value should ideally be very small, and with a properly designed telescope it is. Stray light rejection is difficult to calculate and is best left to a number of computer programs on the market. These range from simple *personal computer* (PC) programs to very elaborate programs that are expensive and difficult to learn how to operate. Breault Research Organization of Tucson, Arizona with their APART and ASAP programs and Lambda Research of Boston, Massachusetts with their GUERAP and the simpler SOAR program offer software designed to analyze stray light rejection.

Laser communication links are generally more jam resistant than *radio frequency* (RF) links due to the much shorter wavelengths but certain design measures can be applied to RF systems to increase robustness. The incorporation of AJ characteristics in RF systems increases their weight, complexity, and cost.

Some of the factors which improve the PSR of a laser communications system are:

- Incorporation of a baffle, which shields the primary at larger off-axis angles.
- Use of internal baffle surfaces of very low reflectance; on the order of a few percent at the highest. Paints and nonpaint finishes are available with very low reflectances.
- The primary mirror is very smoothly polished. The polish is expressed in angstroms-rms. A "super polish" is desired.
- A reimaging group of lenses is used following the telescope, which permits the incorporation of a field stop and a Lyot stop (see Figure 4.13).
- The telescope design is important. Some specialized designs are optimized for off-axis rejection. The off-axis Gregorian is good in this respect since there is no secondary and no spider, and there is a field stop at the focus of the primary which has the effect of eliminating extraneous off-axis light.
- The FOV of the system is small. For a given telescope design, the smaller the field, the smaller the amount of stray light that reaches the detector. The rejection capability is proportional to the angular field area. For example, the rejection of a 100-μrad FOV would be approximately 100 times better (20 dB) than for a 1-mrad FOV. This is a strong function and an adjustable field stop will be discussed later when tracking across the solar disk is covered.

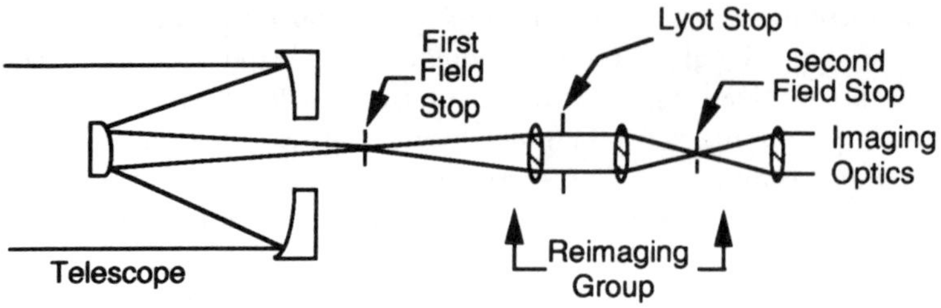

Figure 4.13 Reimaging group for stray light rejection.

Table 4.2 indicates the results of analysis for the Laser Crosslink Subsystem with and without a tertiary fold mirror and for a second-generation off-axis design (unobscured): Note at only 0.25-deg off-axis (barely outside of the FOV) the rejection is greater than six orders of magnitude (10^{-6}). This is a very desirable trait of laser communication systems and not attainable with RF antennas. The effect of polish of the telescope mirror can be dramatic in the rejection of off-axis energy. Table 4.3 covers angles up to 20 deg, was calculated using SOAR for a "well baffled" Cassegrain telescope front end with reimaging as discussed earlier, and illustrates the effect of polish. The 120Å surface roughness corresponds to a diamond-turned (not polished) aluminum primary mirror, a low-cost option. The 30Å value corresponds to a vacuum-deposited beryllium mirror. The 10Å value approximates a metallic mirror (aluminum or beryllium) with a polished electroless nickel coating while the 3Å value is representative of a "super polish" on ultra low expansion (ULE), fused silica or Zerodur.

Table 4.2
Predicted Rejection Performance
1 mrad FOV, values $\log_{10}$, angle in degrees

Angle	*LCS*	*LCS–No Tertiary*	*Off-Axis*
0.0	0.0	0.0	0.0
0.25	−6.66	−6.76	−7.80
2.5	−8.19	−8.53	−9.37
10	−8.20	−9.88	−10.5
15	−8.56	−10.3	−10.9
25	−9.66	−11.1	−11.3
35	−11.6	−11.8	−11.8
42	−11.9	−12.1	−12.1
60	−12.5	−12.8	−12.8
75	−13.3	−13.6	−13.6
85	−13.8	−14.1	−14.1
90	−14.1	−14.4	−14.4

Table 4.3
Point Source Rejection
6-in Telescope, PSR Values $\log_{10}$,
100 μrad FOV, Angle in Degrees

	Primary Mirror Roughness-RMS Angstroms			
Angle	120	30	10	3
1	−7.4	−8.1	−9.2	−10.2
5	−8.1	−10.8	−11.0	−12.9
10	−9.2	10.1	−11.1	−12.1
15	−10.6	−11.3	−12.3	−13.2
20	−11.0	−12.5	−13.8	−14.5

Table 4.3 represents values for clean surfaces. When dust collects on the primary mirror during integration or the mirror collects contaminants on-orbit, the rejection capability degrades (i.e., the values of the PSR increase).

4.4.4 Diffraction-Limited Telescopes

The term "diffraction limited" is generally associated with the Rayleigh quarter-wave criterion, which means that any telescope (or any optical element also) that exhibits a wavefront error equal to or less than 0.075 ($\lambda/13.3$) waves rms can be considered to be diffraction limited. It is referenced to the wavelength of interest.

To achieve diffraction-limited performance in a telescope after it has been installed in a satellite, launched, and on-orbit for years while under the influence of temperature swings, thermal gradients, and material instabilities is no easy task. The design must be exact, the materials chosen must be suitable for the task, the manufacture and alignment must be perfect, and finally, the environment in which the telescope must operate needs to be well understood and controlled to appropriate limits. Remember, the telescope forms the transmit beam and any errors in the figure will degrade the beam quality, thus weakening the received signal at the companion terminal. For some telescopes, a window across the aperture may be desirable for rejection of sunlight by use of an appropriate solar-rejection coating. The direct impingement of the sun on the primary can cause severe distortion with resulting transmit signal loss. Resistance to a laser weapon threat may also be increased. Analysis is required to determine whether these are problems or not.

4.4.5 Telescope Materials

Materials suitable for constructing a telescope are listed in Table 4.4 along with a number of their properties. They fall into several categories: (1) glasses, (2) ceramic type materials (3) metals, and (4) composites which may include metals. Each has

Table 4.4
Telescope Material Properties

Materials	*Density* *gm/cm²*	*Expansion Coefficient* $10^{-6}/°K$	*Thermal Conduction* *W/m-°K*	*Elastic Modulus* *E (GP)a*	*Micro yield Stress* *MYS(MPa)*	*Specific Heat* *Cp (J/Kg°K)*
Beryllium*	1.85	11.4	193	287	34	1820
6061-T6* Aluminum	2.70	23.5	175	69	124	960
ULE	2.20	0.03	1.13	67.7	n/a	766
Glass (BK-7)	2.53	7.1	1.12	81	n/a	880
Fused silica	2.02	0.56	.37	73	n/a	741
Zerodur	2.53	0.05	1.46	91	n/a	821
Invar 36*	8.03	0.54	13.8	145	7	502
Titanium*	4.43	8.8	7.2	110	482	565
SXA(Al)	2.91	13	125	117	117	795
Graphite Epoxy*	1.78	0.02	35	93	138	921

*Commonly used for structure.

its own advantages and shortcomings. The first-generation crosslink telescope is built with an egg-crated ULE primary mirror and a ULE secondary. The tube is constructed of a very thin sheet of Invar (low coefficient of thermal expansion steel) to control the thermal expansion of the barrel and the primary to secondary spacing. These tolerances are extremely critical in maintaining focus. A second-generation crosslink telescope has been constructed of beryllium including the primary and secondary mirrors (see Figure 4.14).

The beryllium approach is the lightest to date, although beryllium is expensive. There are concerns about beryllium's long-term diffraction-limited stability with some researchers. Silicon carbide is a newcomer and may see service in the future as the technology matures. Classically, Zerodur, fused silica, and ULE have been used for space astronomy telescopes where diffraction-limited performance was essential. For longer wavelengths in the infrared, beryllium has been a favorite since figure requirements relax in proportion to the wavelength. Invar steel, with its low coefficient of thermal expansion, has been used as a primary to secondary spacer as has Zerodur. Fused silica is intermediate in thermal expansion between glass and ULE or Zerodur. The 2.4m Hubble Telescope, is producing outstanding imagery on-orbit following installation of correction optics by National Aeronautics and Space Administration astronauts. The Hubble's primary to secondary spacers are made of graphite epoxy composite with a near-zero coefficient of thermal expansion.

In the past few years a material of interest for use in mirrors, structure, and gimbals has appeared, sponsored by the University of Arizona's Optical Sciences

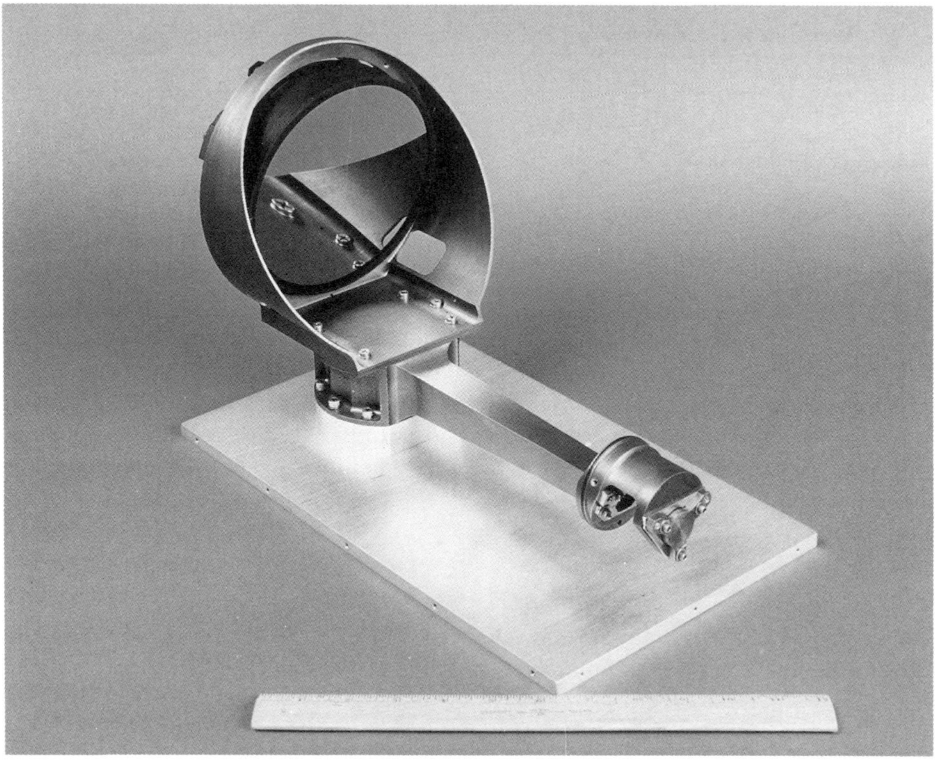

Figure 4.14 Second-generation beryllium telescope. (Courtesy of McDonnell Douglas Aerospace.)

Center [11]. The material, SXA, is a metal matrix composite (aluminum alloy with 30% silicon carbide) that exhibits interesting characteristics, several of which are included in Table 4.2. The coefficient of thermal expansion can be tailored and may be made equal to that of electroless nickel. Aluminum and beryllium mirrors have historically been coated with electroless nickel since these metals do not take the optical polish desired. Techniques, however, are now available for polishing beryllium mirrors. When aluminum is coated with electroless nickel to achieve a good polish, large stresses are set up due to the differences in coefficient of thermal expansion between aluminum and electroless nickel. The result is some loss in figure. The situation is made worse by on-orbit changes due to temperature swings. The tailoring of the SXA mirror material coefficient of thermal expansion to that of electroless nickel results in a good match. As a result, the mirror retains its figure over a greater range of temperatures and creep is minimized. In addition, SXA is stable with time. The use of SXA for structure is also feasible. When the telescope

Table 4.5
Representative Crosslink Telescope Weights Diffraction-limited, Near Infrared, Cassegrain

Telescope Materials	*Aperture (ins)*	*Weight (lbs)*	*Comments*
All beryllium	7.5	5	Engineering model
Cervit mirrors, machined Invar barrel	7.5	17	Flown in aircraft
ULE primary/ secondary, fused silica window, sheet Invar barrel	7.5	10	First generation, space qualified, eight produced
All beryllium off-axis Gregorian	5.5	4	Second generation, one made, tested

structure is made of the same material as the mirrors, this creates an athermal telescope design. Changes in temperature do not change the focal plane position, only the image scale, provided the telescope is mounted near the focal plane and there are minimal temperature gradients.

4.4.6 Telescope Weight

Table 4.5 shows weights for several telescopes the authors have worked with. The sizes are useful for crosslink applications.

4.4.7 Photon Buckets

In certain crosslink applications, high-data-rate communication flows in only one direction (i.e., from the data source to the user). A much lower command and control data rate flows in the opposite direction. As a consequence, an interesting optical approach results. Since high data rates demand large apertures, in general, the *geosynchronous Earth orbit* (GEO) receiver for a low Earth orbit to GEO high-data-rate link should be of large aperture. Large diffraction-limited telescopes are extremely heavy. The question arises, why does the large aperture receiver have to be diffraction limited [12]? The diffraction-limited requirement in crosslinks arises from the need to produce a very narrow transmit beam to maximize the signal at the receiver. Unless the system uses coherent detection, the receiver can be many times the diffraction limit. This "photon bucket" approach yields some interesting results. First, the cost of the receiver telescope is less. Second, the quality of the optics can be considerably reduced. Third, the weight of the telescope and gimbals) can be dramatically reduced, a point of real interest to the satellite builder. Fourth, the risk of producing a large, high-quality telescope is greatly reduced. The quality of the telescope need only be good enough to produce a spot that is small enough to fit a high-speed communication *avalanche photodiode* (APD) detector.

The spot is readily tracked on a larger quadrant APD. One example [4] of a photon bucket is a high-data-rate receiver (multigigabit per second) that is a 24-in aperture, two-axis gimbaled telescope that weighs a mere 95 pounds, about the weight of a 12-in aperture diffraction-limited gimbaled telescope. The telescope and gimbals can be fabricated of either beryllium or the newer material SXA.

4.5 IMAGING OPTICS

The *imaging optics* (IO) are central to the operation of the terminal and they incorporate a myriad of functions. Most of the elements are refractive; that is, light passes through them and is shaped or redirected by them as opposed to reflective elements where light is redirected by reflection. The exceptions to this refractive rule are beamsteerers, which are typically steerable mirrors.

The authors are familiar with at least 17 functions which the imaging optics perform in a crosslink terminal design. Table 4.6 outlines these functions. While not all system designs use all functions (i.e., some short-range system designs wouldn't), for most long-range crosslinks all 17 functions are necessary. Each of these functional responsibilities will be discussed in detail below.

Table 4.6
Imaging Optics (IO) Functions

Transmit Functions	
(1)	Route transmitter signal to telescope
(2)	Perform point-ahead
(3)	Perform beam shaping
(4)	Isolate transmit from receive path
(5)	Monitor transmit power
(6)	Interface to the transmitter
(7)	Provide focus capability
Receive Functions	
(8)	Route received signal to detector
(9)	Filter out background for detector
(10)	Split wide and narrow field detector paths
(11)	Interface to the receiver
(12)	Adjust field stop to track across the sun
Common Functions	
(13)	High-speed fine steer of transmit and receive paths
(14)	Align boresight between transmit and receive
(15)	Provide redundancy and perform switching (mechanisms)
(16)	Provide common structure for mounting
(17)	Provide thermal control of optics

4.5.1 Signal Routing

The IO provides the path for and shaping of the beam from the transmitter to the telescope. The lenses and beamsteerers in the path are generally designed to be of as small diameter as feasible to reduce size and weight. The afocal magnification necessary to expand the beam to transmit diameter is performed by the telescope. With care, this path can be designed to provide a high transmission to conserve laser power. The path is shown in Figure 4.15. Also included is an example of the afocal magnification accomplished when implemented from the transmitter to the narrow transmit beam.

4.5.2 Point-Ahead

The point-ahead function is necessary due to the finite speed of light. Although light travels in a vacuum at 186,000 statute miles per second (3×10^8 m/sec), the round trip from one terminal to another is still finite in time so that the receive terminal has moved and is no longer at the angle at which it was observed. This is analogous to a hunter "leading" a duck in his or her gunsight so that the time of flight of the pellets will coincide with the forward distance traveled by the duck over the same time period. Figure 4.16 illustrates this effect and includes the formula for calculating the point-ahead angular value. The point-ahead angle may be as much as 100 μrad for some orbital cases, but is usually less. The means for implementing this angle in the transmit beam is to use a beamsteerer in the transmitter path before it merges with the receive path to form the common path. Adjustment of this beamsteerer produces the required point-ahead angle. This beamsteerer may also be used to coalign the transmit and receive boresights. Implementation requires two axes and in certain orbital arrangements changes

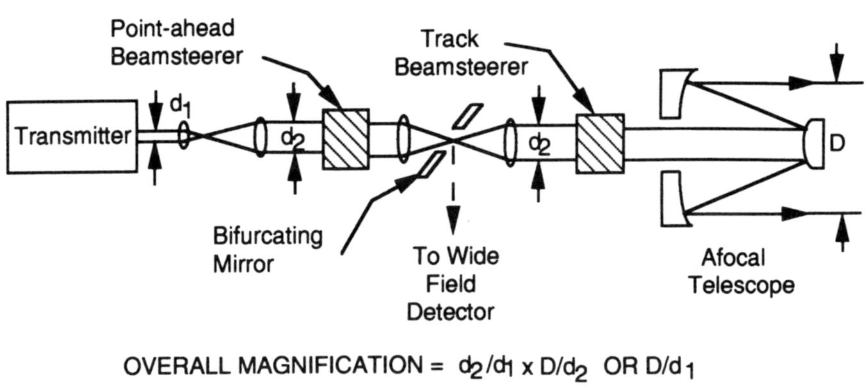

Figure 4.15 Transmitter to telescope routing.

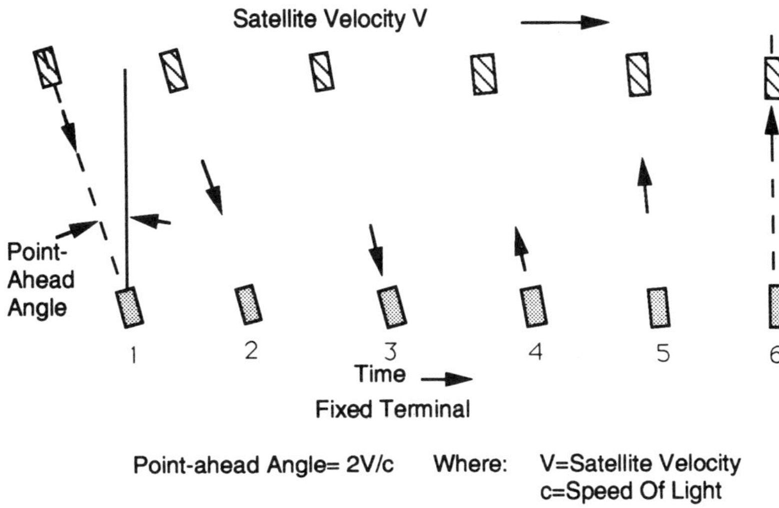

Figure 4.16 Point-ahead function.

continuously. The first-generation LCS, mounted on a spinning satellite, required continuous high-speed computation and implementation of the point-ahead angle. Figure 4.16 also illustrates the arrangement of the point-ahead in the IO.

4.5.3 Beam Shaping

The beam that emanates from a diode laser may be elliptical in cross section. This is due to the fact that single-mode diode dimensions are not equal along and across the diode stripe. This ellipticity varies with diodes but may be three to one (3:1) or so. Since the transmit beam from the telescope generally is circular in cross section, it is necessary to circularize the laser output. This is accomplished either by anamorphic prisms or cylindrical lenses [12]. Figure 4.17 illustrates both approaches. The designer generally makes a selection based on the needs of the system.

4.5.4 Isolation

Given a detector that is sensitive to nanowatts and a laser that produces watts, the backscatter of the laser signal could easily be enough to saturate the detector. To protect the detector from the transmitter, an isolation of watts/sensitivity is required. A quantitative example of this is a laser that produces 1W of peak optical power and a detector that is sensitive to 1 nW. An isolation of 1×10^{-9} is required

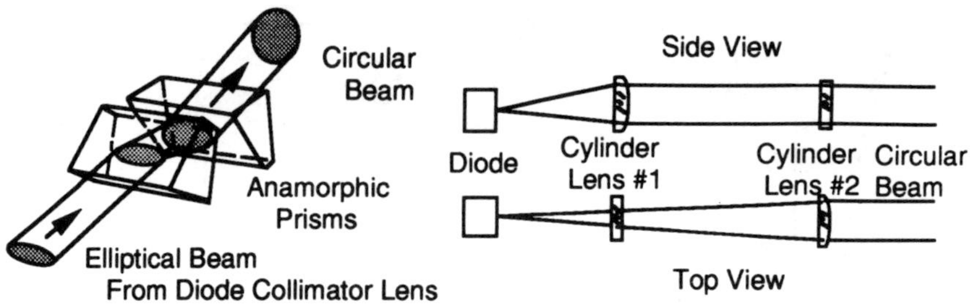

Figure 4.17 Beam circularization approaches.

to achieve equal signal levels. In practice, an additional 10 to 20 dB of isolation would be required to guarantee no interference from the transmitter into the receiver. Isolation is accomplished in a variety of way. Three ways the imaging optics provide isolation are given below.

Spatial isolation—In space, separate transmit and receive apertures provide a high degree of isolation. This is satisfactory for short-range terminals where a large transmit aperture is not required (wide beam divergence).

Polarization isolation—This technique is used where the full telescope aperture is required to form a narrow transmit beam (see Figure 4.18 for a illustration of how the technique works). The LCS uses this isolation approach. The system uses only one wavelength of light (1.064 μm) (i.e., only one laser wavelength is needed).

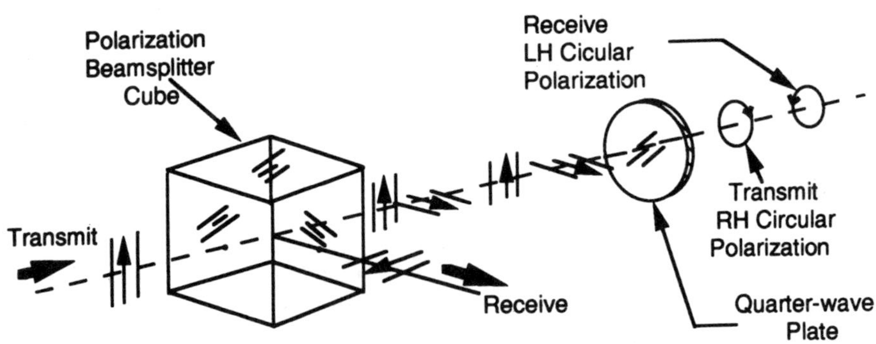

Figure 4.18 Polarization isolation.

Wavelength isolation—This is the easiest approach in terms of meeting stringent isolation requirements. It is used where the full aperture may or may not be needed for a narrow transmit beam. It is possible to achieve 110 to 120 dB of isolation with high-performance optical components. The second-generation crosslink uses this approach. Two separate laser wavelengths are required (see Figure 4.19).

4.5.5 Power Monitoring

The status of the laser can be determined on-orbit by monitoring the laser power. A small portion of the laser power is "picked off" using a beamsplitter. Typically, 1% might be used.

4.5.6 Transmitter Interfacing

The IO interfaces with the transmitter. The IO may contain beamshaping optics such as circularization optics. A first stage of afocal magnification may also be performed at this point. A mechanical mount for the laser may be provided. Where the laser is mounted on separate structure, as in the first-generation LCS, an interface steering optics function is required. Flexure of the laser mount can cause the laser beam to translate or misalign angularly. The interface steering assembly corrects the laser location and angle by picking off laser power with beamsplitters, which direct a small portion of the energy to detectors (quadrant types). By sensing the errors, the steering corrections are made by galvanometer beamsteerers. Where the laser is mounted on the IO baseplate, the steering correction is not required if attention is paid to the laser beam behavior and the environment controlled.

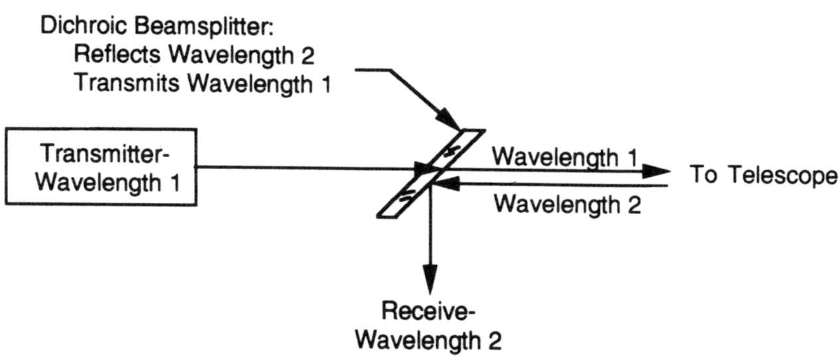

Figure 4.19 Dichroic isolation.

4.5.7 Active Focus

Crosslink designers are split on this one. There is a school of thought that says that a focus capability is not necessary, that if the optics and the environment are understood well enough and the design is proper, no focus change is needed for the life of the terminal. There is also the other school, which maintains that there will always be some unforeseen change that will require that the terminal transmit beam be refocused. This is the designer's choice. In the first-generation LCS, a focus capability was originally included in the design but later dropped. Subsequently, the vacuum test environment resulted in a defocus due to slight bending of some fold mirrors due to the coatings drying out. At this point, the focus ability would have been useful. The mirror coating problem was solved but the thought still lurks in the designers' minds that something could happen due to launch vibration and so forth. An example of this type of problem would be a shift in the primary to secondary spacing of a Cassegrain telescope with it's few micrometer tolerance. The focus is implemented in the laser beam near the laser where the axial movement of a lens a few micrometers at a time is sufficient to refocus the transmit beam. The sharpness of focus can be determined from telemetry of the signal level at the companion terminal as the focus is slowly adjusted. In an early engineering model terminal, a beamsteerer was used with a mechanism that converted rotary motion to axial motion. The position pickoff of the beamsteerer was used to quantify the motion. The focus adjustment need not include the receive path, unless it is a coherent system, since the margin for defocus here is much greater than it is for the transmit path.

4.5.8 Signal Routing

The IO provides the path for the received signal from the telescope to the detector. The path, illustrated in Figure 4.20, includes fold mirrors for redirection and lenses for providing the proper scaling for the detector image. In addition, the path provides a properly collimated (afocal) space for the narrowband detector in front of the detector.

4.5.9 Filtering Out Stray Light

The background for a receiver can vary from a star field to the sun. It is necessary to dramatically reduce the background for a direct-detection laser communications receiver if the faint signal is expected to be detected. For example, reflected sunlight from Earth covers a broad band, running from ultraviolet through the near infrared. The transmitter signal will be extremely narrow, on the order of a fraction of an angstrom to a few hundred angstroms. To provide background rejection, a narrowband filter is used in front of the detector. The filter is sensitive to angle of

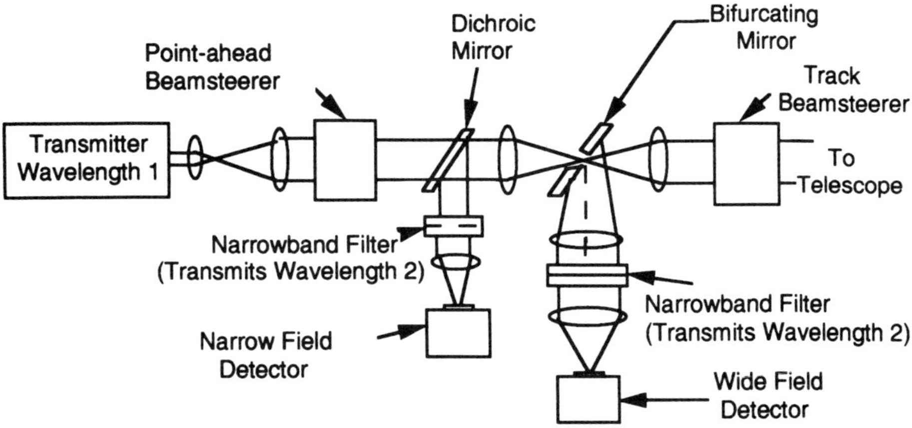

Figure 4.20 Telescope to detector routing.

incidence so it must be situated in an afocal or collimated space, a condition created by lenses. Figure 4.20 shows the position of the components. The filter transmits in the "bandpass" but rejects "out of band." The rejection or blocking of the out-of-band energy must be great enough to avoid still passing enough energy to saturate the detector with "nonsignal." Transmission of some filters at several wavelengths of interest is shown as Figure 4.21. Filters are of several types, the most common type being the Fabry-Perot dielectric. This type of narrowband filter can provide outstanding dichroic combining/separation of wavelengths [14], a property used in the design of multiple diode (dichroic combiner) lasers discussed in Chapter 8.

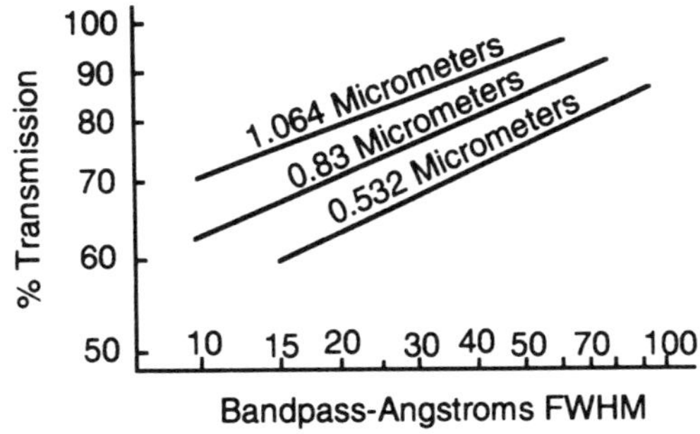

Figure 4.21 Narrowband filter transmission.

4.5.10 Receive Viewfield Splitting

Separate wide and narrow field detectors are frequently required. A narrow field is needed to minimize background while a wide field is required to facilitate acquisition in a reasonable length of time. Several approaches may be used but the most common and conserving of laser power is the use of a bifurcating mirror. This mirror has a hole in its center that is sized to the narrow field detector. During acquisition, the companion terminal will most likely not be on the boresight so the received signal is routed to the wide field detector, as shown in Figure 4.20. As soon as acquisition occurs, the system reduces the pointing error and as the image approaches the boresight of the system, it falls through the hole in the bifurcating mirror onto the narrow field detector, also shown in Figure 4.20. If very high data rates (> 100 Mbps) are not required, the narrow field detector can act both as fine track detector as well as the communications detector by judicious handling of the signal.

4.5.11 Detector Interfacing

As with the transmitter, the IO provides the interface to the detector(s), including any required optics to provide the appropriate focal length(s). In addition, also as with the transmitter and if so constructed, the IO can provide the baseplate upon which the detector(s) are mounted.

4.5.12 Field Stop Adjustment

A requirement for the first-generation LCS was the ability to track and communicate with the Sun in the background [1]. This is a very stringent requirement. Even 60-GHz crosslinks suffer degradation when the Sun is in the background and data rate capability decreases. To solve this problem, a unique approach was used. The LCS already used pulse interval modulation (high peak power Nd:YAG laser) so that the effect of background was reduced since only the background during the pulse interval is added to the signal at the detector. This is not enough, however, so an additional dual approach was adopted. A very narrow band optical filter was used (about 20Å) and an adjustable field stop included. The adjustable field stop is illustrated in Figure 4.22. The field stop is open during acquisition and lower background tracking and communication. This is shown in the left illustration of Figure 4.22. When the LCS line of sight approaches the Sun, the field stop is closed by a drive mechanism. The narrow field/communication detector continues to track the centered spot but the background is dramatically reduced (a couple orders of magnitude). The addition of a mechanism is obviously not desirable but with a requirement to track/communicate across the solar disk, the addition appears justifiable. It is even possible to track and communicate across the solar disk with a laser-diode-based system, but at a much reduced data rate [15].

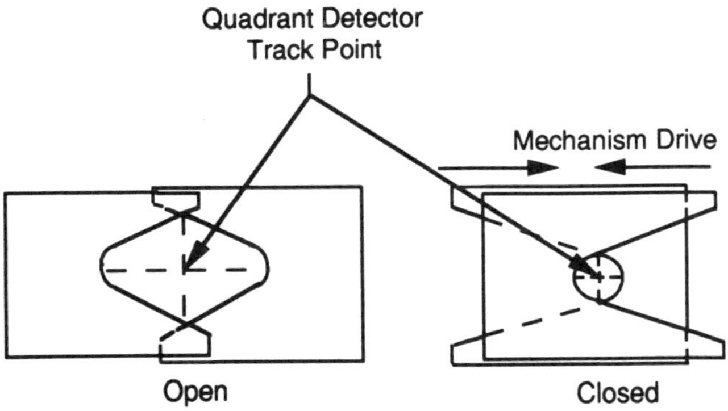

Figure 4.22 Adjustable field stop.

4.5.13 Beamsteering

When very narrow beams on the order of 10 μrad or so are used, any disturbance of the satellite platform can cause mispointing of the transmit beam. If this is severe enough the companion terminal will lose track and the link will be broken. The mispointing of the line of sight is observed by the detector signal, which records any deviation as an error. Unfortunately, the gimbaled telescope is not fast enough to respond to the error. To respond rapidly, a track beamsteerer is incorporated in the common path of the transmitter and the receiver. This mechanism location in the IO is shown in Figure 4.23. The element, driven in two axes, is actually a mirror. An error in the received signal sensed by the detector is processed and acts on the track beamsteerer in a direction that nulls the error. The speed of response of this mechanism is critical and must be greater than the satellite disturbance frequency in order to compensate for satellite-induced pointing errors. Early steerers were

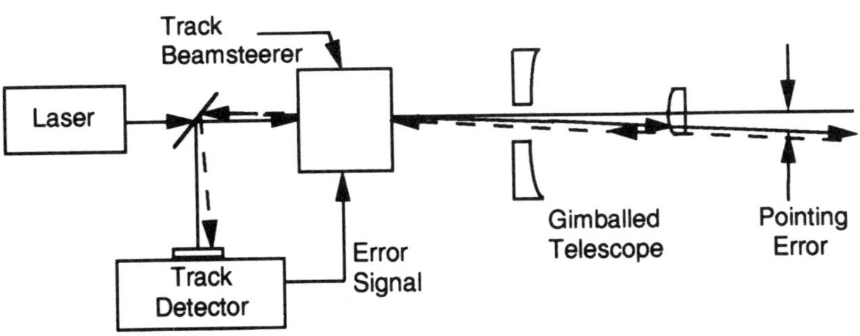

Figure 4.23 High-speed track beamsteerer.

capable of a couple of hundred hertz response, but units today possess bandwidths of 1,000 Hz or more for the sizes required in the IO.

4.5.14 Boresight Alignment

Again, with a 10-μrad beam divergence, it is not only necessary that both the laser and detector be pointed at the companion terminal, it is necessary that the transmit and receive paths be exactly coaligned. A method that has been used for a number of terminals is one which is quite rigorous. If the terminal is allowed to illuminate a corner cube, the received signal will be returned exactly along the same line of sight, a unique property of corner cubes. The telescope need not be very accurately pointed at the corner cube since a corner cube has a fairly wide angle of acceptance. The corner cube error in this return can be reduced to a fraction of the beam divergence with precision optical manufacturing. This setup is illustrated in Figure 4.24. A requirement of this arrangement is that the signal, greatly attenuated on the detector, must still be of sufficient strength to be detected. This requires an accurate analysis of the transmission of this path. There must also be a mounting location for the corner cube on the satellite and gimbal access to the corner cube line of sight. It is obvious that the terminal is not linked to its companion terminal while this boresight alignment is going on. A possible alignment scheme that can be accomplished while tracking and communicating involves dithering the transmit signal in a known manner and tying the signal changes at the companion terminal to the dither directions. In this fashion, the optimum pointing for the transmit beam and its relationship to the received boresight may be established. This technique is known as "closed loop boresight alignment" and was considered for the LCS. There is also a continuous boresight alignment scheme that involves use of a charge coupled device for the track detector [13].

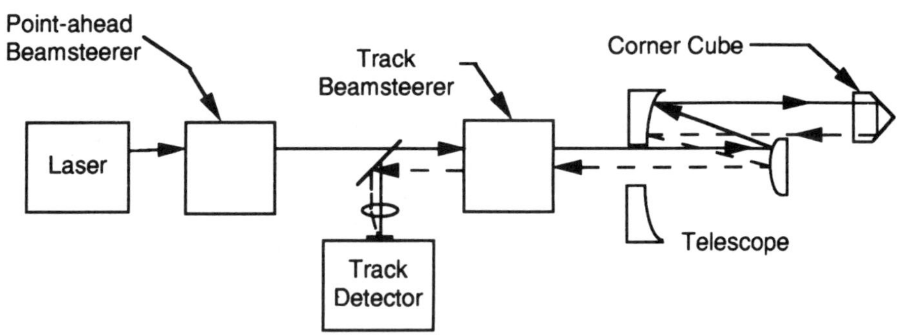

Figure 4.24 Boresight alignment.

4.5.15 Redundancy and Switching

Another major function of the IO is accommodating redundancy and switching. For critical items like point-ahead and track beamsteerers, lasers, detectors, and so forth, redundant units may be required, based on failure rates and the life required of the terminal. There are some techniques that can be used to provide redundancy which don't necessitate use of mechanisms. For example, two lasers can be independently operated if polarization is not a functional requirement in the system. Diode and solid-state lasers are strongly linearly polarized. The individual lasers are combined using a polarization beamsplitter cube in much the fashion that polarization isolation is used and shown in Figure 4.18. Generally, however, mechanisms are used that switch paths between required redundant elements. The mechanism itself, of course, has its own failure rate and this must be taken into account when calculating reliability. Redundant paths for many functions require duplicate optics, adding weight and complexity to the IO. An interesting form of redundancy has recently been pursued with the second-generation terminals. The terminals are small and light and consume very modest power so as a consequence, three terminals form a satellite communications suite with a common electronics where only two terminals are operating at any one time. Although there is some redundancy built into each terminal, the biggest improvement in reliability results from the addition of the third terminal.

4.5.16 Common Mounting Structure

Second-generation systems are being designed to place the transmitter and receiver on a common baseplate with the imaging optics. This is very conserving of weight and space and makes it easier to control the environment. The alignment of this arrangement is also easier to perform initially as well as maintain. The first-generation LCS used a separate assembly approach and as a consequence is much more complex and weighs a great deal more than the second-generation terminal (317 lb versus 80 lb). Of course, the LCS incorporates a lot of redundant assemblies and components. Selection of a baseplate material is subject to the same trades as selecting materials for the gimbaled telescope although the tolerances are not as demanding. It must be remembered that an angular misalignment on the baseplate where the beam diameter is small will be divided by the afocal magnification when referenced to the far field (as transmitted by the telescope). Translations in the IO, however, are magnified at the telescope (i.e., multiplied by the afocal magnification).

4.5.17 Thermal Control

Motions are such with an IO that incorporates the transmitter and receiver that on-orbit environmental control, primarily temperature, is required. It is also required during test to establish that the IO meets its requirements. The systems the authors are familiar with have IO temperature tolerances on the order of $\pm 5°C$ centered about a point near room temperature (20°C). Much of this control is desirably accomplished passively, but some active control will probably be needed. A thermal analysis by a thermodynamicist, which includes orbital considerations, is required to establish what is needed. Thermal control means can include multilayer insulation, coatings, shutters, heat pipes, heat sinks, location on or integration in the satellite structure, and of course, heaters.

4.6 BUDGET ALLOCATIONS / INTERFACES

Budgets based on allocations form an important part in the design process. By breaking down requirements, the various contributors to error can be identified and each contributor assigned its share of the error. Control can therefore be exercised over the design, manufacturing, and assembly process to assure that the final results meet the specifications. This approach is also useful in initially structuring the system/assembly specifications, which may be sent to manufacturing for estimating or out to other companies for bid. Several of the important budgets are discussed.

4.6.1 Alignment Budget

This budget details the alignment requirements of each optomechanical assembly/component used in the system. These may include the IO, the pointing element, the transmitter, the receiver, mounting to the satellite, and even the navigational pointing accuracies that form a part of the pointing and acquisition sequence. Errors are systematic or random (subject to statistical processes). The budget can be applicable to design, manufacturing, assembly, test, and final integration of the terminal into the satellite. The alignment generally includes location and angular tolerances.

4.6.2 Wavefront Quality Budget

This budget details the quality of each element and its alignment into optical assemblies such as the telescope. Physical errors are assigned to each contributor and these in turn are expressed as wavefront errors. The wavefront error is readily converted to link margin loss, an overall system parameter. Overall wavefront error is split and allocated among the various assemblies that comprise the system (e.g., telescope, IO, transmitter). The portion of error allocated to the IO reflects error resulting from a great number of elements in the optical path. As a consequence, lenses, fold mirrors, polarization cubes, dichroics, and all of the other optical elements of the IO must exhibit extremely high quality performance if the system is to demonstrate diffraction-limited performance. Many of the elements of the LCS were required to have wavefront errors no greater than one-hundredth wave rms. This is extremely stringent optical quality and it takes a very good optical fabrication shop to produce these elements. The errors in the telescope are fewer but still number perhaps a dozen error sources including the primary figure, secondary figure, tertiary and fold mirror accuracies, primary to secondary spacing, primary to secondary centration, and primary to secondary tilt. There are more when one considers the effect of temperature gradients on the telescope and its elements. The errors (in decimal portions of a wave, e.g., 0.05 waves) are frequently root sum squared (rss'd) to determine overall error. This is simply the square root of the sum of the individual contributors squared. This may not be accurate in some instances, but provides the designer with a good tool that can help identify major contributors. Again, statistics play a role in this budgeting process. One consideration is that the overall error will always be greater than the largest contributing error. This has an interesting implication. One cannot achieve diffraction-limited performance even with a perfect telescope if there is one element in the transmit path that is not diffraction limited.

4.6.3 Transmission Budget

An important factor in the design of a crosslink is the transmission of the transmit and receive paths. Obviously, the transmission should be as high as possible. To achieve high transmission, we resort to two approaches. First, we use as few elements as possible in the path while accomplishing what is necessary. Second, elements with as high an individual transmission as possible are used. This invariably requires tailored coatings. It is not satisfactory to use simple coatings. As an example, let's say we have 10 lenses in our path. With simple lenses, there are then 20 surfaces. If we use a conventional *magnesium fluoride* (MgF_2) *antireflection* (AR) coating on each surface with a 2% loss at each surface, we would have a an overall transmission of 0.98^{20} or 67%. Contrast this with the same number of lenses with a V-coating, which is designed for a specific wavelength. The V-coating has a maximum reflectance of 0.25% or a transmission of .9975. Our overall transmission in this case is 0.9975^{20} or 95%, a dramatic improvement! As a rule of thumb nearly twenty years ago, an average loss per element was assumed to be 1% and the overall transmission equal to 0.99^{n}, where n was the number of elements. Mirror coatings reflect, to coin a phrase, a similar situation. A standard aluminum mirror coating reflects about 87% at 830 nm. This can drop at large angles of incidence. If we used this coating in our system of about ten mirrors we would have a transmission of 0.87^{10} or 24%. Using high-reflectance multilayer dielectric coatings of 99.3% reflectance, we would have an overall transmission of 0.993^{10} or 93%, again a dramatic improvement. The transmission budget is not trivial, if done properly. Each coating and element in the path must be tabulated with its

Table 4.7
Example Transmission Budget

	Transmit Channel	*Wide Field Receive Channel*	*Narrow Field Receive Channel*
Beginning of life	0.85	0.73	0.78
End of life	0.64	0.55	0.60

characteristics noted for each of the wavelengths of interest. These coatings must be capable of withstanding years on-orbit where there is no humidity, the temperature varies, and energetic radiation is constantly bombarding them.

As part of the transmission budget, consideration must be given to end-of-life transmission. The coatings will degrade somewhat with time due to radiation. For systems the authors are familiar with, a budget has been prepared and allocates a certain amount of loss to each element due to radiation of the coating, the substrate, aging, and so forth. When this is accomplished, the as-manufactured beginning of life system performance must reflect this degradation. In other words, the transmission of a new system prior to launch must be considerably higher than that expected after a number of years on-orbit. Both the LCS and the second-generation terminal specifications reflect this factor. An example of this type of budget follows in Table 4.7.

4.6.4 Interface Control

A form of budgetary control is an interface control document (ICD). In essence, this document elaborates on the conditions that must be met to achieve the desired performance. It is frequently established between adjacent assemblies and is most important in any application. Frequently, different organizations manufacture adjacent assemblies that must play together and the ICD is a necessary instrument for control. When one organization changes its design, the ICD will reflect this change and alert the interfacing organization that a change has been made. The ICD is particularly important in the integration of the laser communication system in the satellite. It ensures that the system will provide a "good ride" and not jeopardize its performance. What is demanded of the satellite in the way of support is specified. On the other hand, the ICD assures the satellite provider that the system will do nothing to jeopardize the vehicle or its mission.

4.6.5 Weight Budget

The weight of the system should be continuously monitored throughout its development cycle. It would seem that weight tends to increase as development proceeds and a conscientious effort to reduce weight must be pursued. Implementing a weight reduction program has proven beneficial in laser communication system development in the past. This was evident in the second-generation system development. Exponential increases in reliable laser diode power facilitated weight reduction by reducing the number of laser diodes in each of the redundant multiple diode combiner lasers of the terminal. This weight reduction was reflected in the three terminals that comprise a laser crosslink communications suite for a geosynchronous satellite, so a sizable weight saving resulted. There are occasions where a weight reduction such as this could enable use of a small launch vehicle with large cost savings or alternatively, an improvement in the satellite performance by carrying additional fuel and so forth.

4.7 THE USE OF OPTICAL DESIGN PROGRAMS

An essential part of the optical design effort on a terminal is optical design program analysis. These programs, of which there are quite a few, vary a great deal in cost although some of the simpler, less expensive programs can go along way in getting answers that validate a conceptual design. Programs are currently available that perform well on PCs. These are available in cost from a hundred dollars or less to several thousands of dollars. High-capability programs such as Code V and ACCOS V are very expensive to lease and operate although they are capable of fully analyzing and optimizing an entire optical system that contains almost any type of element. An interesting comment should be made regarding the optimization of the system. For a number of years, optical programs have been capable of this function, a task most other programs don't attempt to the authors' knowledge. Given a specified optical system and optimization constraints, the program will change the design to achieve the end the designer desires. Even the inexpensive programs perform this optimization.

References

[1] Deadrick, R. B. and Deckelman, W. F. "Laser Crosslink Subsystem-an Overview," *SPIE OE-LASE '92*, Los Angeles, CA, Vol. 1635, Jan. 13–24, 1992.

[2] Dreisewerd, D. W., et al. "A Twice Synchronous Range GEO to GEO Optical Intersatellite Link," *IEEE Military Communications Conf. MILCOM '92*, San Diego, CA, Vol. 1, p. 177, October 1992.

[3] Dreisewerd, D. W., et al. "A GEO to GEO High Data Rate Optical Crosslink Approach," *IEEE Military Communications Conf. MILCOM '92*, San Diego, CA, Vol. 1, p. 183, October 1992.

[4] Dreisewerd, D. W., et al. "A High Data Rate LEO to GEO Optical Crosslink Design," *IEEE Military Communications Conf. MILCOM '92*, San Diego, CA, Vol. 3, p. 1163, October 1992.

[5] Lambert, S. G., et al. "Short-Range Multi-Terminal Satellite Crosslink Communication," *IEEE Military Communications Conf. MILCOM '92*, San Diego, CA, Vol. 3, p. 1170, October 1992.

[6] Swanson, E. A., and Bondurant, R. S. "Using Fiber Optics to Simplify Free-Space Lasercom Systems," *SPIE OE-LASE '90*, Los Angeles, CA, Vol. 1218, Jan. 15–17, 1990.
[7] Klein, B. J., and Degnan, J. J. "Optical Antenna Gain 1: Transmitting Antennas," *Applied Optics*, Vol. 13(9), Sept. 1974.
[8] Schroeder, D. J. *Astronomical Optics*, San Diego, CA: Academic Press, 1987.
[9] Kingslake, R. *Optical System Design*, New York, NY: Academic Press, 1983.
[10] Kingslake, R. *Lens Design Fundamentals*, San Francisco, CA: Academic Press, 1978.
[11] Vukobratovich, D. "Ultra-Lightweight Optics for Laser Communications," *SPIE OE-LASE '90*, Los Angeles, CA, Vol. 1218, 18, Jan. 15–17, 1990.
[12] Ross, M., Editor. *Laser Applications*, Vol. 1, New York, NY: Academic Press, 1971, pp. 283–285.
[13] *Optics Guide V*, (Optics Catalog), Melles Griot, Irvine, CA.
[14] Cotton, J. M., and Casey, W. L. "Narrowband Optical Interference Filters," *SPIE OE-LASE '91*, Los Angeles, CA, Vol. 1417, 48, Jan. 21–22, 1991.
[15] Lambert, S. G., Dreimeyer, D. E., and Casey, W. L. "Operation of a Diode-Based Direct Detection Laser Crosslink Through Solar Crossing," *IEEE Military Communications Conf. MILCOM '93*, Boston, MA., October 1993.

Chapter 5

The Communication Receiver

Chapter 3 explored the system design and determined the received signal as a function of signal spatial and temporal characteristics. In this chapter, the required signal is developed as a function of the *signal to noise ratio* (SNR) for the various modulation formats identified in Chapter 3.

In order to standardize the determination of the required signal power of the link, the required optical signal will be referenced to the surface of the detector assembly. Figure 5.1 illustrates this reference point. Referencing the signal to the front of the detector assembly is natural since physical measurements of the required optical signal levels can be made at this "hypothetical photocathode." When referencing the signal to this photocathode surface, any noise contributions that are present downstream must be adjusted appropriately by the gain mechanism that precedes it in the detection process.

Two types of detection processes will be considered for communication system design. First, direct detection will be discussed both in the classical sense and with optical preamplification. The noise terms associated with this process will be determined and an expression for the required signal in terms of the SNR requirement will be given.

Additionally, a coherent receiver process will be discussed. The noise terms associated with this process will also be characterized and as before an expression for the required signal will be given.

5.1 THE DIRECT-DETECTION RECEIVE PROCESS

For optical detection processes using photodetectors, the SNR can be written as

$$\text{SNR} = \frac{(P_s R_d)^2}{N_o B} \tag{5.1}$$

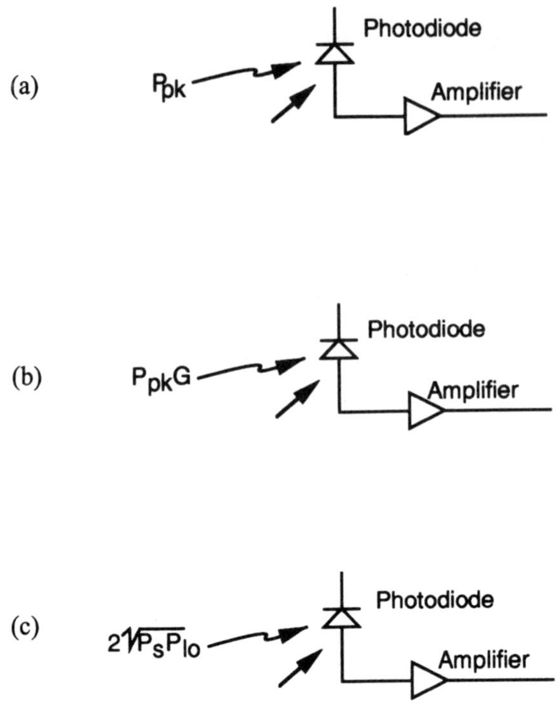

Figure 5.1 Receiver referenced to hypothetical photocathode. (a) Classical direct detection. Noise contribution from background, detector, preamplifier, and signal shot noise. (b) Direct detection with optical preamplifier. Noise contributions from amplified background, detector, preamplifier, amplified signal shot, amplified spontaneous emission, amplified signal spontaneous beat, and spontaneous beat noise. (c) Coherent detection. Noise contributions from background, detector, preamplifier, signal shot, and local oscillator shot noise.

where P_s is the required signal in watts, peak, R_d is the responsivity of the detector in amps per watt, N_o is the noise density in amps squared per hertz, and B is the receiver bandwidth in hertz.

Given the SNR (recall this is determined from the bit error rate and modulation format; see Chapter 3) and the transmitter extinction ratio, the required signal P_s can be determined if the statistics of the noise are found. Hence, the major contributor to the noise process in the communication detection process will now be identified.

For most direct-detection optical receivers, when the signal and noise levels are moderate to large (i.e., large numbers of photon events), the detection noise process can be characterized by simple Gaussian statistics. Using this characterization, the required signal levels can then be determined.

5.1.1 Classical Direct Detection—No Optical Preamplification

The discussion of classical direct-detection receivers begins with a determination of the noise process. In a direct-detection receiver of this nature, one must be concerned with noise sources generated by the detector itself (bulk dark noise), the receiver amplifiers (preamplifier thermal noise), the stellar and celestial bodies as well as reflected and scattered light (background noise), and the signal self-generated noise currents (signal shot noise). In common aperture systems, a transmit signal may also manifest itself as receiver noise due to scattering in the optical systems, which is not efficiently filtered.

All of these noise sources can contribute to the total noise process in varying amounts depending upon the link geometry, the optical design, the receiver configuration, and the data bandwidth. Each will be discussed and characterized below.

5.1.1.1 Background Noise Sources

Direct current (dc) background light incident upon the photodetector manifests itself as noise due to the shot noise process inherent in photodetection. It can be characterized as a dc signal that contributes to the *root mean square* (rms) noise in the receiver. The background noise power (in square amps) can be defined in terms of a noise current density and is given by the expression

$$i_b^2 = 2qFI_bB \tag{5.2}$$

where q is the quantum charge density, B is the receiver bandwidth, and F is the front-end noise factor of the receiver. The value of F is given by

$$F = M \cdot k_{\text{eff}} \cdot (1 - k_{\text{eff}}) \cdot (2 - 1/M) \tag{5.3}$$

where k_{eff} is the ionization coefficient of the photodiode and M is the avalanche gain of the detector. In the general case where *p-intrinsic* (PIN) photodiodes are used, M becomes ideally unity yielding a value for F of 1.

The term I_b in (5.2) is the average received background current and is given by

$$I_b = P_bR_d \tag{5.4}$$

where P_b is the received background power on the detector and R_d is the detector responsivity, as before. The value of P_b can be comprised of many factors. P_b can consist of stellar and celestial background radiant fluxes. It can also consist of scattered light from the receive optics or telescope structure, or even scattered transmitted light energy. In the simplest forms, however, each can be reduced to a spectral radiance or irradiance value.

Now P_b can be obtained from the following expression

$$P_b = H_b \cdot \Omega_{\text{fov}} \cdot L_{\text{rec}} \cdot A_{\text{rec}} \cdot B_{\text{filter}} \tag{5.5}$$

for radiant targets whose angular subtense is larger than the receive detector viewfield and as

$$P_b = N_b \cdot L_{\text{rec}} \cdot A_{\text{rec}} \cdot B_{\text{filter}} \tag{5.6}$$

when the background noise source can be represented as a point source and is fully captured within the receiver viewfield.

The terms H_b and N_b in (5.5) and (5.6) are the background radiance and irradiance energy densities of large angular sources and point sources, respectively. The term H_b is expressed in terms of $\text{W/m}^2\text{/sr/Å}$ and N_b is expressed in terms of $\text{W/m}^2\text{/Å}$. As can be seen, the value of P_b is a function of the receiver solid angle, Ω_{fov}, (for radiant sources), the effective receive area, A_{rec}, the transmission losses of the receive optics, L_{rec}, and the receive filter bandpass width, B_{filter}. The solid angle can be determined by the approximation for small angles as

$$\Omega_{\text{fov}} \approx \frac{\pi}{4} \cdot (\theta_{\text{rec}})^2 \tag{5.7}$$

where θ_{rec} is the receive detector viewfield diameter, and the receiver area can be expressed in terms of the aperture diameter, D_{aper}, and obscuration diameter, D_{obs}, as

$$A_{\text{rec}} = \left[1 - \left(\frac{D_{\text{obs}}}{D_{\text{aper}}}\right)^2\right] \cdot D_{\text{aper}}^2 \tag{5.8}$$

Figures 5.2 through 5.12 show values of H_b and N_b using blackbody radiation and stellar magnitudes to determine the radiances and irradiances of stellar objects [1–3]. Table 5.1 gives values of H_b and N_b for specific stellar sources at wavelengths of interest for laser communication systems.

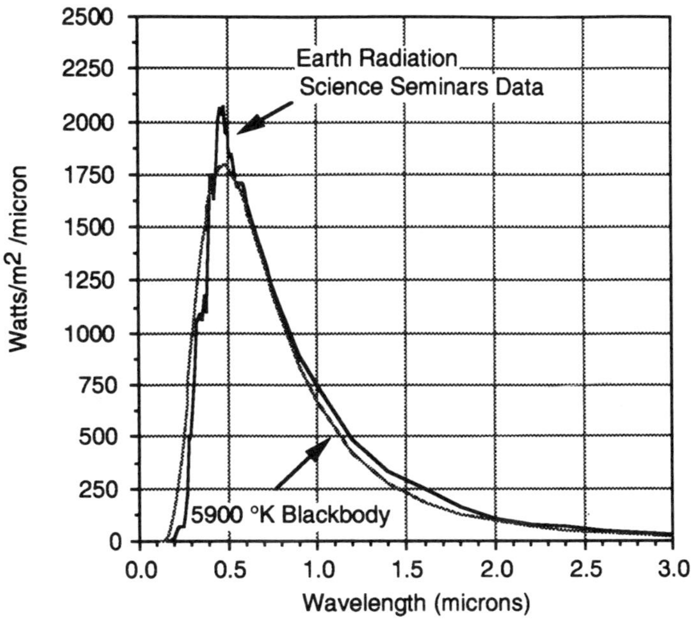

Figure 5.2 Solar spectral irradiance above Earth's atmosphere. (After [1].)

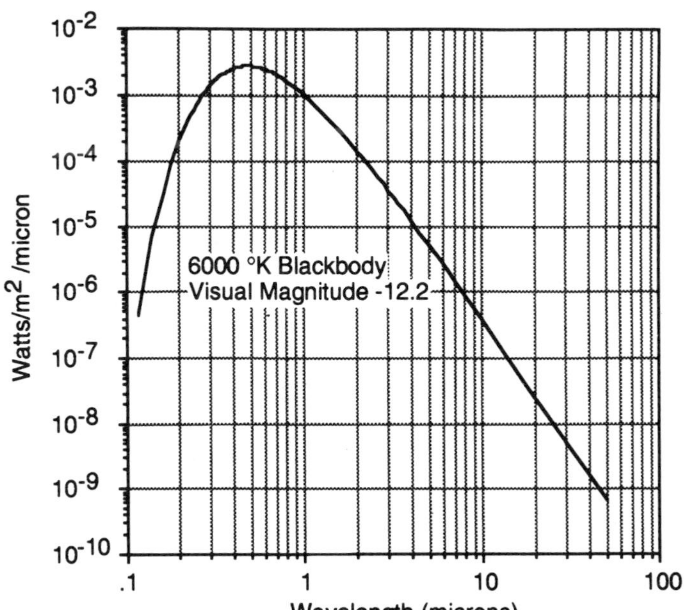

Figure 5.3 Lunar spectral irradiance above Earth's atmosphere.

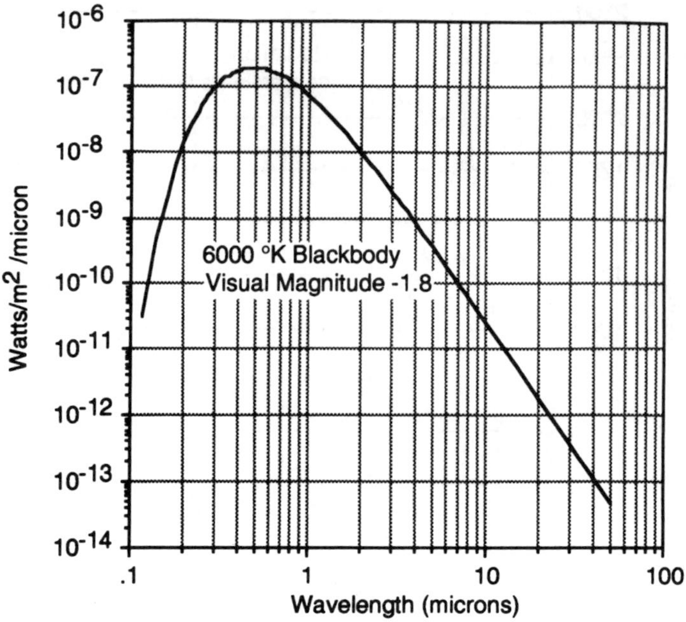

Figure 5.4 Irradiance of Mercury.

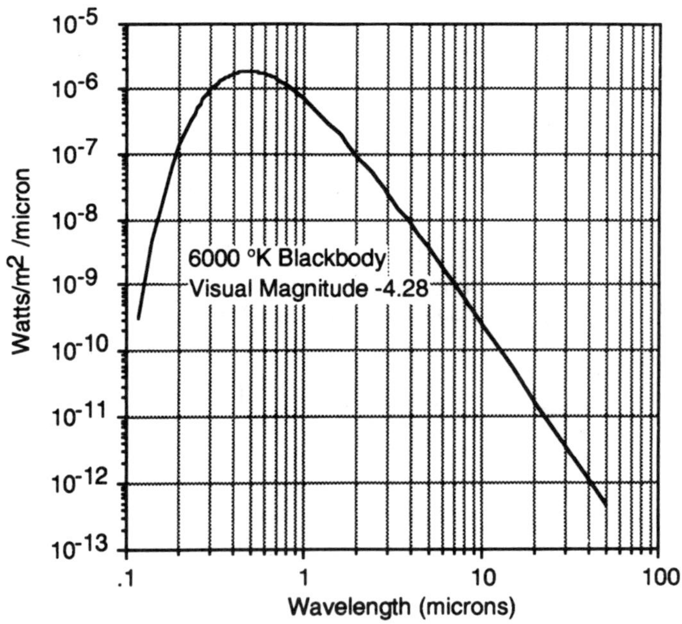

Figure 5.5 Irradiance of Venus.

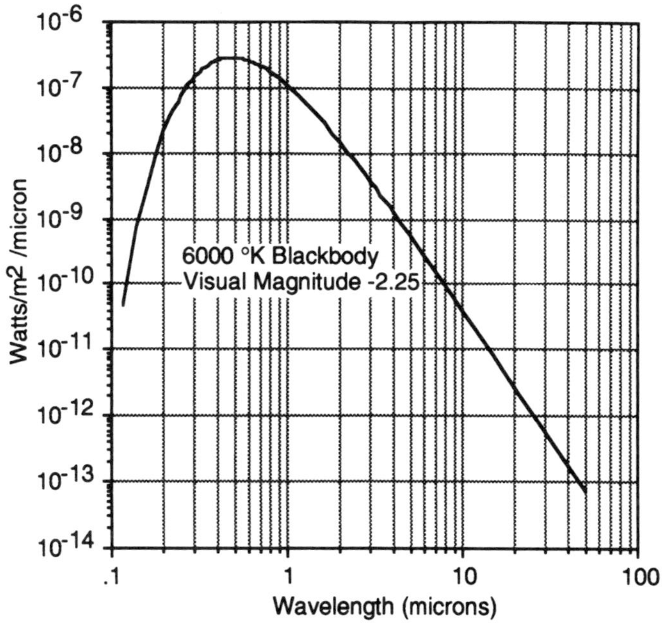

Figure 5.6 Irradiance of Mars or Jupiter.

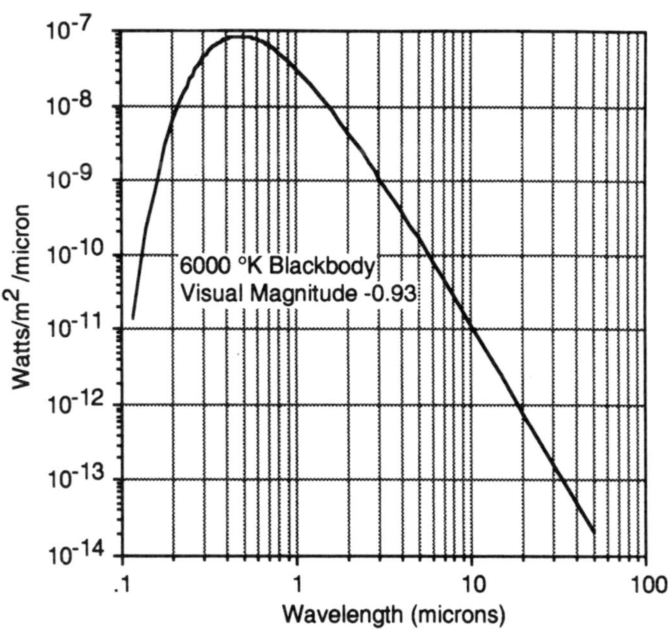

Figure 5.7 Irradiance of Saturn.

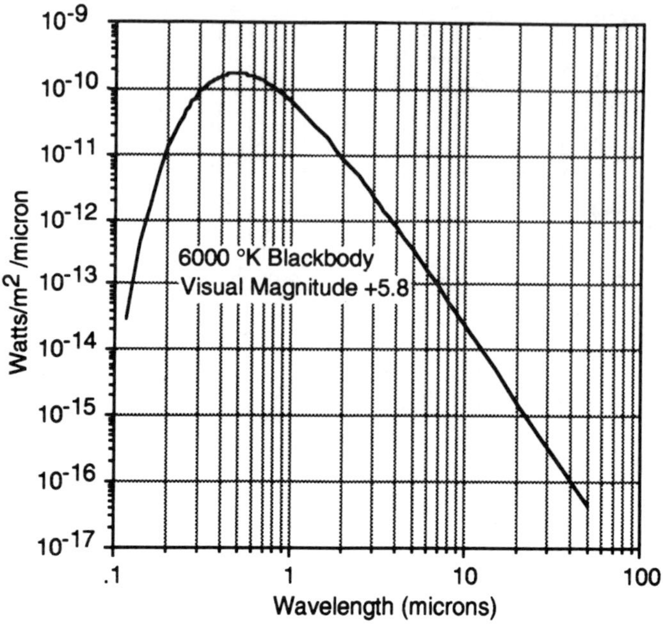

Figure 5.8 Irradiance of Uranus.

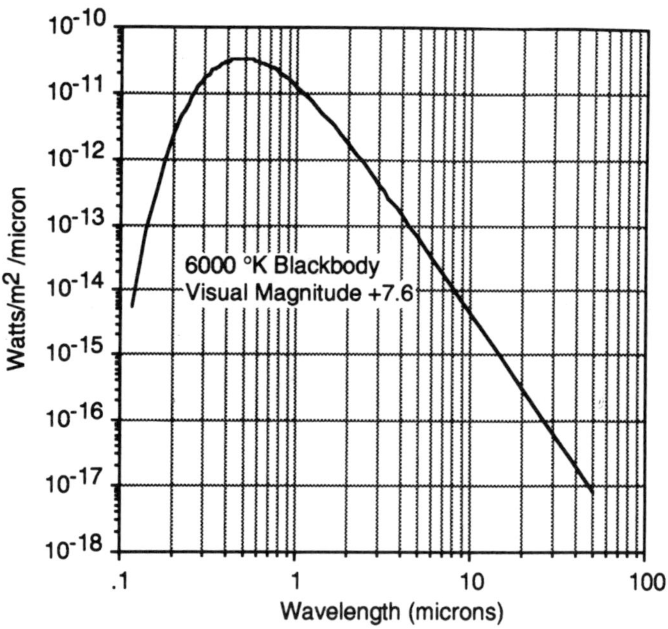

Figure 5.9 Irradiance of Neptune.

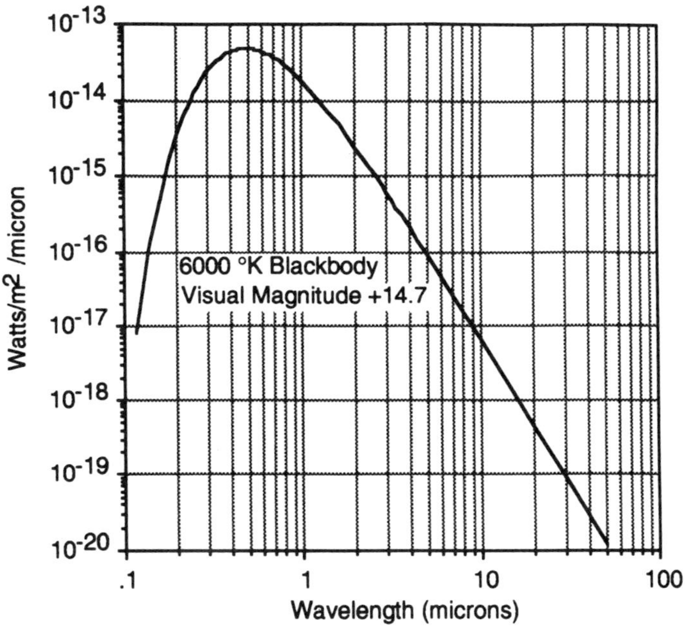

Figure 5.10 Irradiance of Pluto.

5.1.1.2 Detector Dark Noise

Bulk detector noise power, also known as dark noise since it is the amount of noise present when the detector is in the dark, in the photodetection process can be expressed in rms amps squared by the following equation.

$$i_d^2 = 2qFBI_{\text{dm}} + 2qBI_u \tag{5.9}$$

where F, q, and B are as before and I_{du} and I_{dm} are the unmultiplied and multiplied dc dark noise currents. In the case of a PIN photodetector, the dc noise is attributed to only the unmultiplied current (i.e., the multiplied current is zero). In this case, (5.9) reduces to

$$i_d^2 = 2qBI_u \tag{5.10}$$

for the PIN photodiode.

The dark noise is normally dependent upon the operating temperature of the photodiode and its physical volume. The noise increases for high temperatures and

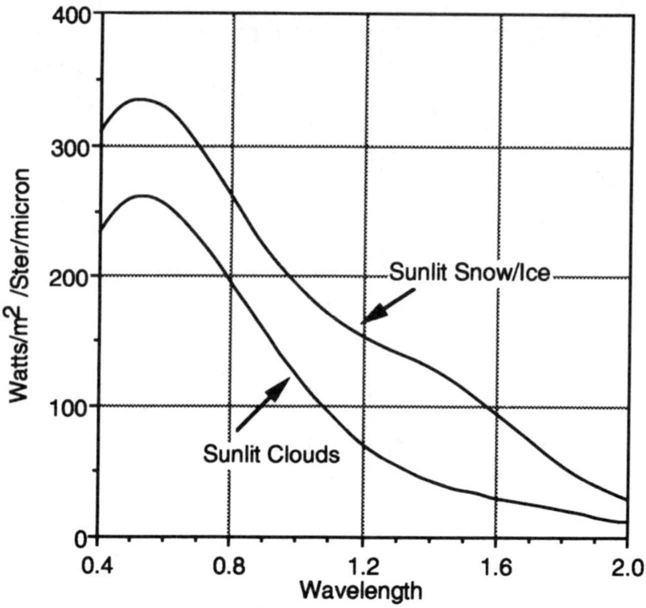

Figure 5.11 Radiance of sunlit clouds and sunlit snow and ice. (After [2].)

for larger physical dimensions. Thus, dark noise can be reduced by cooling the detector and by reducing the physical size of the detector. Cooling is not normally used in optical communication systems for two reasons. First, the testing of optical systems below the laboratory dewpoint requires special handling and purging of the assemblies with dry air or, in most applications, dry nitrogen. Condensation is not tolerated in optical testing. This can be a very expensive when considering production recurring costs. Secondly, the dark noise contributions, except for very large area devices, is normally not the dominant noise source.

Reducing the effective dark area has been explored, but for a different reason than reducing the effective dark noise. Reduction in physical size also makes the photodiode less susceptible to stray ionizing radiation sources. Both natural and manmade radiation sources require the designer to use prudent shielding of the photosensitive device. A technique employing guard rings, which effectively electrically isolates the excess detector material outside the receiver viewfield, reduces the effective dark noise volume of the device. A photodiode with a guard ring is depicted in the schematic in Figure 5.13.

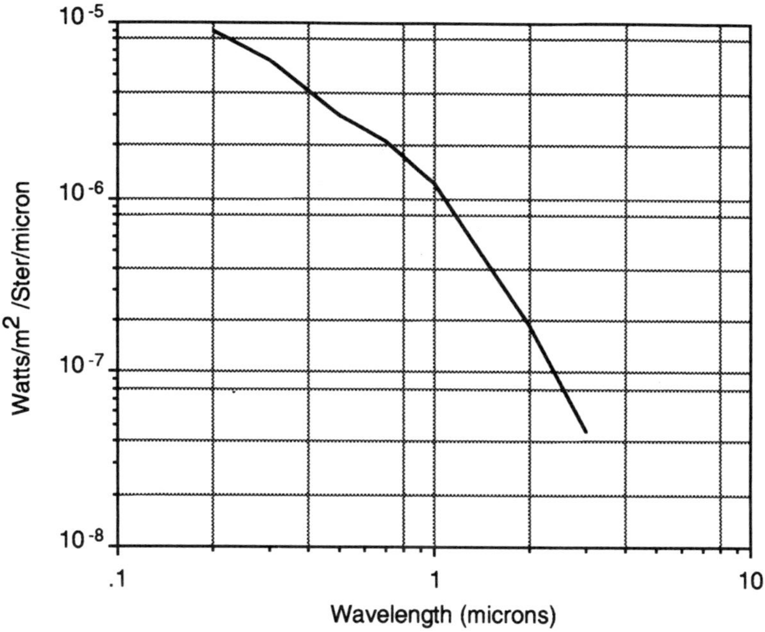

Figure 5.12 Radiance of combined stellar background. (After [3].)

Table 5.1
Background Values for Wavelengths of Interest for Laser Communications

	Irradiance ($W / m^2 / \mu m$)					
Wavelength (μm)	Solar	Lunar	Mercury	Venus	Mars/Jupiter	Saturn
0.53	1842	.0027	1.8×10^{-7}	1.8×10^{-6}	2.8×10^{-7}	8.4×10^{-8}
0.85	940	.0015	9.5×10^{-8}	9.0×10^{-7}	1.5×10^{-7}	4.6×10^{-8}
1.06	748	.001	7.2×10^{-8}	7.1×10^{-7}	1.1×10^{-7}	3.2×10^{-8}
1.3	411	.00054	3.7×10^{-8}	3.6×10^{-7}	5.6×10^{-8}	1.7×10^{-8}
1.5	204	.00024	1.7×10^{-8}	1.6×10^{-7}	2.5×10^{-8}	7.5×10^{-9}

	Radiance ($W/m^2/Sr/\mu m$)		
Wavelength (μm)	Sunlit Clouds	Sunlit Snow and Ice	Starfield
0.53	245	330	3.0×10^{-6}
0.85	180	220	1.4×10^{-6}
1.06	120	190	1.1×10^{-6}
1.3	50	140	6.0×10^{-7}
1.5	40	100	4.0×10^{-7}

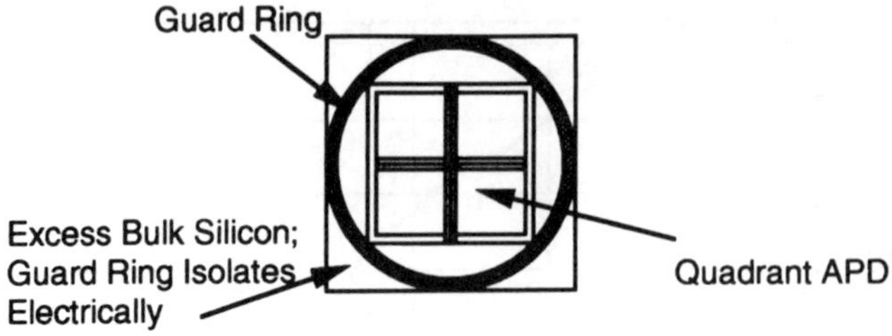

Figure 5.13 Quadrant detector with guard ring.

5.1.1.3 Preamplifier Noise

A third noise contributor to the detection process involves the characterization of the preamplifier used to increase the small signal levels present at the detector output leads. This preamplification is necessary to amplify the signal to manageable levels for detection.

The physical effects of the preamplifier can be characterized, in most cases, as the classical thermal noise of the feedback resistor in an amplifier circuit and is expressed as

$$i_{\text{pa}}^2 = \frac{4kTB}{R_f \cdot M^2} \tag{5.11}$$

where k is Boltzman's constant in J/°K, T is the effective noise temperature of the feedback resistor in °K, M is the avalanche gain of the APD detector, and R_f is the value in ohms of the feedback resistor. Recall, M exists here because of the referencing of the signal to the surface of the photodetector.

Several different types or configurations for preamplifiers have been explored by Personick and Smith [4] in their work with detector preamplifiers for fiber-optic applications. These include *field effect transistor* (FET) and bipolar amplifiers. Personick and Smith state "at low bit rates the FET is superior whereas at high bit rates the bipolar device produces superior performance." The crossover is given as near 50 Mbps within a factor of two. Silicon FET high-frequency operation is also stated to be "limited by the finite gain of the first stage."

The basic noise density of a FET front end is given from [4] as

$$i_{\text{pa FET}}^2 = \left[\frac{4kT}{R_f}\left(1 + \frac{\Gamma_{\text{pa}}}{g_m R_f}\right) + 2qI_{\text{gate}}\right] \cdot \frac{I_2 B}{M^2} + 4kT\Gamma_{\text{pa}}\left(\frac{2\pi C_T}{g_m}\right)\frac{I_3 B^3}{M^2} \tag{5.12}$$

where k is Boltzman's constant, q the elemental electron charge in coulombs, T the thermal temperature of the feedback resistor in °K, R_f the feedback resistance in ohms, g_m the transconductance of the amplifier in mhos, I_{gate} the current of the FET gate, B the receiver bandwidth in Hz, M the avalanche gain of the detector (this relates the noise current to the detector surface), Γ_{pa} a factor (about 0.7 for silicon and 1.1 for gallium arsenide), C_T the total capacitance of the receiver circuit in farads and I_2 and I_3 nonideal receiver filtering factors.

Similarly, the bipolar device noise current density can be expressed as

$$i^2_{\text{pa BIPOLAR}} = \frac{2qI_{\text{base}}I_2B}{M^2} + \frac{2qI_{\text{col}}}{g_mM^2} \cdot \left[\left(\frac{1}{R_f} + \frac{1}{r_{b'e}}\right)^2 I_2B + 2\pi C_T^2B^3I_3\right] + \frac{4kTr_{bb}}{M^2}\left[\frac{BI_2}{R_f^2} + (2\pi)^2(C_d + C_{s'})^2B^3I_3\right] \quad (5.13)$$

where terms common with (5.12) are as stated above and I_{base} is the base current in amperes, I_c is the collector current also expressed in amperes, $r_{b'e}$ the base to emitter dynamic resistance given in ohms, $r_{b'b}$ the dynamic base to base resistance also in ohms, and C_d and C_s the drain and source capacitances given in farads.

While ideally the preamplifier noise will be minimized if the feedback resistance becomes infinite, most systems are operated at finite R_f (i.e., $R_f \ll \infty$) giving rise to preamplifier noises that are dominated by thermal noise. However, at high frequencies, the preamplifier noise can be dominated by the capacitance effects. Van Moui [5] has measured preamplifier noise performance out to gigahertz bandwidth. His results are shown in Figure 5.14.

One technique [4] for minimizing the thermal noise effects in preamplifier circuits is to employ a high impedance (or integrating) front end amplifier using either a FET or a bipolar device. The capacitance for high bandwidth receivers is reduced by selecting low capacitance, high frequency devices. R_f is made large and the receiver effectively integrates the signal received.

The high-impedance design, while in theory a very low noise device, suffers potential drawbacks. The high-impedance design requires the incorporation of an equalizer, which is hard to realize ideally. Parasitic capacitances would require individual equalization and the resultant circuit would be difficult and costly to manufacture. A second detrimental effect results from the high-impedance amplifier's reduced dynamic range.

Perhaps the most common approach to front-end preamplification results from the use of a transimpedance amplifier [4]. The transimpedance amplifier is characterized by current-to-voltage conversion. The transimpedance design has high bandwidth and wider dynamic range when compared to the high-impedance approach. The noise performance, when optimized, approaches that of the high-impedance design. Figure 5.15 schematically illustrates the transimpedance design.

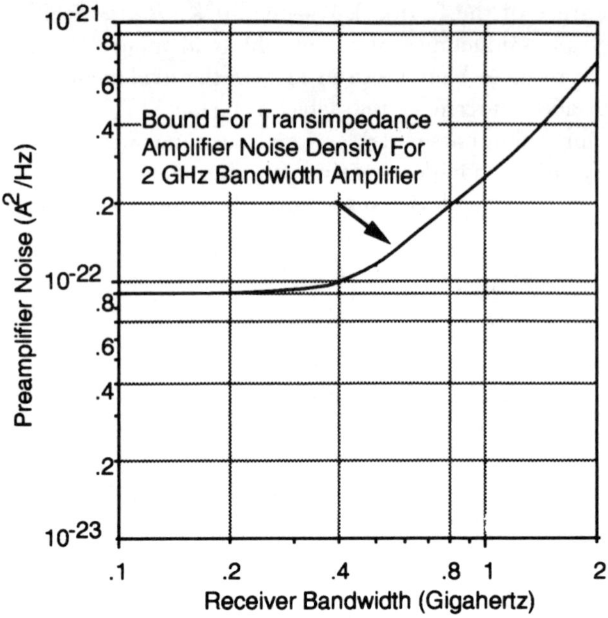

Figure 5.14 Typical transimpedance amplifier noise for high-bandwidth amplifiers. (After [5].)

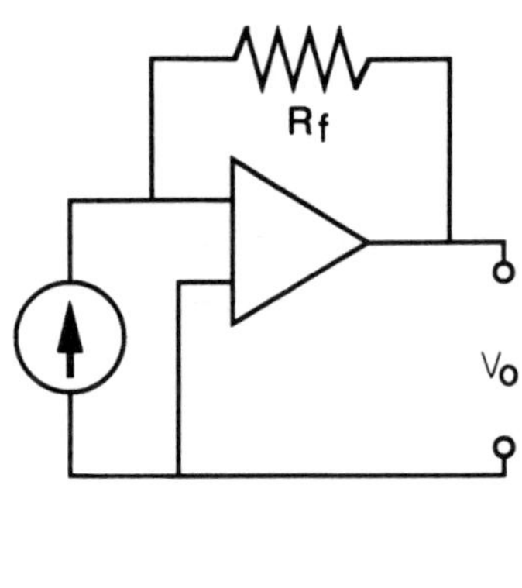

Figure 5.15 Transimpedance amplifier. (After [4].)

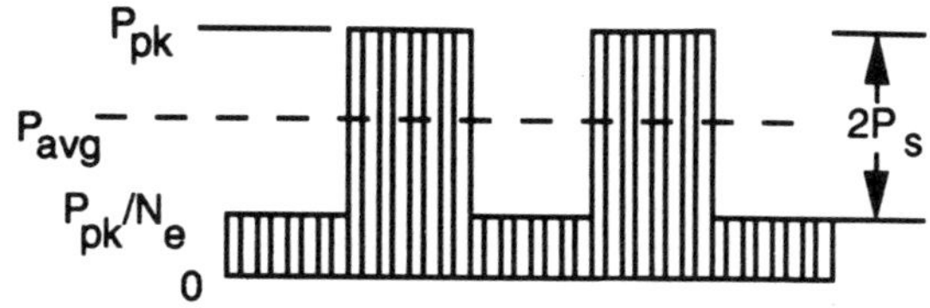

Figure 5.16 Definition of P_{pk}, P_{avg}, P_s, and N_e.

5.1.1.4 Signal Shot Noise

The noise contributions resulting from shot noise can be expressed in terms of the average signal current as

$$i_{ss}^2 = 2qFI_{avg}B \tag{5.14}$$

where q, B, and F are as before and I_{avg} is given by

$$I_{avg} = P_{avg}R_d \tag{5.15}$$

where R_d is the responsivity of the photodetector and P_{avg} is the average signal power over the bit period. For signals with finite extinction ratios, ($N_e \ll \infty$), P_{avg} is given as

$$P_{avg} = \frac{P_{pk}(1 + 1/N_e)}{2} \tag{5.16}$$

where P_{pk} is the peak received level and can be determined from Figure 5.16.

5.1.1.5 Total Noise in Classical Direct-Detection Systems

The total noise is the quadrature sum of all the noise densities and is given by

$$N_oB = i_d^2 + i_{pa}^2 + i_b^2 + i_{ss}^2 \tag{5.17}$$

Now that the noise sources have been determined and characterized, the signal to noise ratio can be expressed in terms of total signal power to total noise power. The SNR from (5.1) can be written as

$$\text{SNR} = \frac{(P_sR_d)^2}{N_oB} = \frac{I_s^2}{i_d^2 + i_{pa}^2 + i_b^2 + i_{ss}^2} \tag{5.18}$$

Substituting the values in (5.15) and (5.16) and writing the received signal levels, I_s, in terms of the peak laser power and extinction ratio yields

$$\text{SNR} = \frac{I_{pk}^2(1 - 1/N_e)^2}{4 \cdot \left[\left(i_d^2 + i_{\text{pa}}^2 + i_b^2\right) + qFBI_{\text{pk}}(1 + 1/N_e)\right]} \tag{5.19}$$

Solving the quadratic in I_{pk} and letting SNR = z yields

$$I_{\text{pk}} = \frac{2qFBz(1 + 1/N_e)}{(1 - 1/N_e)^2} + \sqrt{\left[\frac{2qFBz(1 + 1/N_e)}{(1 - 1/N_e)^2}\right]^2 + \frac{4 \cdot \left(i_d^2 + i_{\text{pa}}^2 + i_b^2\right)}{(1 - 1/N_e)^2} z} \tag{5.20}$$

which can then be related to the peak optical power incident upon the photodetector surface as

$$P_{\text{pk}} = \frac{I_{\text{pk}}}{R_d} \tag{5.21}$$

5.1.2 Direct Detection With Optical Preamplification

Using optical preamplification of the received signal in a direct-detection laser communication system has enormous potential. Recent work done in the fiber-optic industry [6–9] to advance the state of the art in optical amplifiers as signal boosters and as receiver front ends has spilled over into the free-space laser communication arena.

Optical amplifiers can be configured using various techniques, such as semiconductor broad area amplifiers, solid-state rod and slab amplifiers, and fiber-optic amplifiers. Each offers the potential of high front-end receiver gain. However, each also brings with it excess baggage in the form of *amplified stimulated emission* (ASE). This ASE manifests itself as excess noise in the receiver process. The ASE is present even when signal is not, and is broadband. Fortunately, optical filtering techniques can minimize the ASE contributions to overall system noise, providing the amplified signal is narrowband.

Optical preamplification requires identification and characterization of several new noise sources not present in classical direct detection. Among these are the increased signal shot noise of the amplified signal, the shot noise associated with the ASE, the beat noise between the signal and ASE, the beat noise between the ASE components, and the excess amplified intensity noise.

The noise sources previously defined in Section 5.1.1 are also present, but their effects are minimized by the enormous shot noise contribution of the amplified signal and ASE.

Defining the amplification process of the signal as, ideally, G, the received signal is then characterized by

$$I_s = \frac{P_{\text{pk}}}{2}\left(1 - \frac{1}{N_e}\right)GL_A R_d \tag{5.22}$$

where P_{pk} and N_e are as given previously and L_A is the loss associated with coupling the signal into the amplifier medium, the so-called insertion loss. In fiber based amplifier systems, the amplification gain, G, is a function of the amplifier length and the fiber pump power.

Assuming only the amplified components are critical noise contributors (true for sufficiently large GL_A), we begin a discussion of the noise characteristics of optical preamplified signals. From [6] we derive the noise current densities from the major noise contributors.

5.1.2.1 Amplified Signal Shot Noise

The amplified signal contributes to the overall signal shot noise at the receiver. It can be expressed as

$$\begin{aligned} i^2_{\text{ss}_A} &= 2qFBP_{\text{avg}}GR_d L_A \\ i^2_{\text{ss}_A} &= qFBGR_d L_A P_{\text{pk}}\left(1 + \frac{1}{N_e}\right) \end{aligned} \tag{5.23}$$

where all terms are as previously defined.

5.1.2.2 Amplified Background Noise

The optical background energy, previously characterized, is also affected directly by the gain of the amplifier. This amplified background signal contributes to the noise process as

$$i^2_{\text{ba}} = 2qFBGR_d L_A P_b \tag{5.24}$$

where again the terms are as previously defined.

5.1.2.3 Amplified Stimulated Emission

From [5], the shot noise contribution from ASE can be expressed as

$$i^2_{ss_A} = 2qFB(G-1)R_d P_{sp} \Delta \nu_{ASE} \tag{5.25}$$

where P_{sp} is the total spontaneous power emitted from the amplifier at the detector and $\Delta \nu_{ASE}$ a measure of the amount of the total ASE, which is within the passband of the receiver filter.

5.1.2.4 Amplified Signal—Spontaneous Beat Noise

The signal components interact with the ASE, producing beat notes that are within the passband of the receiver filter. The contribution to the noise process due to beat notes can be expressed as

$$\begin{aligned} i^2_{\text{sig-spon}} &= 4L_A R_d^2 \left(P_{sp}(G-1)P_{avg}G\right)B \\ i^2_{\text{sig-spon}} &= 2L_A R_d^2 \left(P_{sp}(G-1)\cdot P_{pk}\left(1+\frac{1}{N_e}\right)\cdot G\right)B \end{aligned} \tag{5.26}$$

5.1.2.5 Amplified Spontaneous—Spontaneous Beat Noise

Likewise, the ASE signal components interact with each other to form beat notes. The noise contributions of the ASE beat note noise is given as

$$i^2_{\text{spon-spon}} = 4R_d^2\left[P_{sp}(G-1)\right]^2 \Delta \nu_{\text{sp-sp}} B \tag{5.27}$$

where $\Delta \nu_{\text{sp-sp}}$ is the measure of the percent of ASE beat noise that is within the passband of the receiver filter.

5.1.2.6 Amplified Source Relative Intensity Noise

The amplified *relative intensity noise* (RIN) of the laser source also contributes to the total noise process and is given by

$$i^2_{\text{RIN}_A} = \left(R_d G L_A P_{avg}\right)^2 \cdot \text{RIN} \cdot B \tag{5.28}$$

where RIN is a measure of the relative intensity noise of the transmit source in units of hertz^{-1}.

5.1.2.7 Signal To Noise Ratio

The SNR of optically amplified signals can be expressed as

$$\mathrm{SNR} = \frac{(I_s)^2}{i_{\mathrm{tot}}^2} = \frac{\left(P_{\mathrm{pk}}\left(1 + \dfrac{1}{N_e}\right) G L_A R_d\right)^2}{4\left(i_{\mathrm{ss}_A}^2 + i_{b_A}^2 + i_{\text{sig-spon}}^2 + i_{\text{spon-spon}}^2 + i_{\mathrm{ASE}}^2 + i_{\mathrm{RIN}_A}^2\right)} \tag{5.29}$$

As can be inferred from (5.23–5.27), the dominate noise sources under high gain conditions are the signal shot noise and the signal-spontaneous beat noise. From this then, (5.28) reduces to

$$\mathrm{SNR} = \frac{\left[P_{\mathrm{pk}}\left(1 - \dfrac{1}{N_e}\right) G L_A R_d\right]^2}{4qFBGL_A P_{\mathrm{pk}}(1 + 1/N_e) R_d + 8 L_A R_d^2 P_{\mathrm{sp}}(G - 1) P_{pk}(1 + 1/N_e)\mathrm{GB}} \tag{5.30}$$

If the signal shot noise dominates, the expression for SNR further reduces to

$$\mathrm{SNR} = \frac{P_{\mathrm{pk}} G L_A R_d \left(1 - \dfrac{1}{N_e}\right)^2}{4qFB(1 + 1/N_e)} \tag{5.31}$$

and following directly, P_{pk} is given as

$$P_{\mathrm{pk}} = \frac{4qFBz(1 + 1/N_e)}{G L_A R_d (1 - 1/N_e)^2} \tag{5.32}$$

where $z = \mathrm{SNR}$.

If the noise is dominated by signal-ASE beat noise and G is large, the SNR can be expressed as

$$\mathrm{SNR} = \frac{P_{\mathrm{pk}} L_A \left(1 - \dfrac{1}{N_e}\right)^2}{8 P_{\mathrm{sp}}(1 + 1/N_e) B} \tag{5.33}$$

where the expression for P_{pk} is given as

$$P_{\mathrm{pk}} = \frac{8 P_{\mathrm{sp}}(1 + 1/N_e) zB}{(1 - 1/N_e)^2 L_A} \tag{5.34}$$

5.2 THE COHERENT-DETECTION PROCESS

Recall that coherent detection is characterized by the mixing of the received signal wavefront with a *local oscillator* (LO) signal wavefront. The LO provides gain to the signal and is represented as

$$I_s = 2\sqrt{P_s P_{\mathrm{lo}}}\, R_d \tag{5.35}$$

where P_s is the received signal power in watts, P_{lo} is the local oscillator power also in watts, and R_d is the detector responsivity in amps/W.

This gain process is the dominant front-end gain and acts exclusively on the signal. Because the background energy is not phase or frequency-coherent with the LO source, no background gain is realized. The extra signal power incident upon the detector, however, does contribute to the optical detector noise process as shot noise. The receiver noise statistics for coherent detection of optical waveforms is generally dominated by this signal shot noise of the local oscillator.

The noise density of a coherent receiver can be represented by

$$i_{\mathrm{coh}}^2 = i_d^2 + i_{\mathrm{pa}}^2 + i_b^2 + i_{\mathrm{ss}}^2 + i_{\mathrm{lo}}^2 \tag{5.36}$$

where i_{ss} is the rms signal shot noise current, i_b is the rms background noise current, i_d the rms detector dark noise current, i_{pa} the rms preamplifier noise current, and i_{lo} the rms local oscillator shot noise current. The terms i_{ss}, i_b, i_d, and i_{pa} are as previously defined for the direct-detection receiver. The local oscillator noise power is given as

$$i_{\mathrm{lo}}^2 = 2qFBP_{\mathrm{lo}} R_d \tag{5.37}$$

where again P_{lo} is the local oscillator power incident upon the detector, B is the receiver bandwidth, and R_d is the detector responsivity.

When the LO power is made sufficiently large, the LO shot noise dominates. When this occurs, the SNR can be written as

$$\mathrm{SNR} = \frac{I_s^2}{N_o B} = \frac{4P_s P_{\mathrm{lo}} R_d^2}{2qFBP_{\mathrm{lo}} R_d} = \frac{2P_s R_d}{qFB} \tag{5.38}$$

Letting SNR $= z$, as before, then solving for P_s yields

$$P_s = \frac{2zqFB}{R_d} \tag{5.39}$$

for the LO shot noise limited case. If the LO noise does not dominate, then (5.38)

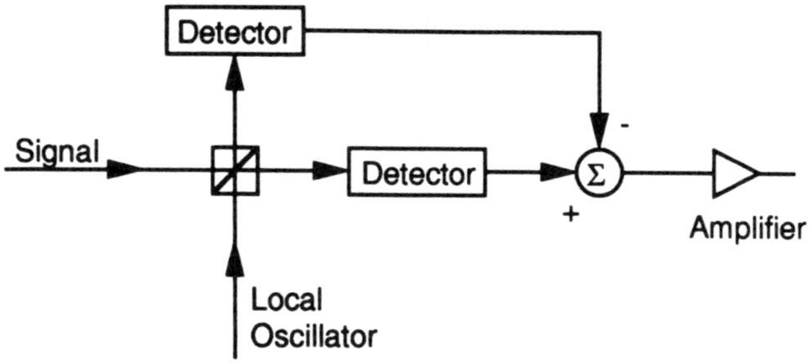

Figure 5.17 Noise cancellation receiver. (After [10].)

must be solved for P_s in terms of all the noise contributors. Of course as before, when using a PIN photodiode the value of F is identically one. The peak signal power required is then given by

$$P_{pk} = \frac{4zqFB}{R_d} \tag{5.40}$$

Additional effects must be considered when a coherent receiver is implemented. These include the intensity fluctuations present in the local oscillator, mixing efficiency of the LO and the signal wavefronts, and phase and frequency noise resulting from finite laser linewidths. Each of these can contribute to degradation in the coherent receiver process.

5.2.1 LO Intensity Fluctuations

Fluctuations in intensity that can be present, especially in semiconductor laser devices, can contribute to excess noise in the receiver. Novel techniques [10] have been developed to reduce the relative intensity noise. This technique uses a dual-detector receiver that effectively cancels the noise due to intensity fluctuations. Figure 5.17 depicts this noise cancellation receiver configuration.

5.2.2 Mixing Efficiency

The efficiency of the mixing process is a strong function of the spatial distribution of the LO and the signal wavefronts. For a diffraction-limited receiver that collects a uniform receive field, the spot intensity on the detector will result in an Airy distribution with a spot size related to the receiver aperture size. This spot size is given as

$$\theta_{spot} \approx \frac{2.44\lambda}{D_{aper}} \tag{5.41}$$

where λ is the wavelength and D_{aper} is the effective aperture diameter.

The LO, through appropriate optical techniques, can take on spatial distributions that range from uniform in intensity to an Airy pattern that directly matches the received wavefront distribution on the detector.

When the detector is in the focal plane, the incident angle of the received energy, θ_{ff}, becomes a spatial displacement d, given by

$$d = f_d(\theta_{ff} M_{lo} + \theta_{lo}) \tag{5.42}$$

where f_d is the focal length of the detector focusing lens, M_{lo} is the magnification between local and far-field space, and θ_{lo} is the angle between the local oscillator and the receive optical axis.

Figure 5.18 illustrates the mixing efficiency as a function of far-field angle error for a focal length less than 1m. From this it can be seen that the optimal LO intensity distribution for a diffraction-limited receiver is a matched Airy disk. This is true for small angle errors but is less efficient at large angle errors.

5.2.3 Phase and Frequency Noise

Another contributor to the overall performance efficiency of a coherent-detection system is the amount of phase and frequency noise inherent in the transmit source. Lasers have a finite linewidth (i.e., a finite frequency purity). Most solid-state lasers are very nearly pure in their emission spectrum when proper control is used. Linewidths of a few kilohertz are not uncommon. On the other hand, semiconductor devices tend to have linewidths of megahertz in width. Techniques to improve stability using spectral feedback techniques (distributed Bragg reflectors and etalons) to control the frequency of the semiconductor devices have been successful in reducing linewidth to kilohertz levels.

The phase and frequency noise present in the coherent-detection process contributes to the detection process as a function of the linewidth. The noise density can be given from [11] as

$$N_o = \frac{B_L}{2\pi} \tag{5.43}$$

where B_L is a measure of the laser linewidth in hertz.

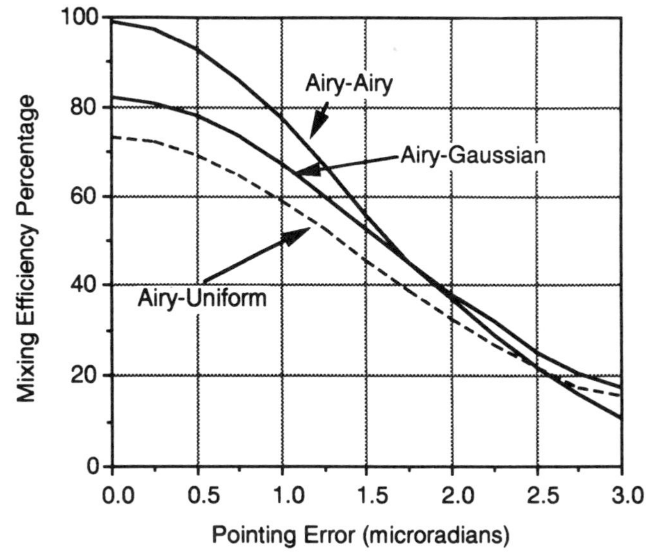

Figure 5.18 Mixing efficiency.

The ratio of data rate to laser linewidth becomes the discriminating factor. Thus, higher bandwidth systems can accommodate laser sources with wider linewidths; however, narrow linewidths still provide the best performance. In addition, some modulation formats are more susceptible than others.

References

[1] Hall, J. B., Jr. "Earth Radiation Science Seminars," NASA Conference Publication 2239, October 1982, Hampton, VA.

[2] Wolf, W. L., Editor. *Handbook of Military Infrared Technology*, Washington, DC: Office of Naval Research, Dept. of The Navy, 1965.

[3] *Electro-Optics Handbook*, Lancaster, PA: RCA Corporation, 1974.

[4] Personick, S. D., Smith, R. G., and Kressel, H. (ed.), *Semiconductor Devices For Optical Communication*, Ch. 4—Receiver Design For Optical Fiber Communications, New York, NY: Springer-Verlag, 1980.

[5] Van Muoi, T. V. "Receiver Design for High-Speed Optical-Fiber Systems," *Journal of Lightwave Technology*, Vol. LT-2, No. 3, June 1984.

[6] Laming, R. I., and Payne, D. N. "Noise Characteristics of Erbium-Doped Fiber Amplifier Pumped at 980 nm," *IEEE Photonics Technology Letters*, Vol. 2, No. 6, June 1990.

[7] Tonguz, O. K., and Kazovsky, L. G. "Theory of Direct Detection Lightwave Receiver Using Optical Amplifiers," *Journal of Lightwave Technology*, Vol. 9, No. 2, Feb. 1991.

[8] Aoki, Y. "Properties of Fiber Raman Amplifiers and Their Applicability to Digital Optical Communication Systems," *Journal of Lightwave Technology*, Vol. 6, No. 7, July 1988.

[9] Olshansky, R. "Noise Figure for Erbium-doped Optical Fibre Amplifiers," *Electronics Letters*, Vol. 24, No. 22, 27, Oct. 1988.

[10] Chan, V.W.S., and Kaufmann, J. E. "Coherent Optical Intersatellite Crosslink Systems," *Proc. of the Society of Photo-Instrumentation Engineers*, Vol. 988, 1988.

[11] Salz, J. "Modulation and Detection for Coherent Lightwave Communications," *IEEE Communications Magazine*, Vol. 24, No. 6, June 1986.

Chapter 6

The Acquisition Receiver

The acquisition process is defined by the ability to search the initial uncertainty region, attempting to detect the signal transmitted from a companion satellite terminal. It is this detection process that forms the basis for the acquisition receiver. Requirements that define this acquisition link for laser communication systems include acquisition time, uncertainty area, channel noise, and detection criteria. From these requirements, receiver characteristics are formulated and a required *signal to noise* ratio (SNR) is determined.

Several techniques have been explored and used in developing a laser communication system acquisition link. In Chapter 3, these techniques were explored in detail. In this chapter, the receiver required signal expressions are developed for several detection techniques. Receiver expressions are developed for pulse and tone-modulated signal detection using conventional *quadrant avalanche photodiodes* (QAPDs) and for *p-intrinsic photodiodes* (QPINs). Integrating receiver systems are analyzed and expressions for required signals are also developed for this class of device.

6.1 PULSE DETECTION ACQUISITION RECEIVERS

Perhaps the simplest of all acquisition receivers is the pulse detection acquisition receiver. With this technique, the receive waveform is viewed over a detection window (or bandwidth) and compared in amplitude to a set threshold level. If the signal in the receive bandwidth exceeds the threshold limit, the signal is detected. The threshold level is set based on criteria established by the required probability of detection and false alarm rate.

As discussed in Chapter 3, the detection probability flows from the allocation of required acquisition probability, search area coverage probability, and scan repetition. Similarly, the false alarm rate is derived from the acquisition time and

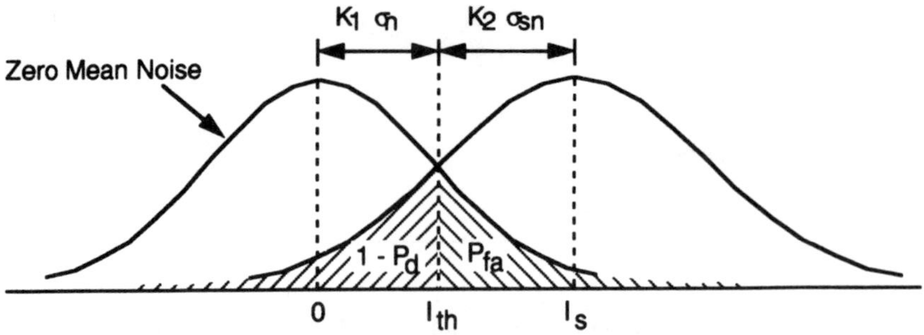

Figure 6.1 Classical Gaussian detection process. P_{fa} = false alarm probability; I_{th} = detection threshold current; P_d = detection probability; I_s = signal current.

scan repetition. Recall that false alarms must be "serviced," thereby adding additional time to the acquisition process. The number of false alarms per second (i.e., the false alarm rate) must then be kept to a minimum in order to keep acquisition times reasonable.

The detection process for zero mean Gaussian noise is illustrated by Figure 6.1. From classical detection theory [1], the required signal can be written as

$$I_{pk} = K_1 \sigma_n + K_2 \sigma_{sn} \tag{6.1}$$

where σ_n is the *root mean square* (rms) noise when only noise is present and σ_{sn} is the rms noise when signal plus noise is present. (Recall that signal shot noise increases the rms noise when signal is present.) For Gaussian noise statistics, an acceptable assumption under high noise conditions, the values of K_1 and K_2 are determined from

$$P_d = \frac{1}{\sqrt{2\pi}} \int_{-\infty}^{K_2} e^{\frac{-x^2}{2}} dx \tag{6.2}$$

$$P_{fa} = \frac{1}{\sqrt{2\pi}} \int_{K_1}^{\infty} e^{\frac{-x^2}{2}} dx \tag{6.3}$$

where P_d and P_{fa} are the probability of detection and probability of false alarm, respectively.

Typical values of K_1 and K_2 are given in Table 6.1 for various detection probabilities and false alarm probabilities. These values are adapted from [2].

6.1.1 Noise in the Pulsed Acquisition Process

The rms noise values are derived similarly to Chapter 5 and are repeated here for both QPINs and QAPDs. Note that for most all acquisition detectors, the initial

Table 6.1
Values of K_1 and K_2 For P_{fa} and P_d

False Alarm Probability (P_{fa})	K_1	*Probability of Detection* (P_d)	K_2
10^{-2}	2.32635	0.999	3.09023
10^{-3}	3.09023	0.990	2.32635
10^{-4}	3.71902	0.900	1.34076
10^{-5}	4.26489	0.850	1.03643
10^{-6}	4.75342	0.800	0.84162
10^{-7}	5.19934	0.750	0.67449
10^{-8}	5.61200	0.700	0.52440
10^{-9}	5.99781	0.650	0.38532
10^{-10}	6.36134	0.600	0.25335

detection will occur off-axis; therefore separate threshold comparators are used for each quadrant. Figure 6.2 illustrates this.

6.1.1.1 Background Noise

For QAPD detectors, the background noise current is expressed as

$$i_b^2 = 2qFBR_dP_B \tag{6.4}$$

where q is the quantum charge, F the noise factor of the dctection process calculated from $F = Mk + (1 - k)(2 - 1/M)$, M the avalanche gain of the QAPD,

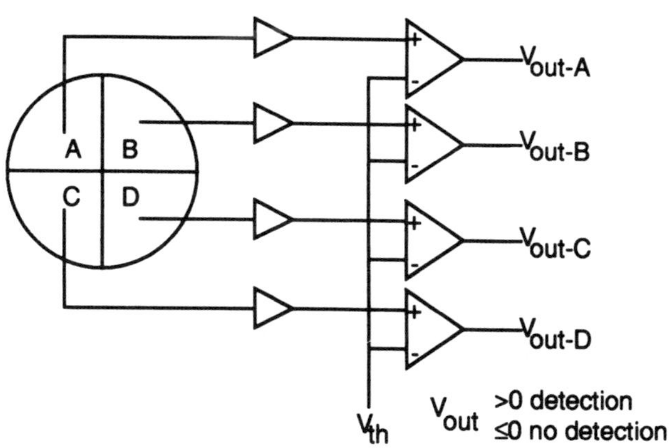

Figure 6.2 Acquisition detector detection scheme.

R_d the responsivity of the detector, B the electrical bandwidth, and P_B as below.

$$P_B = \frac{\pi}{4} \cdot \left(\frac{\text{FOV}}{4}\right)^2 \cdot \frac{\pi}{4} \cdot D_{\text{aper}}^2 \cdot \left[1 - \left(\frac{D_{\text{obsc}}}{D_{\text{aper}}}\right)^2\right] \cdot L_{\text{optics}} \cdot B_{\text{filter}} \tag{6.5}$$

and where FOV/4 is used since only a single quadrant's background noise contributes to the total noise in the quadrant threshold detection. As in Chapter 5, the total *field of view* (FOV) is given by FOV, the aperture diameter by D_{aper}, the obscuration diameter by D_{obsc}, the receiver optics transmission by L_{optics}, and the optical filter bandwidth by B_{filter}.

For QPIN detectors, the background noise current is given as

$$i_b^2 = 2qBR_dP_B \tag{6.6}$$

where the values of q, B, R_d, and P_B are as before.

6.1.1.2 Detector Dark Noise

Detector dark current noise is given for QAPD detectors as

$$i_d^2 = 2qBF\left(\frac{I_{\text{dm}}}{4}\right) + 2qB\left(\frac{I_{\text{du}}}{4}\right) \tag{6.7}$$

where q, F, and B are as before and I_{du} and I_{dm} are the total unmultiplied and total multiplied dark currents, respectively. Note that once again, the factor of 4 is used to divide the dark current into single quadrant values.

Likewise for QPIN detectors, the expression for dark current noise is given by

$$i_d^2 = 2qB\left(\frac{I_d}{4}\right) \tag{6.8}$$

where q and B are as above and I_d represents the total dark current.

6.1.1.3 Preamplifier Noise

Preamplifier noise currents are defined for QAPD detectors and QPIN detectors by the following expressions.

$$i_{\text{pa}}^2 = \frac{(\text{NEI})^2 \cdot B}{M^2} \tag{6.9}$$

is valid for QAPD detectors, and

$$i_{\text{pa}}^2 = (\text{NEI})^2 \cdot B \tag{6.10}$$

is valid for QPIN detectors, where NEI is the noise equivalent current of the preamplifier expression in amps per root hertz and M in (6.9) is the avalanche gain of the QAPD. Chapter 5 defines the preamplifier noise currents in greater detail. The reader is referred to this chapter for additional information on preamplifier noise current.

6.1.1.4 Signal Shot Noise

The final noise component to consider is the signal shot noise and it is expressed for QAPD detectors as

$$i_{\text{ss}}^2 = 2qFBR_dP_{\text{pk}} \tag{6.11}$$

where q, F, B, and R_d are as above and I_{pk} is the peak signal current out of the detector.

Similarly for QPIN detectors, the shot noise can be expressed as

$$i_{\text{ss}}^2 = 2qBR_dP_{\text{pk}} \tag{6.12}$$

where all parameters are as above.

6.1.2 The Pulsed Acquisition Required Signal

With the noise terms established, the values of σ_n and σ_{sn} can be written as

$$\sigma_n^2 = i_b^2 + i_d^2 + i_{\text{pa}}^2 \tag{6.13}$$

$$\sigma_{\text{sn}}^2 = \sigma_n^2 + i_{\text{ss}}^2 \tag{6.14}$$

Using (6.13) and (6.14) , (6.1) can be rewritten as

$$I_{\text{pk}} = K_1\sigma_n + K_2 \cdot \sqrt{\sigma_n^2 + 2qFBI_{\text{pk}}} \tag{6.15}$$

Squaring both sides and solving the quadratic yields a solution for I_{pk} of

$$I_{\text{pk}} = \left(K_1\sigma_n + qFBK_2^2\right) + \left[\left(K_1\sigma_n + qFBK_2^2\right)^2 + \left(K_2^2 - K_1^2\right) \cdot \sigma_n^2\right]^{1/2} \tag{6.16}$$

which can be divided by the responsivity to yield the required peak signal power P_{pk} as

$$P_{pk} = \frac{I_{pk}}{R_d} \tag{6.17}$$

Similarly, (6.16) can be rewritten for QPIN receivers as

$$I_{pk} = \left(K_1\sigma_n + qBK_2^2\right) + \left[\left(K_1\sigma_n + qBK_2^2\right)^2 + \left(K_2^2 - K_1^2\right)\cdot\sigma_n^2\right]^{1/2} \tag{6.18}$$

6.2 TONE ACQUISITION RECEIVERS

Two types of tone acquisition receivers are discussed herein. The first involves envelope detection of the tone signal. The second involves criteria established to lock a phase locked loop (PLL) receiver within a specified time.

6.2.1 Tone Acquisition Receiver With Envelope Detection

When considering a tone detection receiver, the signal received by the detection system is either simply Gaussian noise (when no signal is present) or tone plus Gaussian noise (when tone signal is present). From [3], envelope detection of a tone waveform in the presence of Gaussian noise results in detection statistics that are represented by a Rayleigh density function when no tone signal is present and by a Rician density function when tone signal and noise are both present.

Thus, from [3] (and as previously identified in Chapter 3), the probability of detection and false alarm probability for envelope detection of a tone signal are given by

$$P_d = \int_{I_{th}}^{\infty} \frac{x}{\sigma_{sn}^2} e^{-(I_{st}+x)^2/2\sigma_{sn}^2} \cdot I_o\left(\frac{I_{st}+x}{\sigma_{sn}}\right) dx \tag{6.19}$$

$$P_{fa} = e^{-I_{th}^2/2\sigma_n^2} \tag{6.20}$$

where I_{st} is the signal current level σ_n and σ_{sn} are the rms noise variances when no signal is present and when signal is present, I_{th} the detection threshold level, and $I_o(x)$ the modified Bessel function of order zero.

Using (6.19) and (6.20), a solution for I_{st} can be developed as a function of P_d and P_{fa}. Unfortunately, the solution is not closed form and numerical techniques must be used to determine the value of I_{st}.

6.2.2 Tone Acquisition Via Phase Locked Loop Techniques

Another approach to tone acquisition is to phase lock on to the tone signal using conventional PLL techniques. To that end, the detection of the tone signal is dependent upon the ability to lock onto it. The lock time (i.e., the time to detect) drives the scan time in the acquisition process and ultimately the time to acquire in the acquisition process.

With an established acquisition time requirement, the analysis of Chapter 3 is employed to determine a dwell time and subsequent loop lock time. This derived loop lock time is used along with the following [4] to determine a required loop bandwidth.

$$T_{\text{lock}} \approx \frac{4.2\Delta f^2}{B_{\text{loop}}^3} \tag{6.21}$$

where Δf is the frequency offset of the phase locked loop and B_{loop} the loop bandwidth. It should be noted this equation is for good second-order loops operating with sufficient signal to noise ratio (i.e., SNR $\geq$ 10).

Having established the PLL bandwidth, the expression for SNR using QAPD detection can be written as

$$\text{SNR} = \frac{I_{\text{pk}}^2(1 - 1/N_e)^2}{4\sigma_n^2 + 8qFI_{\text{ave}}B_{\text{loop}}} \tag{6.22}$$

where N_e is the extinction ratio for a squarewave tone, I_{ave} the average signal current given by $I_{\text{ave}} = I_{\text{pk}}(1 + 1/N_e)/2$ and the noise terms are as before with the bandwidth of the acquisition process being the loop bandwidth B_{loop}.

After manipulating the equation, the required signal current can be written as

$$I_{\text{pk}} = \frac{2qFB_{\text{loop}}(1 + 1/N_e) \cdot \text{SNR}}{(1 - 1/N_e)^2} + \sqrt{\left[\frac{2qFB_{\text{loop}}(1 + 1/N_e) \cdot \text{SNR}}{(1 - 1/N_e)^2}\right]^2 + \frac{4\sigma_n^2 \cdot \text{SNR}}{(1 - 1/N_e)^2}} \tag{6.23}$$

and the required signal power follows as

$$P_{\text{pk}} = \frac{I_{\text{pk}}}{R_d} \tag{6.24}$$

For QPIN devices, (6.23) becomes

$$I_{\text{pk}} = \frac{2qB_{\text{loop}}(1 + 1/N_e) \cdot \text{SNR}}{(1 - 1/N_e)^2} + \sqrt{\left[\frac{2qB_{\text{loop}}(1 + 1/N_e) \cdot \text{SNR}}{(1 - 1/N_e)^2}\right]^2 + \frac{4\sigma_n^2 \cdot \text{SNR}}{(1 - 1/N_e)^2}} \tag{6.25}$$

6.3 INTEGRATING DEVICE ACQUISITION RECEIVERS

Another commonly explored acquisition receiver implementation uses an N by N array *charge transfer device* (CTD). A CTD is a photosensitive device consisting of an array of sensors (or pixels). Each pixel integrates charge proportional to the optical energy incident upon its photosensitive surface. The accumulated charge can be read to determine the integrated optical power. Two main types of CTD exist. These are the *charge injection device* (CID) and the *charge coupled device* (CCD). The basic difference between the two lies in the method of readout and charge removal.

6.3.1 Charge-Coupled Device Receivers

The CCD, by far the most common CTD in use today, is read by sequential transfer of charge from pixel to pixel along the columns, similar to a bucket brigade. The charge signals are detected at the ends of the columns and are related to the pixel location by the readout clocking sequence. Each row is subsequently processed in this manner. Other more complex readout techniques have also been developed, which sum charges of several pixels (or rows) at the readout locations. Once the CCD is read, the charge value in each pixel is effectively erased and the device begins integrating anew.

The CCD uses two basic types of readout. One type is referred to as "burst mode" and the second type employs opaque transfer registers. In the burst mode, the frame is integrated and read out rapidly a line at a time. The charge is read directly and, as such, no dead zone exists between elements.

For CCDs employing opaque transfer registers, the charge packets are shifted into registers and clocked out while the next frame is integrating. The opaque areas form dead zones between the elements, the size of which are on the same order as the pixel elements.

6.3.2 Charge-Injection Device Receivers

The CID does not transfer charge via a bucket brigade technique. Instead, the charge is transferred to a row and column-connected capacitor. Individual pixels can be addressed without destroying the charge stored in adjacent pixels. Additionally, the charge is not destroyed during the readout process. This allows multiple reads of the same pixel as a function of time.

As a result of direct pixel addressing, schemes have been developed to address a summed superpixel block of 4 by 4 pixels. This feature is extremely useful since it significantly reduces the time required to read an entire array.

6.3.3 Required Signal for Integrating Device Receivers

For acquisition receivers, the entire N by N array must be read to effectively cover the uncertainty viewfield. While large numbers of pixels reduce the background radiation on a per pixel basis, the readout time increases as the ratio of the number of pixels. Tradeoffs must exist between pixel density, readout time (i.e., speed and ultimately readout noise) and background noise levels.

The required signal for CTD devices follows as (6.1), with the noise given by the following.

6.3.3.1 *Background Noise*

The background noise is given by

$$i_{\mathrm{b}}^2 = 2qR_dB\frac{P_B}{N^2} \tag{6.26}$$

where N^2 is the number of individual pixels in the total FOV and P_B is the overall background signal power collected in the total FOV.

6.3.3.2 *Dark Noise*

Similarly, the total device dark noise is reduced by the number of pixels and is given by

$$i_d^2 = 2q\frac{I_d}{N^2}B \tag{6.27}$$

6.3.3.3 *Readout Noise*

The readout noise is given by

$$i_{\mathrm{ro}}^2 = (\mathrm{NEI})^2B \tag{6.28}$$

6.3.3.4 Fixed Pattern Noise

Fixed pattern noise is present due to nonuniformities in pixel sensitivities and noise properties. Post-processing techniques can be employed to reduce the fixed pattern present upon the CCD, but increases the rms noise current by a factor of the square root of 2.

For CCD readout, the total noise can be written as

$$\sigma_n^2 = i_b^2 + i_d^2 + i_{\mathrm{ro}}^2 + i_{\mathrm{fpn}}^2 \tag{6.29}$$

and for CID detectors using superpixel processing,

$$\sigma_n^2 = 16 \cdot \left(i_b^2 + i_d^2 + i_{\mathrm{fpn}}^2\right) + 4 \cdot i_{\mathrm{ro}}^2 \tag{6.30}$$

The required signal then follows the same logic as the QPIN analysis and is given identically as (6.18), namely,

$$I_{\mathrm{pk}} = \left(K_1\sigma_n + qBK_2^2\right) + \left[\left(K_1\sigma_n + qBK_2^2\right)^2 + \left(K_2^2 - K_1^2\right) \cdot \sigma_n^2\right]^{1/2} \tag{6.31}$$

6.4 SUMMARY

The required signal expressions for several detection waveform and receiver types has been presented. The choice of receiver and waveform should result from tradeoffs that occur at the system level and consider both acquisition and tracking system solutions. While CTDs are attractive from a large viewfield staring consideration, with adequate uncertainty areas QAPDs have consistently been the designers receiver of choice. With the development of the 14 by 14 pixel ssCD device for the European Space Agency laser communication experiment [5], the CCD is beginning to play an ever-increasing role in laser communication system designs.

References

[1] Van Trees, H. L. *Detection, Estimation and Modulation Theory—Part I*, New York, NY: John Wiley and Sons, 1968.

[2] Abramowitz, M., and Stegun, I. A. *Handbook of Mathematical Functions*, U.S. Department of Commerce, National Bureau of Standards, 1967.

[3] Zeimer, R. H., and Tranter, W. H. *Principles of Communications—Systems, Modulation and Noise*, Boston, MA: Houghton Mifflin, 1976.

[4] Gardner, F. M. *Phase-Lock Techniques*, New York, NY: John Wiley & Sons, 1966.

[5] Mathur, R. P., Beard, C. I., and Purll, D. J. "Analysis of SILEX Tracking Sensor Performance," *SPIE Proc.—Free Space Laser Communication Technologies II*, Vol. 1218, Jan. 1990.

Chapter 7

The Tracking Receiver

Because of the inherent narrow beam divergences associated with laser crosslinks, cooperative beam tracking must be employed in most system designs. Tracking accurately on the received signal allows return pointing of narrow beams to the companion terminal, thereby maintaining the required linkage. Tracking is therefore an extremely important aspect of any laser communication system.

In Chapter 3, the types of tracking systems available were identified and the merits of each were discussed. Here. the required signal expressions for a variety of free-space tracking systems are presented. The governing requirements of the tracking system are identified and expressions are formulated for both direct detection and coherent-detection tracking using quadrant detector devices. In addition, array devices for tracking are explored using detectors such as charge-coupled devices. The angle to voltage transfer function (i.e., the slope factor) for various spot intensity profiles is additionally defined.

7.1 GOVERNING REQUIREMENTS

The governing requirements on tracking accuracy can be traced to the closed loop tracking budget. The residual tracking error, that is the *root mean square* (rms) tracking budget, becomes the driving requirement on tracking accuracy. Recall from Chapter 3 that, in its simplest form, the expression for tracking jitter is given as

$$\sigma_{\mathrm{rms}} = \frac{1}{\mathrm{SF} \cdot \sqrt{\mathrm{SNR}}} \tag{7.1}$$

where SF is the angular slope factor of the angle to voltage (or current) transfer function expressed in units of per radian, and SNR is the signal to noise ratio in the tracking system bandwidth. The value σ_{rms} has also been referred to as the *noise equivalent angle* (NEA) of the tracking system. From this equation, it is apparent an understanding of the values of both the *signal to noise ratio* (SNR) and *slope factor* (SF) are imperative in determining the value of σ_{rms}.

7.2 REQUIRED SIGNAL DEVELOPMENT

In developing the expressions for required signal, the type of tracking system must be identified. This text will concentrate on tracking system designs that operate on pulsed, analog, and dc waveforms. Both direct detection and coherent tracking systems will be addressed.

7.2.1 Conventional Quadrant Devices

Most laser communication systems employ quadrant, or monopulse, tracking of the received signal. Quadrant detector devices (a detector divided into four equal sectors or quadrants) and quadrant orientations of detectors have been used to develop the tracking signals. Included in these devices have been *quadrant avalanche photodiodes* (QAPDs), *quadrant p-intrinsic* (QPIN) photodiodes, and quadrant image dividing prisms with discrete *avalanche photodiodes* (APDs). In the simplest sense, the received signal is divided into quadrants as shown in Figure 7.1. The quadrant signals are then summed and differenced to yield azimuth and elevation (or X and Y axes) tracking signals. The azimuth and elevation signals are formulated using the following expressions.

$$\frac{A + B - C - D}{A + B + C + D} = \text{azimuth signal} \tag{7.2}$$

$$\frac{A + C - B - D}{A + B + C + D} = \text{elevation signal} \tag{7.3}$$

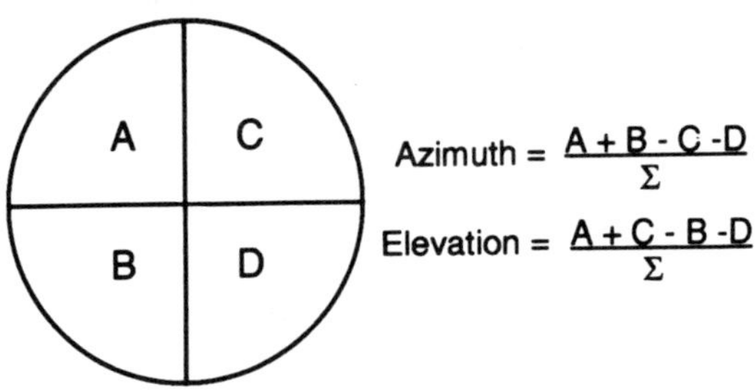

Figure 7.1 Quadrant detector orientation.

Equations (7.2) and (7.3) represent the azimuth and elevation signal when only signal is present. When tracking in the presence of noise, (7.2) and (7.3) are rewritten as

$$\mathrm{AZ} = \frac{A + n_a + B + n_b - C - n_c - D - n_d}{A + B + C + D + n_a + n_b + n_c + n_d} \tag{7.4}$$

$$\mathrm{EL} = \frac{A + n_a + C + n_c - B - n_b - D - n_d}{A + B + C + D + n_a + n_b + n_c + n_d} \tag{7.5}$$

where A, B, C, and D are quadrant signal levels, as before, and n_a, n_b, n_c, and n_d are the quadrant noise levels.

Equations (7.4) and (7.5) can be rewritten for reasonably large SNR levels (i.e., SNR > 10) as

$$\mathrm{AZ} = f(x) + \frac{n_a + n_b - n_c - n_d}{\Sigma} \tag{7.6}$$

$$\mathrm{EL} = f(y) + \frac{n_a + n_c - n_b - n_d}{\Sigma} \tag{7.7}$$

where Σ is the sum of A, B, C, and D, and $f(x)$ and $f(y)$ are the quadrant difference signals $(A + B - C - D)/\Sigma$ and $(A + C - B - D)/\Sigma$.

Now, when each of the noise terms is defined by a zero mean process, the expected values of (7.6) and (7.7) are given as

$$E\{\mathrm{AZ}\} = f(x) \tag{7.8}$$

$$E\{\mathrm{EL}\} = f(y) \tag{7.9}$$

where $E\{\ \}$ denotes the expected value of the bracket function.

The variance of a statistical process is defined by Papoulis [1] as

$$\sigma_x^2 = E\{x^2\} - E\{x\}^2 \tag{7.10}$$

After some mathematical manipulation, the variance of (7.8) follows as

$$\sigma_{\mathrm{AZ}}^2 = E\{\mathrm{AZ}^2\} - E\{\mathrm{AZ}\}^2$$

$$\sigma_{\mathrm{AZ}}^2 = E\{f(x) \cdot f(x)\} + \frac{E\{n_a^2\} + E\{n_b^2\} + E\{n_c^2\} + E\{n_d^2\}}{\Sigma^2} - E\{f(x)\} \cdot E\{f(x)\}$$

$$\sigma_{\mathrm{AZ}}^2 = \frac{E\{n_a^2\} + E\{n_b^2\} + E\{n_c^2\} + E\{n_d^2\}}{\Sigma^2} \tag{7.11}$$

where the random variables $f(x)$ are statistically independent. Noting that the expected values of the noise terms are ideally their variances (since noise process is zero mean), (7.11) reduces to

$$\sigma_{\mathrm{AZ}}^2 = \frac{\sigma_a^2 + \sigma_b^2 + \sigma_c^2 + \sigma_d^2}{\Sigma^2} \tag{7.12}$$

Because all the variances above result from an identical noise process, they are equal and the resultant becomes

$$\sigma_{\mathrm{AZ}}^2 = \frac{4\sigma_n^2}{\Sigma^2} \tag{7.13}$$

where σ_n^2 is the per quadrant noise variance. The rms noise is then given by

$$\sigma_{\mathrm{AZ}} = \frac{2\sigma_n}{\Sigma} \tag{7.14}$$

Similarly,

$$\sigma_{\mathrm{EL}} = \frac{2\sigma_n}{\Sigma} \tag{7.15}$$

Examining (7.14) and (7.15) closely shows the numerators are exactly the square root of the total noise power (sum of all quadrants) and the denominators are the square root of the total signal power. Thus, (7.14) or (7.15) can be written as the square root of the inverse of the SNR, or,

$$\sigma_{\mathrm{rms}} = \frac{1}{\sqrt{\mathrm{SNR}}} \tag{7.16}$$

Equation (7.16) defines the rms noise voltage (or current). Converting to angular coordinates requires multiplying (7.16) by the voltage to angle transfer function. The resultant is identically (7.1), namely

$$\sigma_{\mathrm{rms}} = \frac{1}{\mathrm{SF} \cdot \sqrt{\mathrm{SNR}}} \tag{7.17}$$

The SNR is given as it was previously in Chapter 5 and is expressed as

$$\mathrm{SNR} = \frac{I_{\mathrm{pk}}^2 (1 - 1/N_e)^2}{4 \cdot \left[i_{\mathrm{track}}^2 + qFBI_{\mathrm{pk}}(1 + 1/N_e) \right]} \tag{7.18}$$

where I_{pk} is the peak signal current, N_e is the transmitter extinction ratio, F the detector excess noise factor, and B the track loop bandwidth. This expression holds for all pulsed and tone-modulated direct-detection tracking links that use conventional quadrant devices and techniques. When employing a square law tracker, the SNR is modified as in Chapter 3.

The noise current in (7.18) is given as

$$i_{\text{track}}^2 = i_b^2 + i_d^2 + i_{\text{pa}}^2 \tag{7.19}$$

where i_b is the background noise current, i_d is the detector dark noise current, and i_{pa} is the preamplifier noise current. These currents are defined in detail in Chapter 5 and are repeated here for reference for both QAPD and QPIN detectors.

For QAPD or quadrant orientations of APD detectors,

$$i_b^2 = 2qFBR_dP_B \tag{7.20}$$

$$i_d^2 = 2qBI_{\text{du}} + 2qBFI_{\text{dm}} \tag{7.21}$$

$$i_{\text{pa}}^2 = 4 \cdot (\text{NEI})^2/M^2 \tag{7.22}$$

where q is the quantum charge, F is the detector excess noise factor, B the track loop bandwidth, R_d the detector responsivity, P_B the background signal level, I_{du} the total unmultiplied detector dark current, I_{dm} the total multiplied detector dark current, M the avalanche gain of the quadrant avalanche photodiode, and NEI the noise equivalent current of the quadrant preamplifiers.

For PIN photodiodes, (7.21), (7.22), and (7.23) reduce to

$$i_b^2 = 2qBR_dP_B \tag{7.23}$$

$$i_d^2 = 2qBI_{\text{du}} \tag{7.24}$$

$$i_{\text{pa}}^2 = 4 \cdot (\text{NEI})^2 \tag{7.25}$$

Solving (7.18) for SNR and equating to (7.19) yields

$$\frac{1}{(\text{SF} \cdot \sigma_{\text{rms}})^2} = \frac{I_{\text{pk}}^2(1 - 1/N_e)^2}{4 \cdot \left[i_{\text{track}}^2 + qFBI_{\text{pk}}(1 + 1/N_e)\right]} \tag{7.26}$$

which yields the following when solving for I_{pk} in terms of σ_{rms} and SF:

$$I_{pk} = \frac{2qFBI_{pk}(1 + 1/N_e)}{(\mathrm{SF} \cdot \sigma_{rms})^2 \cdot (1 - 1/N_e)^2} + \sqrt{\left[\frac{2qFBI_{pk}(1 + 1/N_e)}{(\mathrm{SF} \cdot \sigma_{rms})^2 \cdot (1 - 1/N_e)^2}\right]^2 - \frac{4 \cdot i_{track}^2}{(\mathrm{SF} \cdot \sigma_{rms})^2 \cdot (1 - 1/N_e)^2}} \quad (7.27)$$

The peak required signal power follows directly as

$$P_{pk} = \frac{I_{pk}}{R_d} \quad (7.28)$$

Solving for the track required signal is simple once the rms angle error budget is defined and the received image spot and detector characteristics are known.

7.2.2 Integrating Devices

A second class of direct-detection tracking detector devices is known as the integrating device. Typically these devices are characterized by multielement N by N arrays. Devices in this class are the *charge-coupled device* (CCD) and the *charge injection device* (CID). These devices are excellent background discriminators with their small individual pixel viewfields but have the added noise associated with signal readout.

These multielement devices, when used for tracking, are often operated in a subarray readout mode. This offers a smaller Q by Q readout area that can be addressed at a higher rate yielding a much higher tracking bandwidth potential. This is evidenced when one considers a 10-MHz readout rate for a CCD. With 2,048 by 2,048 pixels, each pixel can be, at best, addressed only once every 0.42 seconds (a 2.4-Hz frame rate).

Smaller devices (that is, smaller numbers of pixels), have been explored as tracking devices. Most notable is the device being considered for the *European Space Agency's* (ESA) laser communication experiment [2]. With the smaller number of pixels, in this case 14 by 14, higher frame rates exist. One drawback of smaller pixel devices is the lack of background rejection afforded by the large array pixel devices when using the same small device as an acquisition detector. Careful trades must be performed to determine the optimal size of the custom pixel size devices.

The pixelized devices afford the luxury of having an active area that extends the linearity out to the detector edges. No transfer function (i.e., slope factor) saturation occurs once the spot traverses the image-dividing area as in quadrant devices. This can be useful for systems that require offset tracking of the receive spot. This does, however, require a large number of track-active pixels, which yields a much lower information rate.

Many laser crosslink designs have been postulated using the integrating arrays as trackers [2–9]. Typically, the arrays are used in subarray mode employing a 2 by 2 pixel set for tracking. However, the centroid calculations for received signal can extend to the maximum pixel dimension (i.e., N by N).

In addition to square arrays, other orientations of pixels have also been postulated as possible orientations for tracking. One such orientation involves a hexagonal arrangement of pixels [10].

7.2.2.1 Centroid Calculations

The centroid calculations for a simple 2 by 2 array are identically those developed for the quadrant detectors by (7.2) and (7.3). Extending this to an N by N array, where the centroid is now based upon the summing and differencing of the N by N pixels, yields the centroid locations.

$$\mathrm{AZ} = \frac{\sum\limits_{i}\left(\sum\limits_{j} A_{ij}\right)i}{\sum\limits_{i}\sum\limits_{j} A_{ij}} \tag{7.29}$$

$$\mathrm{EL} = \frac{\sum\limits_{j}\left(\sum\limits_{i} A_{ij}\right)j}{\sum\limits_{i}\sum\limits_{j} A_{ij}} \tag{7.30}$$

where A_{ij} is the pixel signal amplitude at the i, j location defined by Figure 7.2 and the rms noise associated with the centroid calculation follows directly from (7.10).

The azimuth centroid, for the case when a 3 by 3 pixel array with mean background noise subtracted, is given as

$$\mathrm{AZ} = \frac{A_{11} + A_{12} + A_{13} - A_{31} - A_{32} - A_{33}}{A_{11} + A_{12} + A_{13} + A_{21} + A_{22} + A_{23} + A_{31} + A_{32} + A_{33}} \tag{7.31}$$

where A_{ij} is the amplitude of the pixels as defined in Figure 7.2.

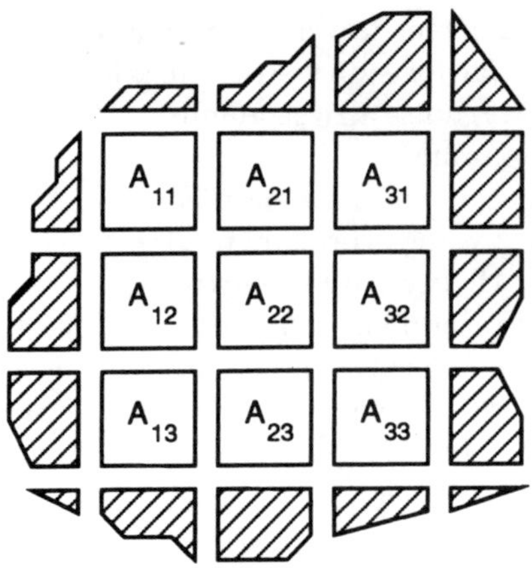

$$\text{AZ} = \frac{A_{11}+A_{12}+A_{13}-A_{31}-A_{32}-A_{33}}{\Sigma} \qquad \text{EL} = \frac{A_{11}+A_{21}+A_{31}-A_{13}-A_{23}-A_{33}}{\Sigma}$$

Figure 7.2 Pixel identification for centroiding.

When noise is present, (7.29) becomes

$$\text{AZ} = \frac{A_{11} + n_{11} + A_{12} + n_{12} + A_{13} + n_{13} - A_{31} - n_{31} - A_{32} - n_{32} - A_{33} - n_{33}}{\Sigma + n_{11} + n_{12} + n_{13} + n_{21} + n_{22} + n_{23} + n_{31} + n_{32} + n_{33}} \tag{7.32}$$

where Σ is the sum of all pixels.

When the sum signal is much larger than the noise, the denominator reduces to the sum and the expected value is given by

$$\text{AZ} = \frac{A_{11} + A_{12} + A_{13} - A_{31} - A_{32} - A_{33}}{\Sigma} \tag{7.33}$$

The expected value of the centroid is identical to (7.29) when the means and variances of the individual pixels are respectively equal.

The mean square value is given by

$$E\{\text{AZ}^2\} = \left(\frac{A_{11} + A_{12} + A_{13} - A_{31} - A_{32} - A_{33}}{\Sigma}\right)^2 + \frac{6n_o^2 - 6\mu^2}{\Sigma^2} \tag{7.34}$$

where μ is the mean of the pixel noise process. The variance following from (7.10) becomes

$$\sigma_{\text{rms}}^2 = \frac{6\sigma_n^2}{\Sigma^2} \tag{7.35}$$

The rms angle error follows as

$$\sigma_{\text{rms}} = \frac{\sqrt{6}\,\sigma_n}{\Sigma} = \frac{\sigma_{\text{total}}}{\Sigma} = \frac{1}{\sqrt{\text{SNR}}} \tag{7.36}$$

Extending this to the general solution for an N by N centroid yields

$$\sigma_{\text{rms}} = \frac{\sqrt{2N}\,\sigma_n}{\Sigma} = \frac{\sigma_{\text{total}}}{\Sigma} = \frac{1}{\sqrt{\text{SNR}}} \tag{7.37}$$

where in each case the total noise is a function of the pixels used in the centroid calculation.

While ideally it would be perfect for these integrating devices to track on continuous wave (CW) signals, the ability to discriminate the tracking signal from point background sources, such as stars, would be extremely difficult, even with spectral rejection filtering. Since these devices operate on the integrated content of the tracking signal, the temporal content of the signal, which is higher in bandwidth than the readout rate is in effect filtered out by the integration process. To overcome this, techniques have been used to impart some temporal content into the track signal.

One such technique involves modulating the tracking signal at a low rate that is within the readout rate of the frames. This usually requires frame sampling the tracking signal at a rate of at least four times its bandwidth. More appropriate would be to sample the signal at at least ten times its bandwidth for proper reconstruction in the track processing electronics.

7.2.2.2 Noise Considerations

Like the quadrant detectors previously discussed, the imaging devices are also subjected to noise sources. The major contributors to the noise sources include detector dark noise current, background noise current, readout noise current, and noise variance due to fixed pattern noise. Each of the noise contributors below is referenced to a single pixel.

Background Noise. The noise contribution to the track process resulting from background is given as

$$i_b^2 = 2qR_dP_BB \tag{7.38}$$

where q is the quantum charge, R_d the detector responsivity, B the readout bandwidth, and P_B the background power per pixel which is given by

$$P_B = H_B \cdot \frac{\pi}{4}(\mathrm{IFOV})^2 \cdot L_{\mathrm{optics}} \cdot A_{\mathrm{aper}} \cdot B_{\mathrm{filter}} \tag{7.39}$$

where H_B is the background radiant flux, IFOV the individual pixel FOV, L_{optics} the optics losses in the receiver system, A_{aper} the effective aperture area, and B_{filter} the bandwidth of the optical rejection filter.

Readout Noise. The noise associated with readout is attributed to the preamplifiers and are bandwidth dependent. The preamplifier noise density is given by

$$i_{\mathrm{ro}}^2 = (\mathrm{NEI})^2 \cdot B \tag{7.40}$$

where NEI is the noise equivalent current given in $\mathrm{A/Hz}^{1/2}$.

Dark Noise. The bulk dark noise is normally a function of pixel size and temperature. The dark current noise is given by

$$i_d^2 = 2qI_dB \tag{7.41}$$

where I_d is the bulk dc dark current per pixel.

Fixed Pattern Noise. Nonuniform background patterns resulting from spatial variations in pixel size, capacitance, and bias charge form variances known as *fixed pattern noise* (FPN). This FPN can be mitigated with additional calculation during post-readout processing, resulting in the FPN being subtracted from the samples. The pattern is stored in memory and subtracted on a pixel by pixel basis. This does result in additional noise since the noise variance of the stored sample is added in quadrature with the signal sample.

Total Noise Power. The total noise power can be written as

$$i_{\mathrm{total}}^2 = 2 \cdot \left(i_b^2 + i_{\mathrm{ro}}^2 + i_d^2 + i_{\mathrm{FPN}}^2\right) \tag{7.42}$$

where the factor of two results from the fixed pattern noise reduction algorithm.

The rms angle error is then given by rewriting (7.35) as

$$\sigma_{\mathrm{rms}} = \frac{\sqrt{2 \cdot \left(i_b^2 + i_{\mathrm{ro}}^2 + i_d^2 + i_{\mathrm{FPN}}^2\right)}}{\Sigma} \tag{7.43}$$

The required signal then follows as in (7.25).

7.3 REQUIRED SIGNAL DEVELOPMENT (COHERENT TRACKING)

Coherent tracking involves the use of a local oscillator to effectively beat the transmitted signal down to an intermediate frequency or to baseband. When tracking using coherent techniques, the required signal is similarly based upon (7.1) but the slope factor can be tailored using various types of local oscillator intensity profiles. In addition, when filling the detector with the tailored local oscillator profile, the mixing of the oscillator and the received intensity profile yields an effective continuum of detectors increasing the resolution of the tracking signal.

The required signal is given again by (7.27); however, the total track noise power, i_{track}^2 is given below with the added inclusion of the local oscillator shot noise.

The noise contributors associated with coherent detection of the signal are similar to the direct-detection noise discussed above with the exception of the local oscillator shot noise contribution. These noise terms were previously discussed in Chapter 5. They are repeated here for completeness

The total noise current is given as

$$i_{\text{track}}^2 = i_b^2 + i_d^2 + i_{\text{pa}}^2 + i_{\text{lo}}^2 \tag{7.44}$$

where again i_b is the background noise current, i_d is the detector dark noise current, i_{pa} is the preamplifier noise current, and i_{lo} the local oscillator shot noise current.

For quadrant PIN detectors

$$i_b^2 = 2qBR_dP_B \tag{7.45}$$

$$i_d^2 = 2qBI_{\text{du}} \tag{7.46}$$

$$i_{\text{pa}}^2 = 4 \cdot (\text{NEI})^2 \tag{7.47}$$

$$i_{\text{lo}}^2 = 2qBR_dP_{\text{lo}} \tag{7.48}$$

where again q is the quantum charge, B the track loop bandwidth, R_d the detector responsivity, P_B the background signal level, I_{du} the total unmultiplied detector dark current, NEI the noise equivalent current of the quadrant preamplifiers, and P_{lo} the local oscillator power.

7.4 SLOPE FACTOR DETERMINATION

As previously addressed, in ratiometric processing, two signals are produced from the four quadrant signals (or more if pixelized), one signal for the azimuth and one for the elevation corrections necessary to recenter the received signal on the detector boresight. For quadrant devices, if the illumination is entirely on the right side of the junction, the output signal is a "1." If the illumination is entirely on the left side of the junction, the output is a "−1." And if the illumination is boresighted, a "0" is output.

If the illumination position is scanned across the junction from left to right, the resultant transfer function will look like the sketch shown in Figure 7.3. In most tracking applications (except for offset tracking), it is desirable to keep the incoming beam as close to the center boresight position as possible. This implies that the azimuth and elevation signals should be near zero in any practical application. The slope factor analysis is based upon the assumption that the curve may be represented by a line tangent to the curve at the origin of the transfer function. This slope factor then becomes a measure of the detector's sensitivity to position (which equates to angle in the far field) changes away from the boresight of the detector.

Normally, a high slope factor indicates the detector is very sensitive to spot movement. Thus, a high slope factor allows the received beam to be tracked very precisely. This however must be tempered by the magnitude of the errors. If the track error magnitude is larger than the linear region, inaccuracies, due to the nonlinearity become prevalent.

For most systems, the linear range region, and therefore spot size, are "customized" to the budgeted tracking accuracy. As a rule of thumb, the spot size should be at least twice the linear range desired. In general, high slope factors (narrow spots) produce narrow linear ranges and large spots produce large linear ranges and small slope factors.

Many detector types, transition types and spot intensity profiles have been studied for the direct-detection quadrant devices. Figure 7.4 shows the layout of the basic quadrant detector. The boundary of the transition region (i.e., the "walls" of

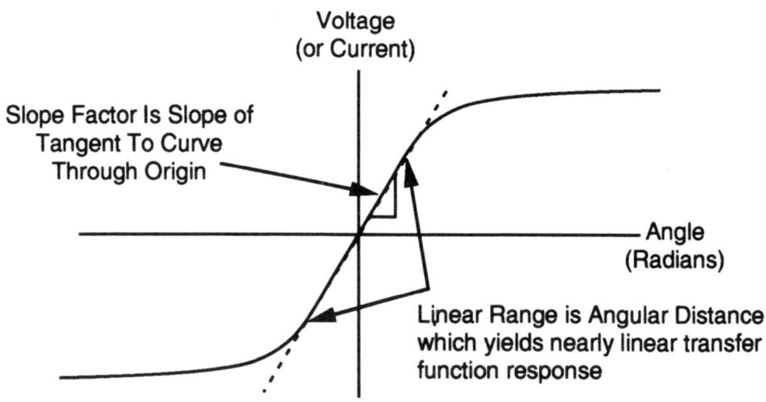

Figure 7.3 Generic transfer function.

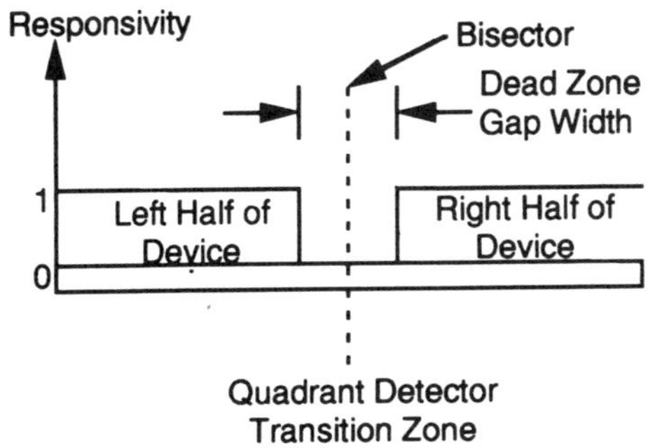

Figure 7.4 Basic quadrant detector with dead zone.

the transition gap) for this basic device is assumed to be parallel. The transition region is defined as the area between the two quadrants. Different transition regions provide different slope factor and linearity characteristics.

7.4.1 Transition Region Types

Several basic transition region types are presented. Each is discussed below.

7.4.1.1 Dead Zone

In dead zone transition region devices, light that falls in the transition region is not detected by any of the quadrants. Light falling in the dead zone is lost. Therefore, minimizing the dead zone size to spot size ratio is important to the system designer. While the system will lose signal in the dead zone region, the slope factor is typically the highest for a dead zone device as compared to other transition types.

7.4.1.2 Shared Linear

A shared linear transition region allows for a more graceful transition between the quadrants. Figure 7.5 depicts sharing linear transition device characteristics. Device types can be fabricated with center responsivities of one, or in the extreme case, zero. For a sharing transition region, the slope factor is limited by the size of the gap.

7.4.1.3 Sinusoidal

Very similar to the shared linear device is the shared sinusoidal transition device. It has a higher maximum slope factor than the shared linear, but like the shared linear is limited by the size of the transition gap. Figure 7.6 illustrates the shared sinusoidal transition region.

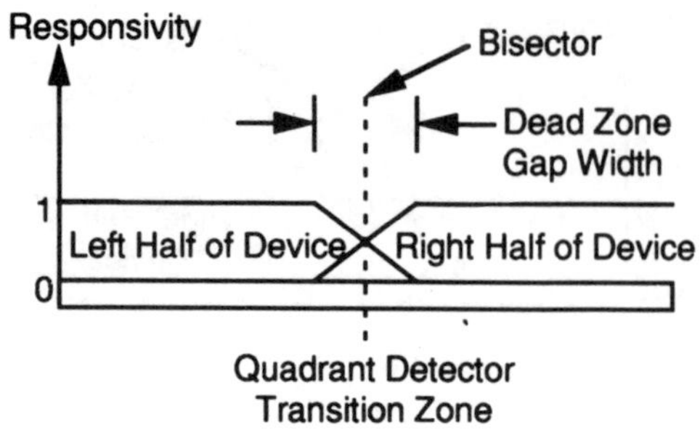

Figure 7.5 Shared linear transition profile.

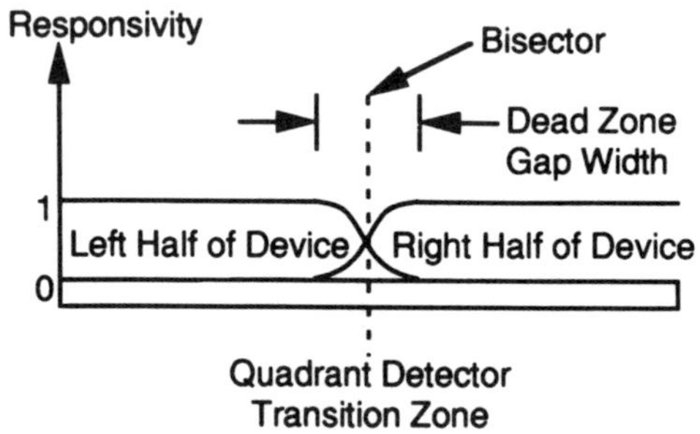

Figure 7.6 Shared sinusoidal transition profile.

7.4.2 Slope Factor

Given the various transition regions, slope factors, which are a function of the transition gap width, have been calculated for the Airy intensity pattern. Figures 7.7 through 7.9 illustrate the slope factors and on-axis efficiencies for these various transition types.

Closed form solutions for the slope at the origin of the transfer function have also been identified [11]. For the Airy pattern, the slope factor can be expressed as

$$\mathrm{SF} = \frac{4.14}{\theta_A} \tag{7.49}$$

in units of per radian where θ_A is the optical spot diameter in radians. This same analysis has been performed for a Gaussian spot intensity profile with the resultant slope factor given as

$$\mathrm{SF} = \frac{1.56}{\theta_A} \tag{7.50}$$

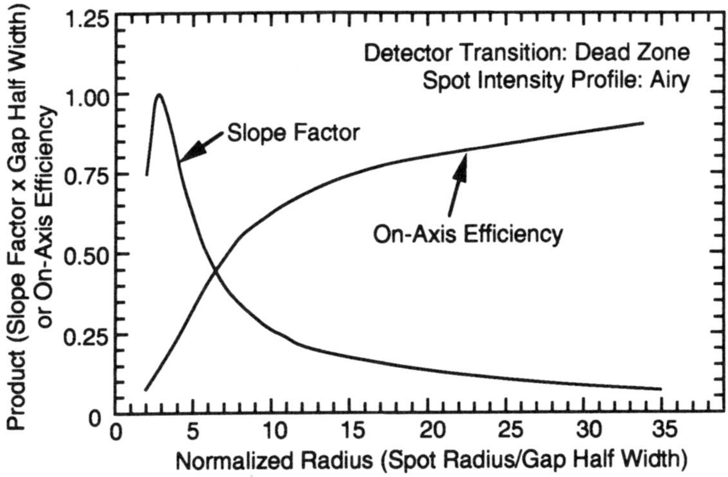

Figure 7.7 Slope factor and efficiency: quadrant detector with dead zone.

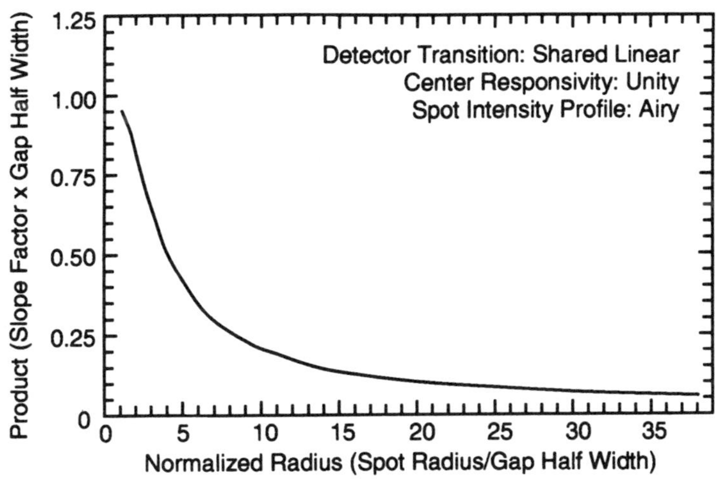

Figure 7.8 Slope factor: quadrant detector with linear transition.

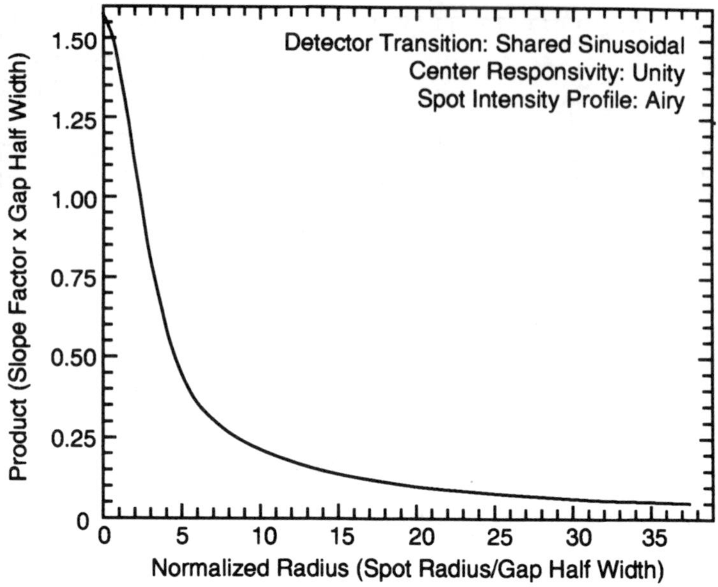

Figure 7.9 Slope factor: quadrant detector with sinusoidal transition.

and for a uniform spot intensity profile as

$$SF = \frac{1.27}{\theta_A} \tag{7.51}$$

Coherent tracking has been considered by Chan and Swanson [12] for various local oscillator intensity profiles and the Airy received spot profile. In general, a radial derivative of the received signal intensity yields the best slope factor performance, although more realizable profiles, such as matched Airy and uniform profiles, also provide good performance.

7.5 SUMMARY

The required signal calculations for both coherent and direct detection have been presented. The signal depends upon the system noise and the required rms tracking error component of the tracking system. Almost all laser crosslink systems to date have employed direct-detection tracking as most tracking systems are a subset of the acquisition receiver system, which for many reasons have remained direct detection.

References

[1] Papoulis, A. "*Random Variables and Stochastic Processes*," New York, NY: McGraw Hill, 1965.

[2] Mecherle, G. S., and Barry, J. D. "Laser Communication Terminals With Automated Video Tracking," *Proc. of SPIE Free Space Laser Communication Technologies*, Los Angeles, CA, Jan. 1988.

[3] Mathur, R. P., and Purll, D. J. "Tracking Sensor Developments For Optical Intersatellite Links," *Conf. Proc. on Optical Space Communications*, Paris, France, April 24–28, 1989.

[4] Wittig, M. and Oppenhauser, G. "The European SILEX Project: Concept, Performances, Status and Planning," *Proc. of SPIE Free Space Communication Technologies II*, Vol. 1218, Jan. 1990.

[5] Clark, D. L., Cosgrove, M. A., Short, R. C., Thielk, M., VanVranken, R., and Thomas, T. "Simultaneous Tracking and Low Data Rate Communications With High Speed Array Detectors for Optical Communications Network Management," *Proc. of SPIE Free Space Communication Technologies II*, Vol. 1218, Jan. 1990.

[6] Mathur, R. P., Beard, C. I. and Purll, D. J. "Analysis of SILEX Tracking Sensor Performance," *Proc. of SPIE Free Space Communication Technologies II*, Vol. 1218, Jan. 1990.

[7] Chen, C. C., Jeganathan, M., and Lesh, J. R. "Spatial Acquisition And Tracking For Deep Space Optical Communication Packages," *Proc. of SPIE Free Space Communication Technologies III*, Vol. 1417, Jan. 1991.

[8] Short, R. C., Cosgrove, M. A., Clark, D. L., and Oleski, P. "Performance of a Demonstration System for Simultaneous Laser Beacon Tracking and Low Data Rate Optical Communications With Multiple Platforms," *Proc. of SPIE Free Space Communication Technologies III*, Vol. 1417, Jan. 1991.

[9] Short, R. C., Cosgrove, M., Clark, D. L., Martino, A., Park, H. and Seery, B. "Acquisition and Tracking Performance Measurement for a High Speed Area Array Detector System," *Proc. of SPIE Free Space Communication Technologies III*, Vol. 1417, Jan. 1991.

[10] Cox, J. A. "Point-Source Location Using Hexagonal Detector Arrays," *Optical Engineering*, Vol. 26, No. 1, Jan. 1987.

[11] Ross, M, Freedman, P., Abernathy, J., Matassov, G., Wolf, J., and Barry, J. D. "Space Optical Communications With the Nd:YAG Laser," *IRE Proc.*, Vol. 66, No. 3, March 1978.

[12] Swanson, E. A., and Chan, V.W.S. "Heterodyne Spatial Tracking System for Optical Space Communication," *IEEE Trans. on Communications*, Vol. COM-34, No. 2, Feb. 1986.

Chapter 8

Key Technologies

8.1 OPTOMECHANICAL TECHNOLOGIES

Laser communication relies on optomechanical devices to perform many of its functions. Pointing is accomplished by a pointing element or gimbal that incorporates both mechanical and optical features. Further, fine track and point-ahead may be implemented with an optomechanical device, the beamsteerer. State changes required of the optics configuration are accomplished with mechanisms that relocate optical elements of the imaging optics to change optical relationships on command. Lastly, the assemblage of lenses, mirrors, and other optics require that they be positioned and held precisely throughout launch and on-orbit for the remainder on the mission.

This section covers these key optomechanical elements and discusses their function and configuration. The first of these devices discussed is the gimbaled pointing element.

8.1.1 Pointing Element Gimbals

In Chapter 4, three types of gimbaled pointing elements were discussed from an optical viewpoint. They share a common design in that each is required to provide pointing, stability and support to an optical element. In the case of the gimbaled telescope, the telescope is pointed by the gimbals. For the gimbaled mirror, a flat mirror is pointed to direct the line of sight. The third approach, the gimbaled package, requires the gimbals to point the entire system optics, leaving only the fixed electronics to complete the system.

Gimbals for the three pointing approaches are basically the same although differences may exist. The gimbals generally share the following characteristics:

- Two axes of motion. Three or more axes may be needed if angular access demands.
- The drives can accelerate the pointing element rapidly enough to follow the required line-of-sight motion. The gimbals work in conjunction with the fine track beamsteerers.
- They incorporate position encoders so that their pointing angles may be controlled precisely.
- They are structurally stiff to support and precisely position the pointing element, not interfere with the drive servo dynamics, and be stiff enough to survive launch. A rule of thumb is that the first structural resonance should be three to ten times the gimbal servo bandwidth with the ten ratio being preferred. This helps avoid structurally induced servo instability.
- For a gimbaled telescope, provide for a Coudé path as discussed in Chapter 4.

Figure 8.1 is a schematic block diagram of a two-axis gimbal.

Gimbal Configurations

There are many configurations for gimbals. Chapter 4 showed several of these, all based on an elevation inside azimuth (az-el) mount with a yoke to mount the elevation (inside) axis. There are many configurations possible, 31 of which are illustrated in Figure 8.2 and described in [1]. Note the very different approaches used to provide gimballing. The Z designation denotes those gimbals that are configured to view the zenith (straight up) and have a varying capability for viewing at the horizon. On the other hand, those gimbals in Figure 8.2 that carry the H designation are those that can view the horizon but are not suitable for zenith viewing due to their singularity at the zenith. The O designation simply stands for others. Two types of beamsteerers are also shown. The first is a pair of single-axis beamsteerers arranged to provide two-axis steering and the second is a two-axis beamsteerer, or fast-steering mirror as they are sometimes known.

A Coudé path was illustrated in Chapter 4. It is possible to provide more than one Coudé path through a gimbal system. If the gimbal system is of the type shown in Figure 8.3, two paths may be Coudéd as illustrated. This approach is limited to an outer axis rotation of about 270 deg so that outer axis rigidity may be retained. Designers should investigate gimbal design as thoroughly as they do other elements of the system since a lighter weight, more balanced, and lower power overall system will result.

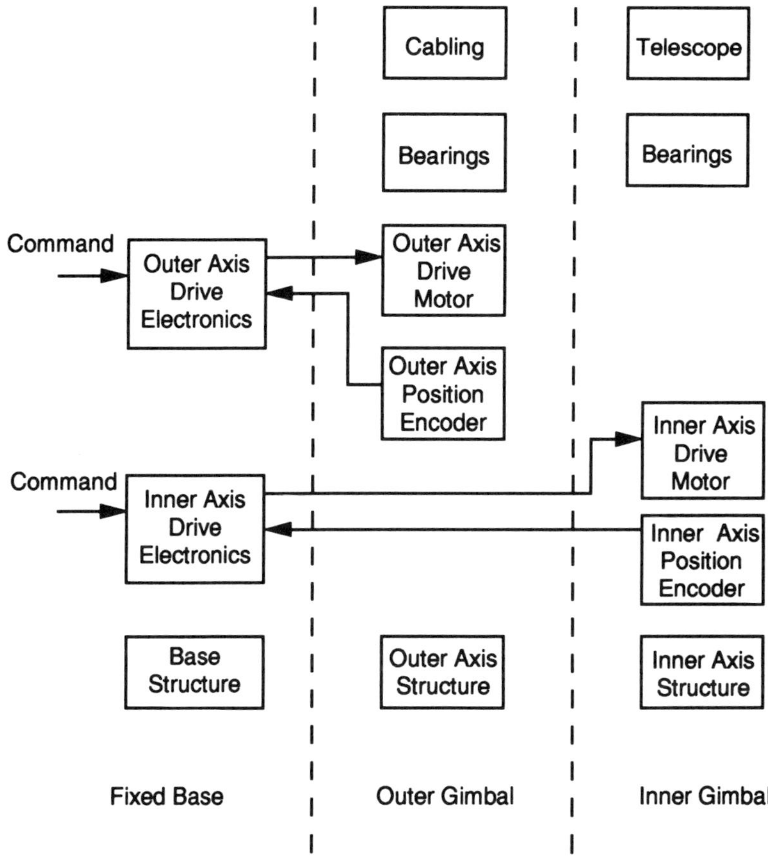

Figure 8.1 Two-axis gimbal block diagram.

The typical location of the components is illustrated in Figure 8.4. On the yoke, the position pickoff and motor are usually mounted on the same side to avoid servo instability, which might result from shaft torsional flexure although there are exceptions to this. The components that comprise a two-axis gimbal will be discussed in the following sections, starting with the drive motors.

Motors. Generally, gimbal motors are direct drive, brushless, *direct current* (dc) devices. Brush-type dc motors don't operate in the vacuum of space since air (and moisture, ideally) is required for the brush/commutator interface. A characteristic of brushless motors, or torquers as they are frequently called, is that since generally no gearing is used, the motors must possess enough torque to accelerate the gimbals

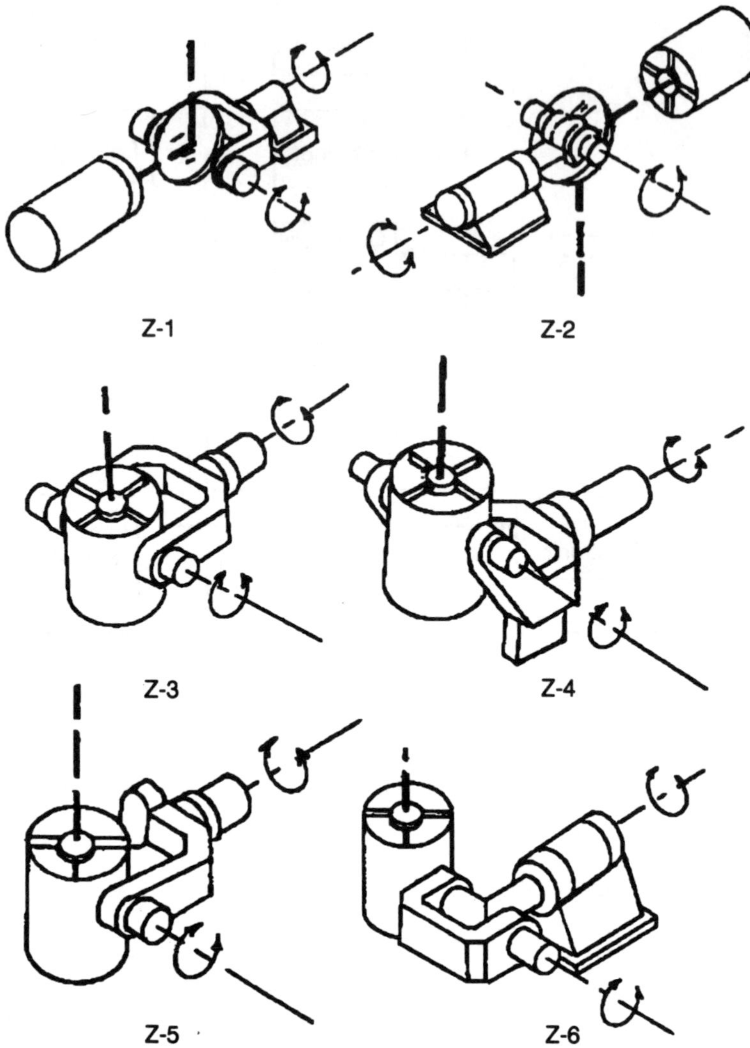

Figure 8.2 Gimbal configurations (31 configurations).

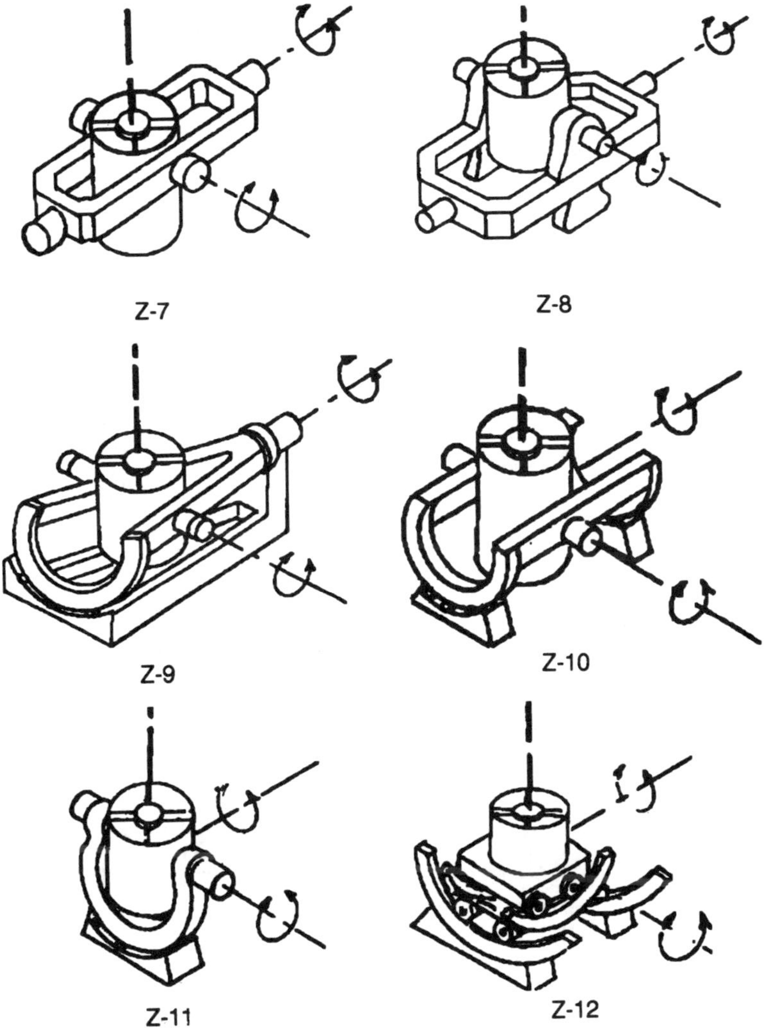

Figure 8.2 (continued)

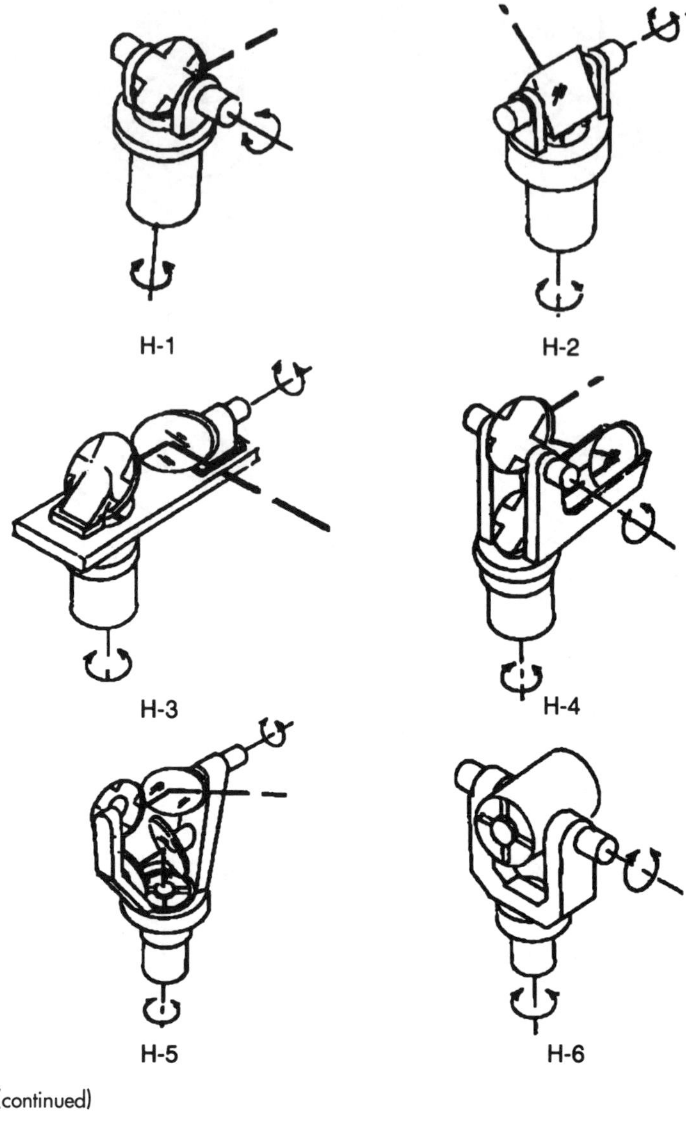

Figure 8.2 (continued)

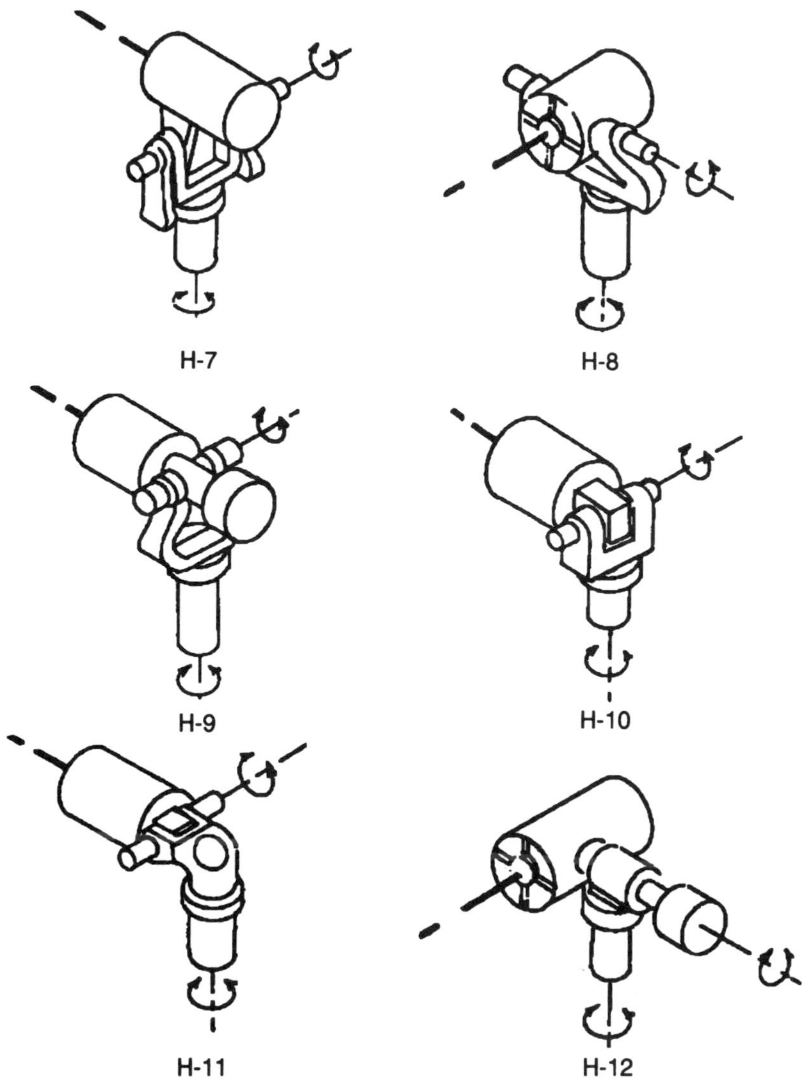

Figure 8.2 (continued)

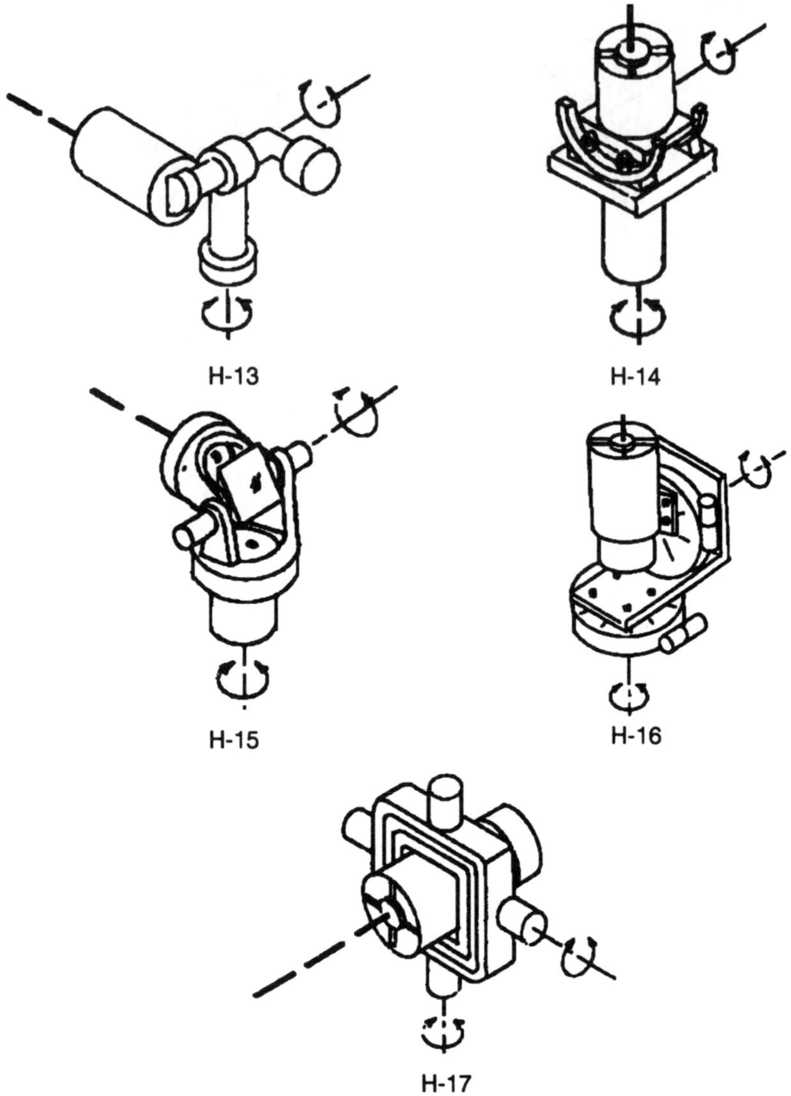

Figure 8.2 (continued)

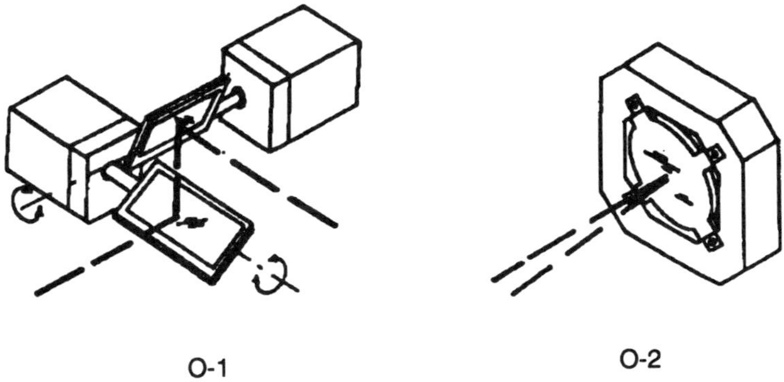

Figure 8.2 (continued)

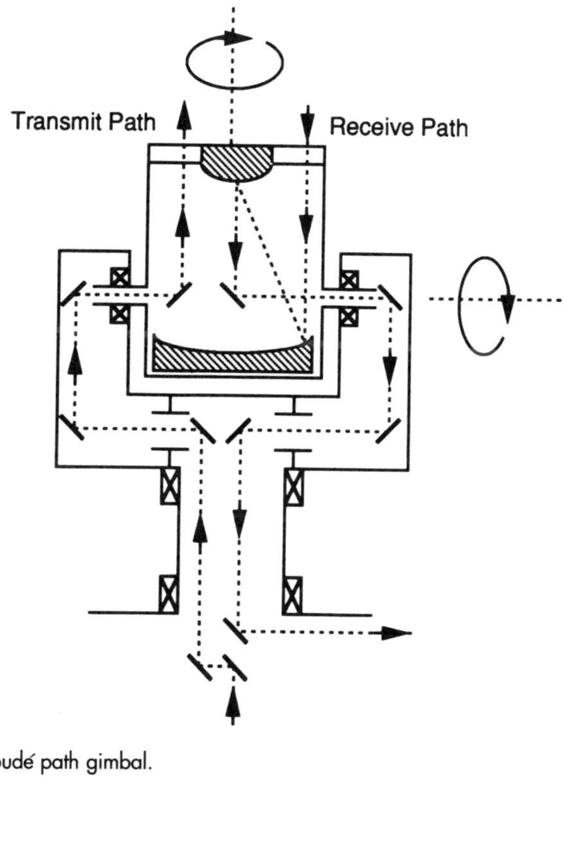

Figure 8.3 Two Coudé path gimbal.

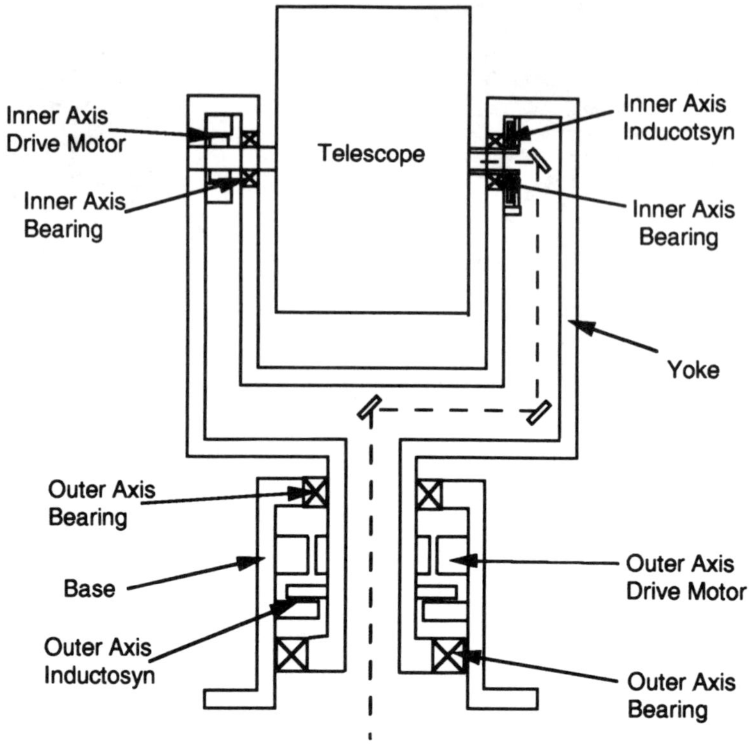

Figure 8.4 Gimbal component locations.

directly. This generally requires large torque and a limited rotation. The limited rotation results from the gimbal access, which is rarely more than 360 deg in either of the two axes.

Torque is proportional to drive current and is linear, which simplifies drive servo design. Since there are no brushes, commutation must be accomplished electronically. For limited rotation (as, for example, in an inner gimbal with a rotation of perhaps 60 deg), it is possible to use a limited rotation motor, again simplifying the design since no commutation is required.

To accommodate the Coudé path of a gimbaled telescope, the central hole of a torquer can be made large compared to the overall diameter. In this manner, there is room to pass a Coudé optical path through the center of the gimbal axis.

Figure 8.5 shows the drive motor circuitry for a two-axis system where a torque motor is used for the outer azimuth axis and a limited rotation motor drives the inner elevation axis. Note that two circuits are used for the torquer in the azimuth axis. This dual circuit accomplishes the commutation necessary for a full

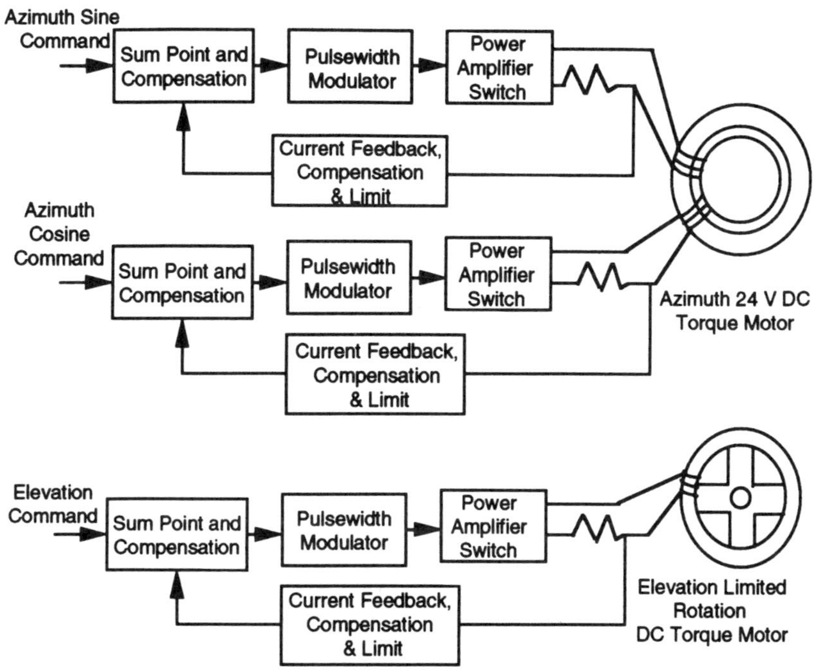

Figure 8.5 Drive motor circuitry.

rotation dc motor. As was mentioned above, no commutation is required for the limited rotation motor for the inner elevation axis. Pulse width modulation is used to drive the motors using pulses rather than a continuous signal, the pulse width being varied to duty cycle the torque. This approach saves power and reduces heating of the motor as well as providing better positioning control.

Some gimbals rely on torque increase by gearing, which, in the broadest sense, covers any drive where there is a rotation ratio between two driven elements and this includes capstan to cable drum torque increases. Conventional gearing may be of a worm, spur, or bevel gear type. Using this torque-increasing approach, the designer can incorporate a smaller, higher speed motor of low torque output to drive the gimbals. Cogging of a pointing element is unacceptable for a very narrow beam system since the cogging could mispoint the beam momentarily, enough to disrupt the received signal at the cooperating terminal. As a result, tooth to tooth action must be smooth. A unique type of gearbox has been used in space by Schaeffer Magnetics of California. This firm has incorporated a gearbox that is backlash free, the Harmonic Drive. The outer gear is flexible and it is squeezed as the inner planetary gear rotates inside it. The mesh between the two gears is broken smoothly at one point in the rotation and then remeshed at another point, the result being a cogless relative rotation between the two gears. The gear ratio is related to the difference in the number of teeth between the two gears. There is a lot of space experience with this drive, which has included long-term space-qualified lubricant. Harmonic Drives drove the astronauts around on the Moon in the Lunar Rover many years ago. A drive suitable for a laser communication system that would use a Harmonic Drive gearbox is illustrated in Figure 8.6. Note that a Coudé path cannot be incorporated into this design since the harmonic drive is a centered unit (i.e., it uses a central shaft). The drive motors can be stepper motors or dc brushless with Inductosyn-position encoders. None of these drives has failed in space in spite of the fact that they have been used on dozens of spacecraft applications. Drives such as these are used on the Hubble Telescope.

Another unique gimbal design is one which offers the promise of very light weight coupled with the capability for a Coudé path, as shown in Figure 8.7. The gimbal, by Sagebrush Technology, uses a drive where multiple steel cables are

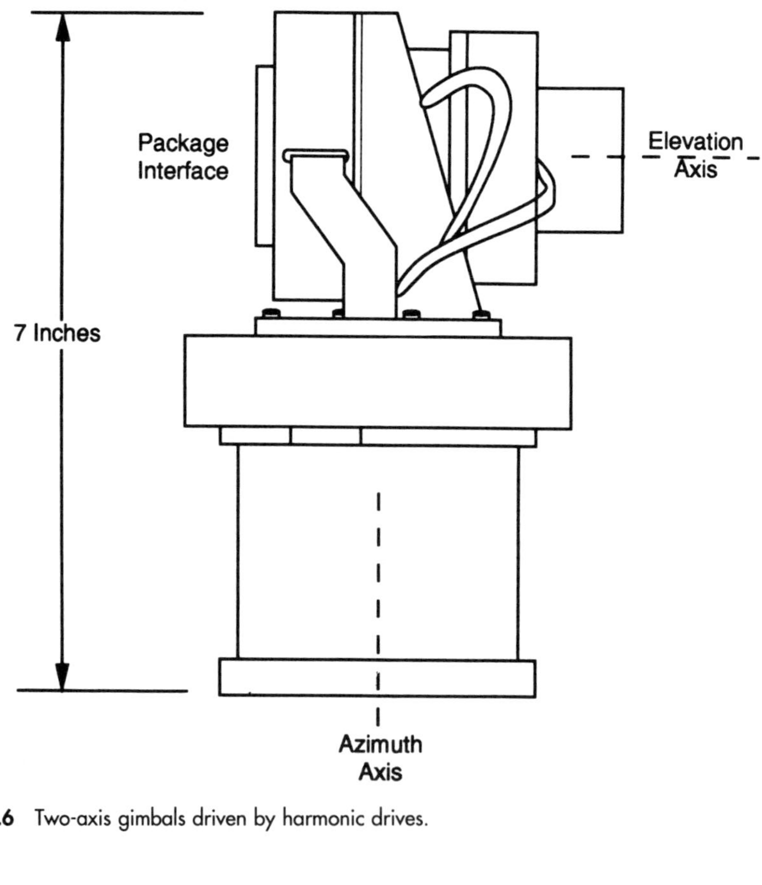

Figure 8.6 Two-axis gimbals driven by harmonic drives.

(a)

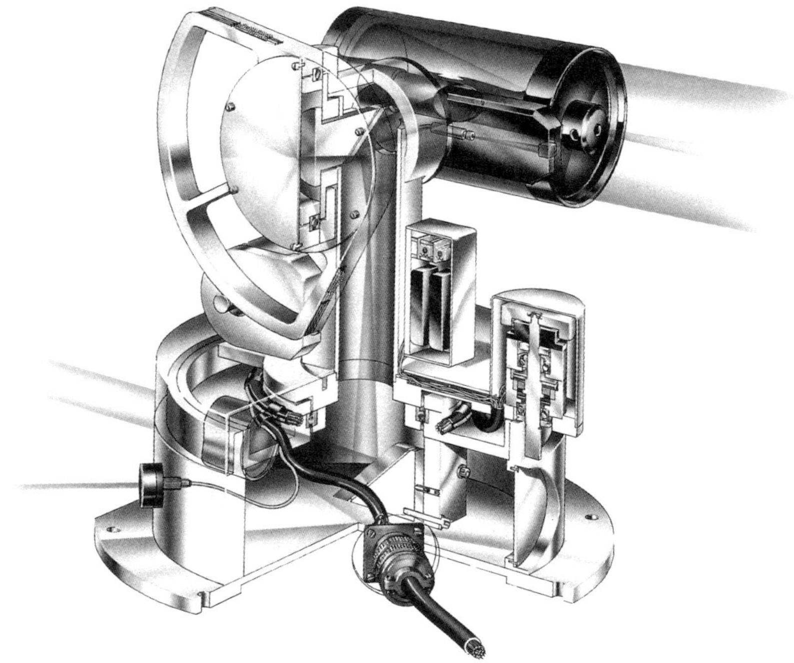

Figure 8.7 (a) RotoLok drive gimbals. (Courtesy of Sagebrush Technologies, Inc.)

wound around a drum and capstan, the capstan being of small diameter and the drum of much larger diameter [2]. The drive is designated the RotoLok drive. The drive motor is a brushless dc type and the position encoder is an incremental optical encoder in one configuration. As with the Harmonic Drive gimbal described above, there is no cogging so it can be used to point very narrow beams on the order of microradians divergence. The torque available at the small motor is multiplied by the ratio of the drum diameter to capstan diameter. The inertia reflected back to the motor is reduced as the square of the gear ratio so the main acceleration load is the motor and not the driven gimbal with its large inertia. For example, if there is a ten to one gear ratio (drive to gimbal rotation ratio of 10:1), then the gimbal inertia reflected back to the drive motor is $(1/10)^2$ of the gimbal inertia the motor sees and is reduced by a factor of 100. There is no backlash in this type of gimbal drive and the stiffness is extremely high. The design can be made of an aluminum matrix composite, such as SXA, which matches the ball bearing's stainless steel coefficient of thermal expansion, a very desirable characteristic where large temper-

(b)

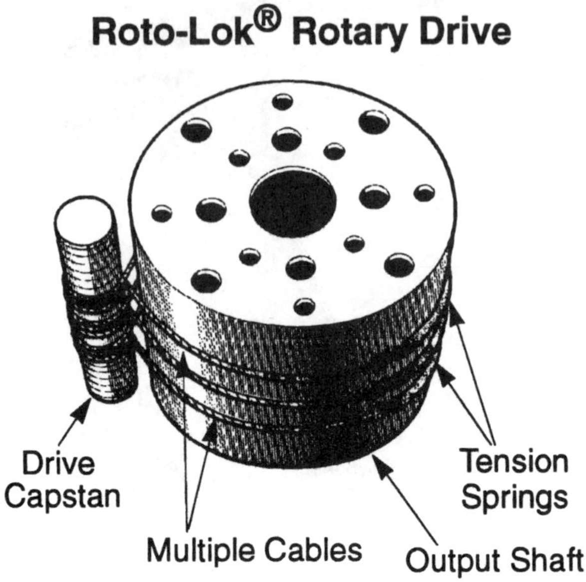

Figure 8.7 (b) RotoLok gimbal drive mechanism. (Courtesy of Sagebrush Technologies, Inc.)

ature excursions are expected on-orbit. An alternate light material is graphite epoxy composite. A gimbal of this material is shown in Figure 8.8.

Figure 8.9 illustrates a direct-drive gimbal for a short-range laser communication system conceptual design [3]. The gimbal, as proposed by Ball Aerospace, weighs a mere 4 lb. Notice the close coupling of the package to the gimbals. Although a Coudé path is not a part of the illustrated design, it could be incorporated into this type of gimbal.

Gimbal weights based on current technology are incorporated in the combined gimbaled telescope curves in Chapter 11.

Position Encoders. The position encoder in a gimbal supplies the angular reference to the system to measure where the pointing element is pointed and provides feedback in a positioning loop where the pointing element is commanded to a position. Several designs of encoders are available to the designer, including the optical encoder, Inductosyn, and resolver. There are others, but these three constitute the most applicable forms of encoder. The authors have seen these three types of position encoders built into gimbals suitable for space laser communications.

The optical encoder operates as shown in Figure 8.10. For reliability, a *light emitting diode* (LED) on the fixed portion of the encoder shines through a spoked

(a)

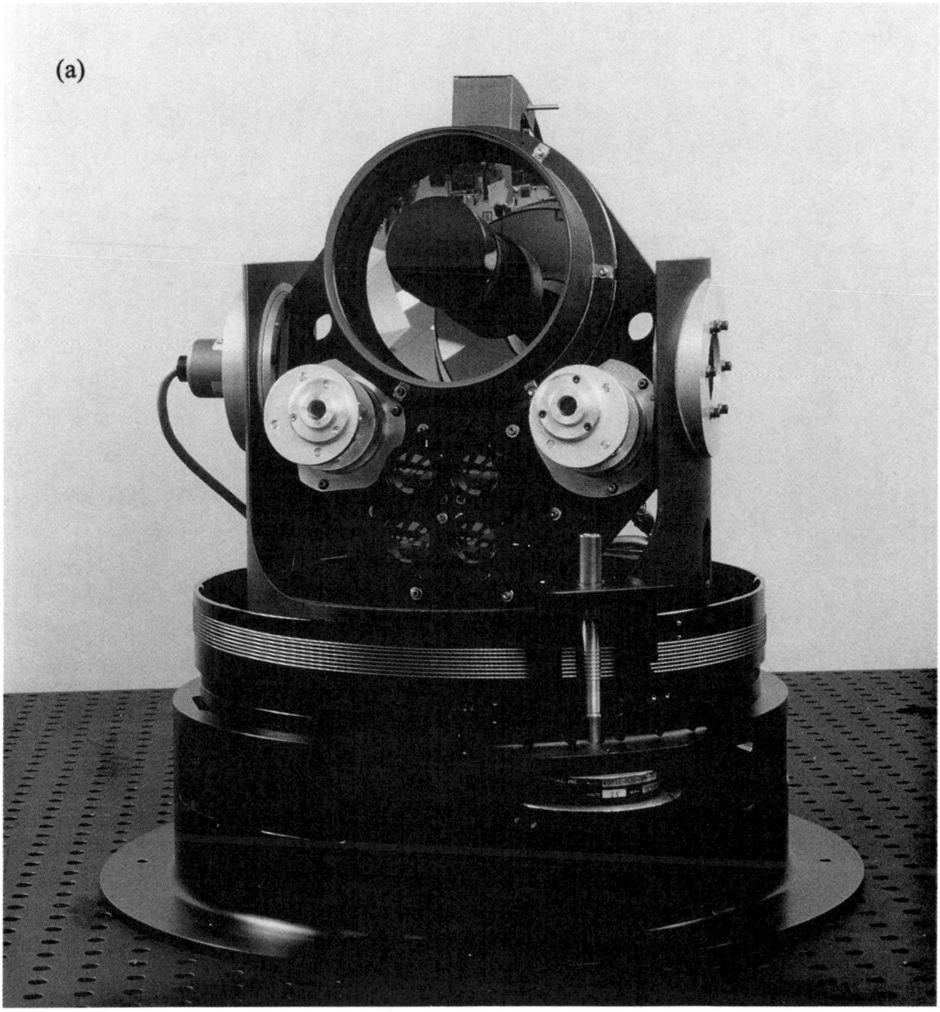

Figure 8.8 (a) Graphite epoxy composite RotoLok gimbals. (Courtesy of Sagebrush Technologies, Inc.)

wheel mounted on the rotating part of the encoder onto a fixed detector. As the spoked wheel rotates relative to the fixed LED, the light is interrupted and the interruption sensed by the detector. The number of times the light is interrupted is a measure of the rotation of the spoked wheel, hence angular position.

There are three types of optical encoders. The optical tachometer senses movement but not its direction. It has only one output. A counter may be used to sense motion provided it is always in one direction. If a clock is used with the

Figure 8.8 (b) Graphite epoxy frame. (Courtesy of Sagebrush Technologies, Inc.)

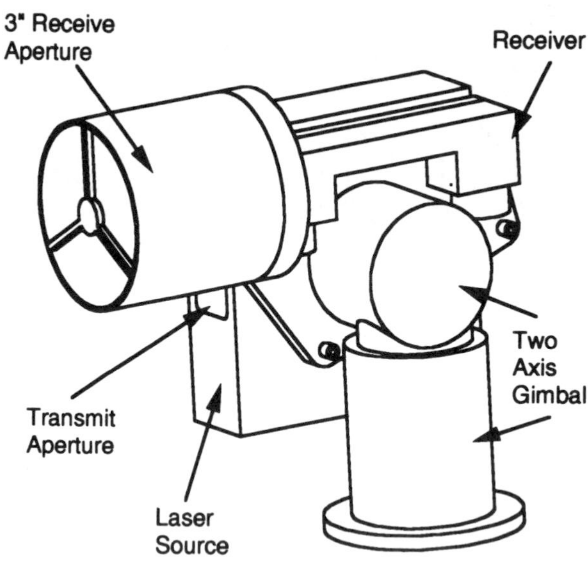

Figure 8.9 Short-range terminal direct-drive gimbals. (Courtesy of McDonnell Douglas Aerospace.)

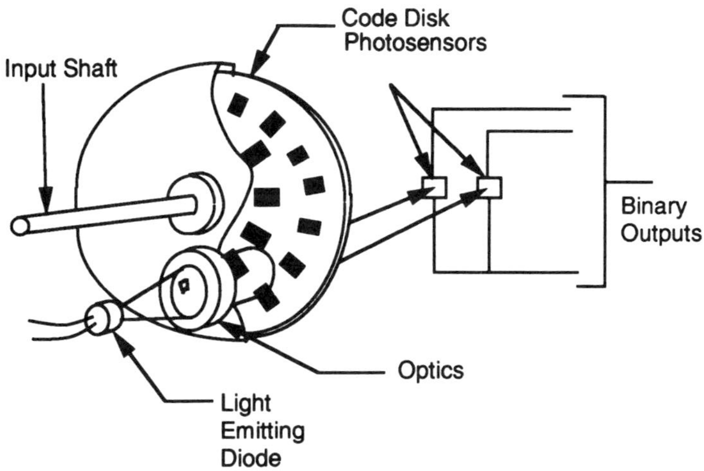

Figure 8.10 Optical encoder operation.

encoder, velocity may be ascertained by noting the number of pulses per unit time. This type of encoder is used for velocity detection and control applications. Figure 8.11 illustrates the binary output of this type of encoder.

The incremental optical encoder can have two or three outputs. With two outputs, it detects motion and direction of the motion. An algebraic counter accumulates plus and minus increments of motion so the total net motion is known. A third output is a zero reference output used to locate a unique position of the encoder. This position defines a known starting reference for the start of an operation or process. For example, in the startup of a gimbal in a laser communication system that uses this type of encoder, the gimbals are driven to a mechanical stop in each axis. This stop represents the zero reference and the counters accumulate the total net position movement that follows. Figure 8.12 shows the output of this type of encoder.

The absolute position optical encoder usually has between six and 20 or so outputs, depending on the accuracy needed. These outputs form a binary representation of the shaft position. There is a unique one-for-one correspondence between

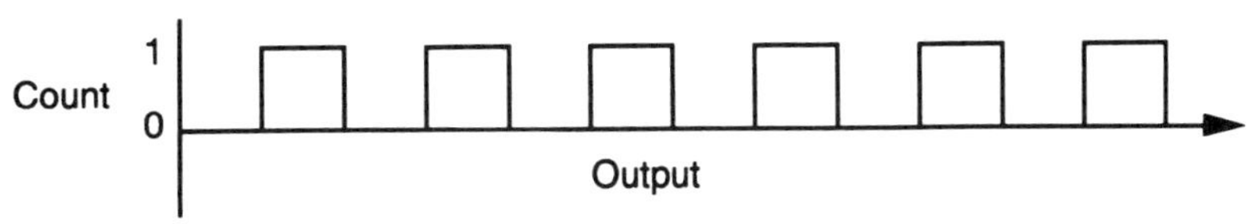

Figure 8.11 Optical tachometer output.

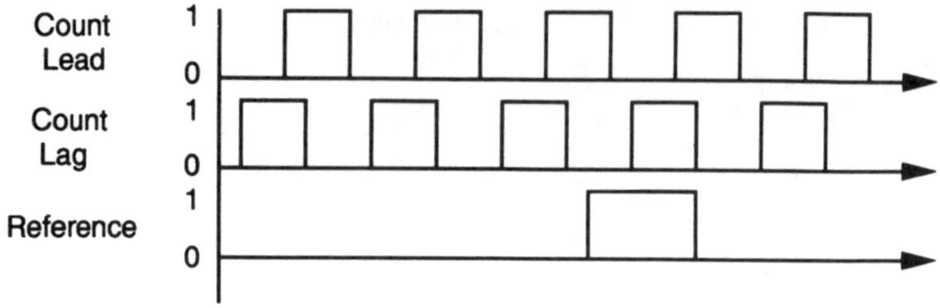

Figure 8.12 Incremental optical encoder output.

the encoder position and the binary output so the encoder is not affected by power loss (i.e., as soon as the power returns, the position is known with the full accuracy of the device). There is no necessity to drive the gimbals to a zero reference as with an incremental encoder. This type of encoder is the heaviest, most complicated, and the most costly of the optical encoders. Figure 8.13 illustrates the output of an absolute optical encoder.

Many encoders are designed for use in hostile environments, including space. Sealed bearings are used with the appropriate space-qualified lubricant. The electronics for encoders are separately mounted in a box.

An Inductosyn-type position encoder is the product of Farrand Controls. The Inductosyn is a precision resolver but unlike the windings around a core, as in a resolver, the Inductosyn consists of printed circuit patterns with parallel hairpin turns repeated along the edge of a disc. As a result, the plates are flat and consume little axial space in a gimbal. Also, redundant "windings" are readily created by use of three plates with patterns printed on the center plate on both sides.

The inductive coupling between sets of windings is used to measure displacement (position). The Inductosyn elements, attached to fixed and moveable plates

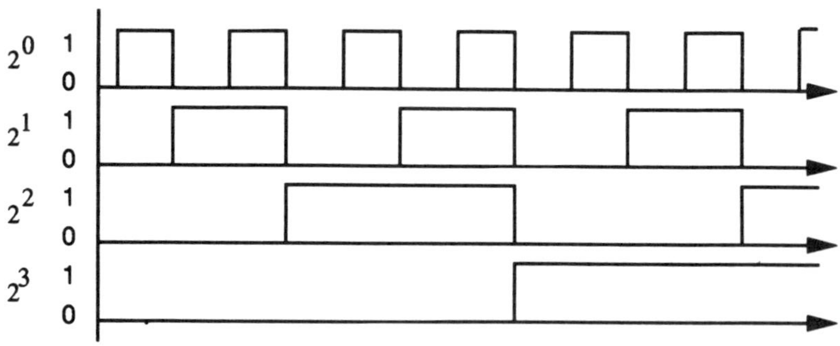

Figure 8.13 Absolute optical encoder output.

with the hairpin loops parallel, are separated by a small air gap. An alternating current flowing in one conductor will induce a voltage in the other that depends on the relative position of the conductors. The induced voltage is a maximum when the loops are directly facing. Passing through zero midway, it rises to a negative maximum at the next facing location. The pitch, P, is the spacing between the hairpin turns. The Inductosyn elements behave as transformer windings with a coupling ratio that is a maximum when the windings are facing. The indsuced voltage is proportional to the cosine of the displacement (position) divided by the pitch. A second winding may be located adjacent to the first but displaced P/4 from it. This induced voltage is proportional to the sine of the displacement divided by the pitch. As the discs rotate relative to each other, the amplitudes of the two outputs change. Phase remains constant while amplitude is a function of relative displacement. There is a unique pair of induced voltages for each position of the plates within one cycle of the pitch P. Thus, by measuring these induced voltages, the accurately known pitch interval may be subdivided with high precision. The air gap between the plates is of the order of 0.005 inches.

The Inductosyn signals are sine waves of 100 mV and may be used directly in servo control loops or may be converted to digital outputs by use of a converter. Digital outputs are at standard *transistor-transistor logic* (TTL) levels to interface directly with computers, numerical controllers, and position readouts. Figure 8.14 illustrates the Inductosyn.

As previously stated, the Inductosyn is actually a resolver that uses printed circuits instead of windings on a magnetic core (as in a resolver). In a resolver, the function is nearly identical to the above description of the Inductosyn. Dual windings provide a high degree of precision, if needed, just as in the Inductosyn. Due to the windings and magnetic laminations, the conventional resolver is bulkier and considerably heavier than the Inductosyn.

Bearings. Bearings are always critical to the operation of space gimbals. A great deal of analysis and testing has been conducted on bearings for space. Computer programs have been developed that characterize the mounting point deflections on the stresses and deflections within ball bearings. A conventional mounting for bearings on a shaft consists of two angular contact bearings mounted at opposite ends of the shaft, arranged back-to-back with one bearing on a flexure to accommodate axial differential expansion due to temperature excursions. Angular contact bearings contain the maximum number of balls within a given bearing envelope and therefore have the greatest load capability.

Of concern is the torque required to drive the bearings, which increases with time on-orbit. Gimbal servo drives require very low frictional torque from the bearings in order to point accurately. As the bearing starting torque increases, the ability of the gimbal to make small pointing changes decreases. Large bearings obviously carry large loads. During launch, forces on the gimbals (and all other

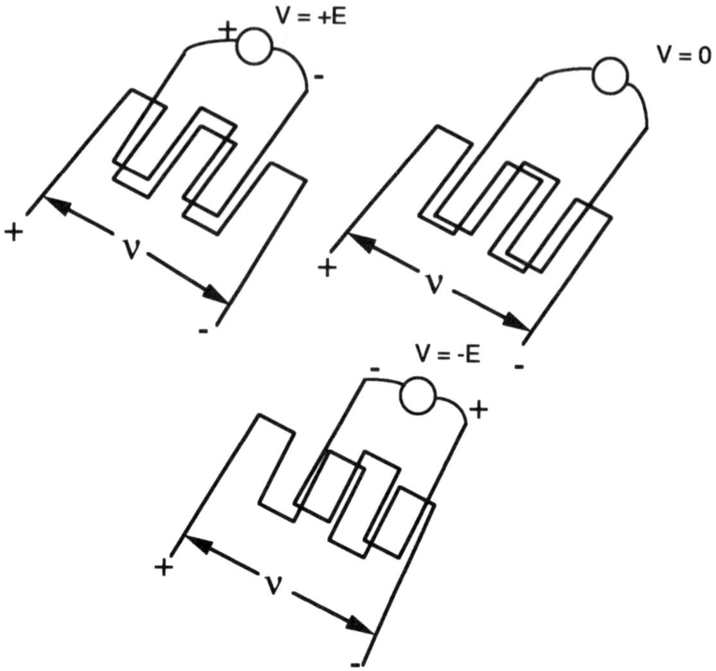

Figure 8.14 Inductosyn operation.

components) can exceed 16 g so that momentary forces over 16 times the mass of the gimbal supported equipment will be seen by the gimbal bearings. Unfortunately, large bearings exhibit larger low-speed friction torque. There is obviously an increase in bearing diameter required for a Coudé path since the optical path passes right through the inner race.

If gimbal axis bearings are constrained as an assembly, and there are materials that are different from the bearings constraining the bearings, then a temperature change on-orbit could provide an exceedingly high stress load to the bearings. Ball Aerospace solved this potential problem in their gimbals by providing an axial diaphragm that preloads the gimbal bearings. Any difference in the expansion of the shaft and the bearing mount is taken up as spring load of the diaphragm, which yields a much lower stress in the bearings than a hard metal constraint would. Another approach to solving this problem is the use of a matched, preloaded bearing pair on one end and a free-floating bearing on the other, which can move axially on the shaft as the shaft differentially expands or contracts relative to the gimbal. Bearings are typically constructed of 440 stainless steel, hardened to a Rockwell_C 58 minimum hardness, and tolerances are usually ABEC-7 or better. To minimize friction, sealed bearings (with a running seal) are not used. On occasion, roller bearings might be used for large, overhung load gimbal designs.

Lubricants. Lubricants for space must not dry out in the space vacuum. For laser communications, they also must not out-gas and contaminate the optics. There are two types of space lubricants, wet and dry. Wet lubricants are the type we all are familiar with. Dry lubricants are more unusual. Dry lubricants require slightly more torque than wet lubricants, so as a consequence laser communications gimbals would normally rely on wet lubricants. Wet-lubricated bearings are also always smoother than dry-lubricated bearings. Another problem with dry-lubricated bearings is that the dry lubricant has a tendency to pile up dams at the turnabout points in an oscillating application. This is especially true for low-angle motions where the ball tracks do not overlap. When a new angular range occurs, high-torque spikes are produced when the balls cross the lubricant piles. The greater tendency for torque noise does not disqualify dry lubricant for oscillating applications, only those where smoothness is important, as in laser communications with its microradian beams. Radiation tolerance is also important in a space lubricant, which must be able to withstand the expected radiation environment without loss of its lubricating properties.

Liquid lubricants are carried internally in trapped cavities surrounding the bearings. The trapping of the lubricant is accomplished by providing close clearances and a torturous path, thereby keeping the lubricant inside the enclosure. In this manner, wet lubricants can be used for space gimbals without loss of lubricant over time and without permitting the lubricant to leak out of the gimbal and contaminate the optics. The vacuum also will boil off the lubricant if its vapor pressure is high; only a small controlled amount is lost with time due to evaporation. As a supplier of custom lubricants, Ball Aerospace has for many years provided space lubricants for their own drives and for others. The wet lubricant is known as VacKote, by Ball, and is a long-chain hydrocarbon oil with a complex metallo-organic additive.

Structures. Gimbal structures have been constructed of a number of materials including beryllium (expensive), titanium, aluminum, and steel. Sheet Invar steel was used for the first-generation gimbaled telescope. A combination of the above materials can often prove to be effective and provide cost savings, although beryllium is always expensive (but always light). One gimbal with which the writers are familiar had a yoke made of aluminum with titanium and steel inserts for the bearings to avoid problems due to differences in coefficient of thermal expansion between the 440 stainless steel bearings and their mountings of aluminum. The azimuth shaft was made of steel with integral bearing mounts. Analysis showed that for the same stiffness, aluminum, titanium, and steel weighed about the same but with steel, less weight was required around the azimuth bearings since no inserts were required. The elevation shaft was made of titanium, which provided a proper coefficient of thermal expansion for direct mounting of the bearings. The azimuth inductosyn housing was made of aluminum and carried the three mounting feet for the gimbal. The choice of aluminum here is a good one since the gimbal would interface with the aluminum structure of the spacecraft. The natural frequency of this gimbal was predicted by NASTRAN modeling to be 98 Hz, a useful value for a servo drive bandwidth of 10 to 15 Hz. NASTRAN, for those unfamiliar with its capabilities, performs finite-element modeling of structures.

Caging. A frequent requirement for a space laser communication gimbal is that it be caged (i.e., it be locked down for launch and released when on-orbit). This prevents the gimbals from slamming into the stops during launch and also can provide additional rigidity. If the gimbaled telescope is perfectly balanced, then theoretically the telescope should never receive a torque due to translation. However, it should be remembered that any imbalance is multiplied by the launch acceleration of perhaps in excess of 16 g so that a slamming torque could be possible.

Note that with the two pulling solenoids, the failure of one solenoid does not prevent the gimbal from being unlocked since the other pulling solenoid can free the gimbal. The caging mechanism is attached to the inner gimbal so that both axes are caged.

Cabling. This is an interesting problem to the designer of a gimbal and no less interesting to the laser communication system designer since the performance of the gimbal can be affected seriously if the approach is not well engineered. Coaxial cables (coax) may be required to cross the gimbals to the inner axis as well as many other conductors. To provide for continuous rotation in an axis, slip rings are used. These have to be custom engineered, in general, since the design varies with the number of conductors, rotation, power levels, redundancy requirements, and so forth. Lubricants are necessary and important for slip rings due to the sliding surfaces. The first-generation crosslink, the laser communication subsystem, used slip rings since the gimbaled telescope was despun.

More commonly, the gimbal access range does not exceed 360 deg, so that cabling techniques can be used. One technique is cable wrap (service loop), where several turns of the cable are wrapped around an axis, for example, the outer or azimuth axis, in an azimuth-elevation configuration. The cable wraps can accommodate more than a full revolution of axis motion without creating excessive windup torque. The cable wrap can be enclosed in a housing if necessary. This approach is useful for passing multiple coax conductors across the gimbals since bending is the preferred mode for the coax and not twisting. The requirement to pass coax cables across the gimbals is the result of adopting a gimbaled package approach and is not a requirement for a gimbaled telescope or gimbaled flat pointing approach. In addition to coax cables, there are many other conductors required to pass across the gimbals for a gimbaled package approach. One such design, which was implemented in hardware, required over 120 conductors crossing the outer gimbal. For such multiple conductors other than coax, a flex capsule approach may be used. Flex lead assemblies or flexible cable capsules have a very desirable characteristic in that they impart a low torque to the gimbal throughout its travel. With 120 conductors, this is a serious concern. The flex capsules, exemplified by Polytwist capsules by Polyscientific, contain their own bearings. The moderate torque imparted to the gimbals by the flex capsules can be handled by propitious design.

As a cable or flex capsule winds up in the gimbal travel, a torque opposing the motion also builds up due to the cabling or capsule. An approach to nullifying this torque buildup is the use of a torque compensator, which is merely a spring so mounted as to provide an opposing torque to the cable or capsule torque.

Stops. It is almost always a requirement that soft stops be provided so that an inadvertent command cannot force the gimbals to bang against the end of their travel. This is easily provided by polyurethane plugs, which cushion the gimbals stop. If the requirement exists to provide a hard stop, as required by an incremental encoder for an initial reference, a spring approach could be used. For use as a reference, the torque applied to the gimbal would not exceed the spring force. When used as a soft stop, the spring would deflect, thereby decelerating the gimbal smoothly while absorbing the gimbal energy.

8.1.2 Beamsteerers

Beamsteerers or steering mirrors are used in laser communication systems as fine, high-speed track devices and also as slow, fairly precise position steerers for the point-ahead function. They are of several forms or types. The driven element is most commonly a mirror that steers the beam. The driver for a mirror may be a galvanometer type or a piezo-electric element. An acousto-optic steerer steers the beam directly as it is transmitted. Each of these will be discussed in turn.

8.1.2.1 Galvanometer Beamsteerers

In the first-generation crosslink system, small galvanometer beamsteerers were used for position and angle steering of the laser as it passed across the interface between the laser and the imaging optics. This was necessary since steering was anticipated at this interface due to thermally induced bending and position shifts between the separately mounted laser and the imaging optics. These steerers did not require position pickoffs since quadrant detectors in the imaging optics sensed the steering of the laser and commanded the steerers to null the shifts in position and angle.

The beamsteerers used for point-ahead and track in the first-generation system were single-axis galvanometer devices and contained capacitive position pickoffs for either loop stability or feedback for point-ahead position. Flex pivots were used as bearings, which cut down friction and extended life. Many design iterations have occurred on galvanometer beamsteerers and the current generation of devices are quite adequate for the job. The current-generation beamsteerer is a two-axis, actuator-driven device. Flex pivots are used to mount the steering mirror. Speed of response is great and devices with a control bandwidth for a 1-in mirror of up to 4 kHz are available, adequate to compensate for any projected spacecraft disturbances.

Beamsteerer accuracy for the track function is not as critical as it is for the point-ahead function, where the point-ahead angle might be commanded open loop. For open loop pointing for a point-ahead angle of, say, 70 μrad maximum, a specification for an accuracy of 0.5% might be called for. This corresponds to an open loop accuracy of 70 divided by 200, or 0.35 μrad in the far field. This is a relatively small portion of the beam divergence, but must be included as an error in the pointing budget. Of course there is an optical magnification factor in the imaging optics so that the throw of the beamsteerer must be greater by this magnification ratio. In its normal configuration, the beamsteerer mirror provides a gain factor of two (steering is greater by a factor of two than mirror motion), which must be included in the calculations for beamsteerer range.

The configuration in the system of the two beamsteerer applications is shown in the block diagram of Figure 8.15. The point-ahead beamsteerer steers the laser (transmitter) beam relative to the track point, while the track beamsteerer provides

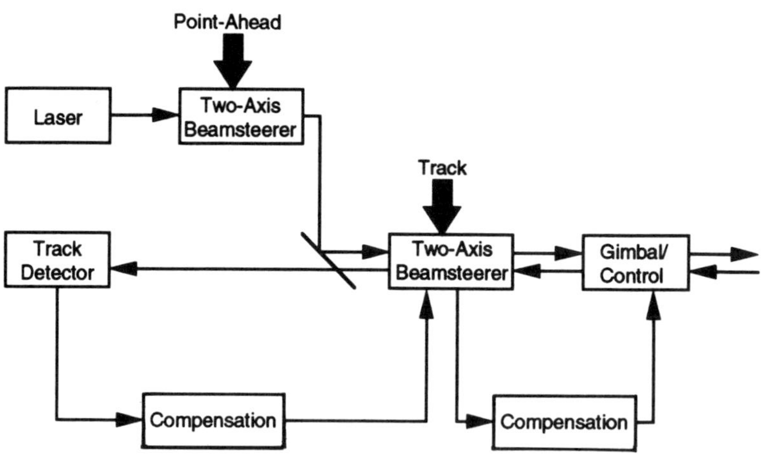

Figure 8.15 Location of beamsteerers in the system.

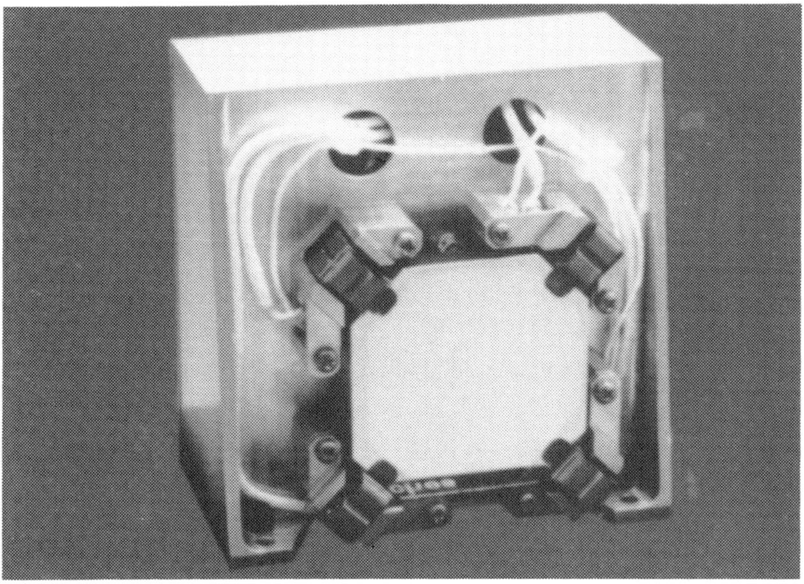

Figure 8.16 Two-axis beamsteerer. (Courtesy of McDonnell Douglas Aerospace.)

the high-speed fine correction of the track point, which the gimbals are unable to do due to bandwidth limitations.

Curves, and the accompanying discussion in Chapter 3, quantitatively describe the raw disturbance spectrum of a typical satellite and the reduction of jitter through the use of a 250-Hz bandwidth track beamsteerer. Figure 8.16 is a photo of a second-generation crosslink two-axis beamsteerer manufactured by General Scanning. Table 8.1 lists the characteristics of two reactionless fast-steering mirrors (beamsteerers) built by Ball Aerospace.

Beamsteerers are customarily placed near a pupil in the system to minimize beam walk (translation) of the transmit beam at the telescope. In the second-generation crosslink, a *lateral photo detector* (LPD) is used to monitor the steering provided by the point-ahead beamsteerer. A portion of the transmit beam is picked off and focused onto the LPD. This provides a measure of the deflection of the transmit beam relative to the boresight.

8.1.2.2 Piezoelectric Beamsteerers

These steerers are widely used in adaptive optics to steer the elements in deformable mirrors. Deflections for this application may be smaller than normally needed for laser communication applications. A position sensor for this approach could pose a

Table 8.1
Reactionless Fast-Steering Mirror Characteristics

	One-Inch Mirror	*Two-Inch Mirror*
Two-axis	Yes	Yes
Tracking accuracy	$<$ 25 nrad	$<$ 80 nrad
Range of travel	+ or -2 degrees	+ or -2 degrees
Control bandwidth	4 kHz	1 kHz
Acceleration	5,000 rad/sec^2	2,600 rad/sec^2
Actuators	Samarium cobalt air core	Samarium cobalt air core
Materials	Beryllium mirror	Beryllium reaction mass
	Stainless steel flexure	Aluminum structure
	Beryllium mirror	Beryllium reaction mass
	Stainless steel flexure	Aluminum structure
Size	3.75-in high by 2.75-in wide by 1.25-in deep	5-in high by 4-in wide by 2-in deep
Weight	9 oz	3 lb

challenge. Early on in laser communications, these steerers were used to drive mirrors in several laboratory systems but were subsequently replaced in flight type systems by the more rugged galvanometer devices.

8.1.2.3 Acousto-optic Beamsteerers

Nonreflective (i.e., transmissive) acousto-optic beamsteerers have been investigated in depth for application to laser communications. Their use results in additional loss in beam quality and transmission beyond galvanometer devices. They also require high voltages to establish the acoustic field in the crystal that performs the deflection of the beam. This is not desirable in space applications. A position sensor would also be needed.

There is the potential for liquid crystal steering although this is quite new and the characteristics of this approach should be investigated in some depth before adoption.

8.1.3 Mechanisms

Excluding beamsteerers, which are mechanisms, there is another group of mechanisms that perform function changes in a laser communication system. The first-generation crosslink employed some thirteen mechanisms to perform the functions of redundancy switching, field stop adjusting, and so forth. This quantity of mechanisms can affect reliability unfavorably so, as a result, there is a great desire on the part of the system designer to minimize the number of mechanisms. The second-generation system only employs two. The functions that might be performed by mechanisms are tabulated in Table 8.2; however, a particular system would likely not have more than a few of these.

Table 8.2
Mechanism System Functions

Adjustable field stop
Adjustable focus
Redundancy switching
Lasers
Beamsteerers
Detectors
Optics Path
Boresight alignment shutter
Caging
Solar protect shutter
Filter changing
Beam divergence
Decentered pupil transmit beam
Driven Risley prisms for steering
Driven polarization elements
Selectable beamsplitter drive

Mechanisms in early lab systems were generally driven by linear solenoids against a spring load. One exception was an active focus device used in the Airborne Flight Test System optics. The drive for the active focus was a galvanometer beamsteerer where the rotary motion was converted to linear motion, which drove a focus lens mounted on a precision slide. The position pickoff of the steerer was used to control the motion of the device. Another example of a linear motion mechanism is the adjustable field-stop device, which was illustrated in Chapter 4. The first-generation crosslink used a number of rotary mechanisms that inserted or removed a component such as a mirror from an optical path to offer a redundant path for the beam. Several of these rotary mechanisms built by Eastman Kodak are shown in Figure 8.17. The drive for each is a stepper motor, which operates as a rotary

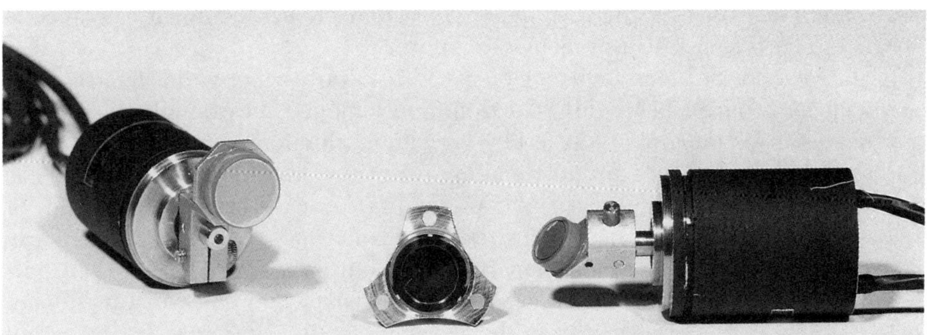

Figure 8.17 Mechanism drives. (Courtesy of McDonnell Douglas Aerospace.)

solenoid. On command, the motor rotates and drives against a hard stop for position accuracy. The stop is located ahead of the required motion for the motor to complete the step so that a continuous torque is applied by the motor at the end of the stop. The only tilt nonrepeatability of the element arises from the runout of the bearings; therefore, precision bearings are used in this application. One feature that is important on-orbit as well as in test is a status sensor in each mechanism, which indicates remotely the state of the mechanism (i.e., its position at a given moment). This status data is telemetered to the ground on command.

8.1.4 Optical Mounts

If a laser communication systems is to operate properly on-orbit, the optical components must be positioned and held in the required position throughout the life of the system. The launch of the system is extremely stressing in that vibration, shock, acceleration, and acoustic noise are present, combined. Components can shift or settle. In addition, the "*g*" force release on leaving Earth will result in bending, the magnitude of which can determine whether or not there is a problem. As a result of these considerations, a great deal of effort is normally put on the structural design (static and dynamic) of the system. Structures are modeled in computer programs such as NASTRAN and ANSYS, and studied thermally by additional programs that evaluate the effects of temperature changes on the alignment of the structure that holds the optical elements.

The baseplate of the *imaging optics* (IO) on which the optical elements are mounted was discussed in Chapter 4. Each component must be positioned on the baseplate within some initial alignment tolerance of three directions of translation and about three axes of rotation. Some of these motions are not critical and fine adjustment need not be provided for them. These positions must be maintained throughout test, launch, and on-orbit to within an additional alignment tolerance beyond initial alignment. This alignment tolerancing is expressed as an alignment budget and was discussed in Chapter 4. A portion of the alignment tolerance is allocated to each component or element mount.

The methods of mounting elements are driven primarily by the element type, the mount for a mirror being different from a lens mount, for example. In general, materials used for mounts are generally beryllium, aluminum, invar, or titanium. Graphite epoxy composites are now making their entry as alternative materials as are metal matrix composites, such as SXA (as previously discussed). Properties of these materials useful for mounts are tabulated in Chapter 4. It is particularly important to pay attention to the properties of mount materials. Elements can shift in position following assembly or even be damaged by improper mount design. They also may not be protected from the rigors of launch or thermal excursion and could effectively be destroyed due to the environment.

Table 8.3
Optical Glass Characteristics

Vendor	*Type*	*Density lb / in³*	*Elastic Modulus ×10⁶ psi*	*CTE min / in °F*	*Thermal Conductivity BTU / ft-hr-°F*
Dow Corning	Fused silica 7940	0.0795	10.6	0.317 @ 32–392	0.80 @ 77
Kimble	Soda Lime CA-2	0.0900	11.1	4.83 @ 32–572	0.60 @ 32
Kimble	Borosilicate KG-33	0.0802	9.5	1.78 @ 32–572	0.65 @ 32
Schott	Borosilicate BK-7	0.0907	11.7	3.94 @ 22–158	0.64 @ 194
Schott	Flint F-2	0.1300	8.4	4.56 @ 22–158	0.45 @ 194
Schott	Lanthanum LaK-9	0.1270	16	3.5 @ 22–158	0.52 @ 194

CTE = coefficient of thermal expansion, BTU = British thermal unit.

The optical materials used in components are generally "glassy" in nature, although metal mirrors are being used more frequently. Glasses are good in compression but very weak in tension, with about a 1,000 lb/in^2 or so stress limit in tension. SXA makes good stable and moderate-cost mirrors, and we expect to see more of this material in optical applications in the future. Characteristics of representative glassy materials used for elements are tabulated in Table 8.3.

The tradeoffs performed to determine the appropriate materials to use for a mount are complex and require many calculations. In a qualitative fashion, certain factors that can be considered are listed in Table 8.4. Note that these considerations are subjective, but at least they provide some insight into the process. For thorough quantitative approaches to mount design, the reader is referred to [4–8].

Adhesives and sealants are used more and more for holding elements in cells. The advantages are numerous, the most important being the isolation of the tender glassy element from the unforgiving characteristics of metal. As mentioned previously, a poorly designed mount can destroy a lens, for example, by cracking or chipping it. Metallic locking rings can cause serious stresses due to thermal expansion differences between metal and glass substrates. It has been said that "optical elements should be held firmly but gently much as one would hold a small bird in your hands." Excessive force on an element can also create stress birefringence, which is disastrous in a system that relies on polarization for isolation between the laser and the receiver.

Adhesives are used to hold an element in a particular position. An important adhesive frequently used in small element mounts and fiber optics is an *ultraviolet* (UV) curing type by Norland. Use of this adhesive permits "active alignment." An element can be positioned relative to a mount and fine-adjusted while the UV curing adhesive is in place but not cured. The position of the element can be observed using a very fine test instrument, such as a microscope or interferometer. When the desired alignment has been achieved, the adhesive is subjected to UV

Table 8.4
Mount Material Considerations

Material	*Advantages*	*Disadvantages*
Aluminum	Excellent machinability. Low cost. High thermal conductivity allows quick thermal equilibrium.	Large coefficient of thermal expansion, which can cause image degradation and maybe lens breakage.
Stainless Steel	Thermal coefficient of expansion close to glass. Good corrosion resistance.	Difficult to machine. Heavy material.
Beryllium	Very light. High stiffness. Thermal coefficient of expansion close to that of glass.	High cost. Difficult to machine. Machining operation is hazardous to the operator unless highly efficient exhaust systems are used.
Titanium	Light. Stable and yields itself to grinding and lapping operations. Thermal coefficent of expansion close to glass.	High cost. Difficult to machine. Low thermal conductivity keeps material from reaching thermal equilibrium quickly.
Invar	Very low coefficient of thermal expansion.	Heavy. Difficult to machine.

light and curing takes place in short order. The alignment will be maintained to a high degree of accuracy after the adhesive sets up.

Elastomeric sealants such as *room temperature vulcanizing* (RTV) silicone rubbers are frequently used to hold mirrors, lenses, prisms, and other larger elements in mounts. Where extremely accurate alignment is required, the element may contact a metallic or plastic (e.g., nylon) pad or surface but be sealed into position using an elastomeric sealant. Table 8.5 lists properties for several of these materials.

Outgassing is of concern with adhesives and sealants so the *National Aeronautics and Space Administration* (NASA) has established requirements for these in NASA Specification SP-R-0022A, which limits the total mass loss of the adhesive or sealant to 1%, or less, and the collected *volatile condensable materials* (VCM) to 0.1% or less after 24 hours at 125°C. A material that exceeds the NASA specification but used only in very small quantities may be acceptable to the designer. Two-part materials must be properly mixed and samples taken during the process and tested to assure that the material will behave as specified during test and flight. The sample testing is also a requirement for any material used in space. Several of the previous references in this subsection also cover use of adhesives and sealants.

Table 8.5
Adhesive and Sealant Characteristics

Material / Manufacturer	*Tensile Strength psi*	*Elongation %*	*Useful Temperature Range °F*	*Thermal Conductivity BTU / hr-ft-°F*	*Total Mass Loss (%)*	*Collected VCM @25°C (%)*	*Vol. Expansion 77–302°F cc / cc-°C*
6-1104 Dow Corning	510	575	−67 to 392	.068	.015	0.65	8.5×10^{-4}
6-1125 Dow Corning	550	520	−67 to 392	.073	.015	0.065	8.5×10^{-4}
93-500 Dow Corning	790	110	−85 to 392	.085	.025	0.05	9.0×10^{-4}
NOA 61 Norland	3,000	38	−112 to 194		2.2	0.01	
RTV 566 General Electric	925	130	−175 to 600		.1	0.02	
RTV 516 General Electric	800	120	−75 to 400	.11	1.23	0.42	27×10^{-5}

psi = Pounds per square inch, BTU = British thermal units, VCM = volatile combustible materials, cc = cubic centimeters.

8.2 LASER TECHNOLOGY

It is axiomatic that the laser is the heart of the laser communication system. When the crosslink designer needs more link margin, it is frequently more advantageous to increase laser power than it is to narrow the transmit beam and increase the collection area by increasing the aperture. An example of this is the need for a 3-dB increase in link margin for a short-range crosslink where the laser source is a single diode. By simply adding another diode, an additional 3 dB is achievable without altering anything else in the system. To accomplish the same 3-dB improvement with the optics significantly increases the size and weight of the system.

The development of the laser source for a system has frequently been the pacing technology in the development of a system. This was never more true than in the *laser crosslink subsystem* (LCS) development, where laser development set the pace for system development. The LCS laser design incorporated two modes, one of which was a high peak power acquisition mode. The laser included an internal modulator, which proved to be extremely challenging in execution. We are jumping ahead, however, and will pick up the LCS laser story later in this section.

A major driver for laser diode development has been commercial need. For example, diode power has steadily increased in recent years, driven primarily by the need for higher powers used in *compact disc* (CD) players and telecommunications (fiber-optic) systems. Similarly, the development of high-power Nd:YAG lasers has been paced by the military for range-finders and by industry for marking applications. Laser communications engineers have taken advantage of these developments to produce smaller, lighter, and longer life systems.

In Chapter 3, the systems approach to laser selection was discussed and a number of candidate lasers introduced. Here we will discuss in more detail the characteristics, advantages, and shortcomings of these candidate lasers. The word laser stands for light amplification by stimulated emission of radiation. The theory of the laser is offered in many texts and write-ups [9–11].

8.2.1 Required Characteristics for Space Lasers

For long-range crosslinks (one synchronous to two synchronous range), a laser is needed that produces a very narrow beam. This requires the laser to operate in the fundamental mode, TEM_{00} (i.e., runs in a single mode and produces a diffraction-limited beam). Operation in this manner produces the minimum beam divergence for the given aperture. Certain other laser communications applications, such as call-up lasers and acquisition or tracking beacon lasers, do not require TEM_{00} mode operation. These nondiffraction-limited lasers are also discussed below.

For space application, the laser must be capable of very long life, often 10 or more years for current satellites. The long life must be demonstrable by accepted accelerated life testing. Spacecraft power is a precious commodity, so it is important that the laser used be highly efficient and require as little power as possible. Additionally, from a power viewpoint, an inefficient laser requires dissipation of the heat produced by the laser, so the thermal control system design is impacted, adding weight and complexity to the spacecraft.

To communicate, it is necessary to modulate the transmit signal. A means of modulating the laser signal must be available. The laser/modulator must also possess the ability to operate at the speed required for signaling. For the application, the laser must also possess the appropriate wavelength and narrowness of wavelength (linewidth). Last, but not least by any means, the laser must be survivable in space. It is not readily possible to repair lasers on-orbit, shuttle retrieve/repair teams notwithstanding. For geosynchronous links, no manned access to synchronous orbits has yet been demonstrated. Where laser life is not adequate or there is a known failure mechanism, a spare (redundant) laser may be required to prevent single-point failures. Ground lasers are air or liquid cooled, an unacceptable approach to cooling in space. Conduction and radiation are the only thermal effects at the designer's disposal for cooling in space. For cold conditions, adding heat is sometimes necessary, but this can be accomplished by electrical resistance strip heaters, which require electrical power.

8.2.2 Types Of Lasers / Trades

There are a great number of laser types; however, the choice for laser communication system designers is quite limited. An example of this limitation is the *helium neon* (HeNe) laser. The HeNe is a gas laser and has a 50,000-hr life before the gas tube fails. Unfortunately, power output of the HeNe is a few (up to six or so) milliwatts, inadequate to close the links for long-range systems. In light of the limitations for many lasers, we will discuss six types that currently hold promise for crosslinks. The six are *carbon dioxide* (CO_2), *neodymium doped yttrium aluminum garnet* (Nd:YAG) and its solid state "relatives," semiconductor diode, diode combiners, *master oscillator power amplifiers* (MOPAs), and lastly, *erbium-doped fiber amplifiers* (EDFAs). Not all of these possess equal capabilities or equal ability to survive in space, but nonetheless all are considered candidates and indeed work is currently being done on each type to adapt them to space laser communication systems. Each will be discussed in turn.

8.2.2.1 CO_2 Laser

This laser was mentioned in Chapter 3 and was one of the first types pursued for laser communications [12, 13]. The reasons were that the wavelength at 10.6 mm penetrated the atmosphere well, the laser works well in a coherent configuration, and the laser was readily available. The laser is a gas laser (i.e., a CO_2 and N_2 gas mixture in a glass tube is excited to lase by electrical excitation). Figure 8.18 shows

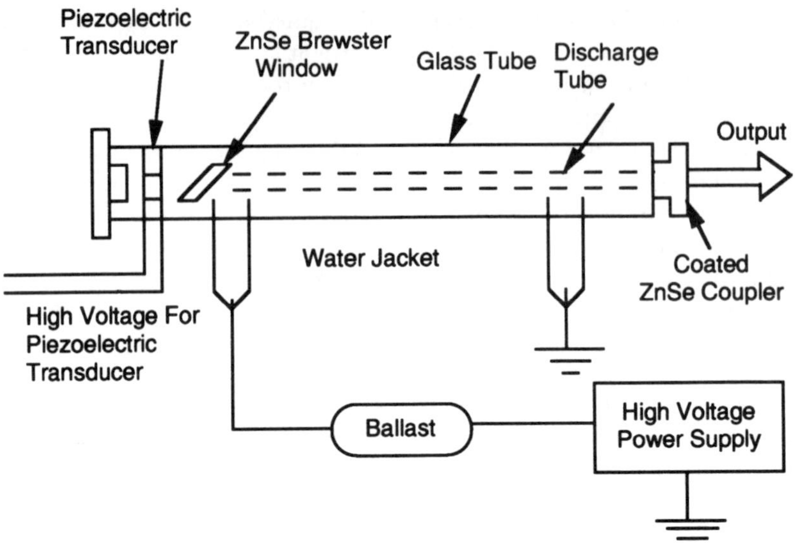

Figure 8.18 Conventional CO_2 laser layout.

the layout of a conventional CO_2 laser producing about 8W. The laser output needed for a coherent system is *continuous wave* (CW). Power available is from 40 to 70 W/m of length. Typical lasers used in ladar (laser radar) systems are about 20W, CW. The gas is cooled by conduction to the tube walls. The power obtainable from the tube is independent of the diameter from about 1.5 to 5 cm. CW power is proportional to tube length of up to 10m. Of course, for space, tube length must be short. The CO_2 laser is used throughout industry for a number of applications in powers up to 25 kW or more. Really large lasers have been studied and built for ground and space weapons, potentially producing output powers up to several megawatts CW. Small CO_2 lasers can be conduction cooled for space. A major shortcoming of the CO_2 laser is tube life. Unfortunately, the tube is self-contaminating after perhaps 20,000 hours of use, the result being that the CO_2 laser is currently not capable of meeting continuous-use long-life requirements on-orbit. For intermittent use, it may be fine. Another issue concerning the use of a CO_2 system is that the detector, for sensitive operation, requires cryogenic cooling, which must be done with a mechanical cryogenerator with a maximum life time of several thousand hours or so, again a life-limited condition. It has been proposed in the past that the cooling be done radiatively but the limitations of this approach bear investigation. The high voltage required for the excitation is also not desirable in space due to possible arcing.

8.2.2.2 Nd:YAG and Its Relatives

The Nd:YAG laser has seen lots of service in laser communications applications and tests. Following the early CO_2 work, the "YAG" laser was pursued as the laser "of choice," to use physicians' terminology. The Air Force in the early seventies funded the development of a Nd:YAG laser communications system for space [14, 15]. Several years later the Air Force selected the Nd:YAG system for an operational crosslink, the LCS that was been mentioned previously in Chapter 1. The original Nd:YAG lasers for space were lamp-pumped and a lot of effort was expended to prolong potassium-rubidium lamp life with the thought in mind of producing a system that would last for the required mission. A lamp life of several thousand hours was achieved, but a separate development to back up the lamp life was the use of another laser onboard the spacecraft that was pumped by the Sun. This sun-pumped laser was breadboarded and operated on the ground, producing the required laser output. The collector for the sunlight was a short barrel two-foot-aperture Cassegrain telescope. The solar image was collected and focused into a solar concentrator, which then pumped the Nd:YAG laser rod.

A major breakthrough for putting the YAG laser into space was the use of semiconductor diodes for pumping the rod at about 810 nm, where the YAG rod exhibits a strong absorption band. There is also a strong pump band at 790 nm but the diodes are not as reliable as diodes at 810 nm. This diode-pumped approach was used in the LCS. The diodes have a potential lifetime of many years, but this required validation. The validation was accomplished by performing accelerated life testing on the pump arrays. This accelerated testing involves running a large sample of the diodes at their rated output but at increased temperature. The arrays must also be carefully controlled throughout manufacture to assure quality. There is a relationship between life and temperature that is generally accepted, the Arrhenius relationship [16]. Using this relationship, the required life has been demonstrated for the diode pump arrays used to pump the LCS laser. The Nd:YAG laser is capable of lasting the required life of currently planned satellites. A feature of diode pumping is that many diodes are used (more than 100) and the failure mode is "gentle" in that a number of diodes may fail before the link is lost.

The Nd:YAG laser is capable of running in many different configurations. These configurations run at different pulse rates and the regime of each is tabulated in Table 8.6. The characteristics of each of the configurations is also included in Table 8.6. As a result of these regimes, the choice of a particular data rate dictates the configuration. Of course, the designer has the option of using an external modulator while running the laser CW, but the ability to achieve high peak power is not available.

The Nd:YAG crystal is triply ionized neodymium in yttrium aluminum garnet ($Y_3Al_5O_{12}$) with approximately 10^{20} Nd^{3+}/cm^3. It lases at a primary wavelength

Table 8.6
Nd:YAG Laser Operating Modes

Excitation	*Modulation*	*Typical Repetition Rate*	*Typical Pulse Length*
CW	None	Continuous	Continuous
CW	Q-switch	10 Hz–50kHz	50–700 ns
CW	Cavity dump	0.2–2 MHz	10–15 ns
CW	Mode-locked	200–500 MHz	30–200 ps

of 1.064 μm. It is a hard, brittle crystal that is clear with a purple tint. It has a thermal conductivity some ten times that of glass, which permits CW operation with good beam quality. Glass has also been doped with Nd and lases well, but the poor conductivity does not make for good beam quality due to uneven heating. Very large lasers used for fusion research with pulse outputs in the thousands of megawatts per pulse are built with Nd-doped glass but are fired in single-shot mode so that heating effects on beam quality are mitigated. The NOVA laser at Lawrence Livermore National Laboratory uses a Nd:glass rod in each of its ten arms to produce 10 kilojoule pulses.

Nd:YAG lasers are generally Q-switched or mode-locked. Cavity dumping is not available in commercial lasers. An exception to the rule of not cavity dumping Nd:YAG lasers was the LCS. The requirements placed on the design of the LCS laser were stringent. The system required a means for creating high peak power pulses for acquisition, but the communications data rate requirements required a cavity-dumped configuration. A requirement was that the terminals had to be able to communicate at full data rate while one terminal crosses the solar disk. A narrow linewidth was needed, which permitted use of a very narrow bandpass filter for solar background rejection. Additionally, the need for redundancy was apparent since no single-point failures were permitted in the LCS.

The LCS met these requirements by use of a Nd:YAG laser that had unusual characteristics and as such posed a real challenge to the laser designers. The resultant laser operated in both a Q-switched mode for acquisition and then in a cavity-dumped mode for communication and tracking. It has been mentioned that the Nd:YAG laser can produce high peak powers. The LCS laser in the Q-switched mode produced short (nanoseconds) pulses at a peak power in excess of a kilowatt. This is advantageous for acquisition since it is easy to detect these high power pulses at the receiver. The peak power for the cavity-dumped mode for communications and tracking was over 40W with approximately 10-ns pulses. Again the high peak power worked in the designer's favor. It may be remembered from Chapter 3 that for a pulse interval modulation format, such as used in the LCS, the background photons are counted only during each pulse slot so the appearance of a high peak pulse in a slot results in a high SNR. The authors believe the LCS to be the only space-qualified Nd:YAG communication laser. The LCS laser is shown in the photograph of Figure 8.19. Much of the internal mounting hardware is beryllium. Much attention must be paid to cleanliness in the manufacture since the power density on the laser optics is so high (87 MW/cm^2), that if contaminants are present on any of the elements, optical damage will result and the laser may cease to function. The optical arrangement of the laser is shown in Figure 8.20. The intracavity modulator shown in Figure 8.20 had been a source of difficulty during portions of the LCS program. The development of a voltage bias with time was puzzling, but the designers believed they had found the cause at the time the program was terminated in 1993.

It is interesting to note that by enhancing this type laser, it may run CW and produce close to a watt of output power. A coherent system can be designed around a CW Nd:YAG laser. Such as system has a number of advantages over a coherent laser diode systems [17].

The Nd:YAG laser was used in the *Air Flight Test System* (AFTS). The configuration was lamp-pumped and a frequency doubler (FD) was used to achieve second harmonic generation with an output wavelength of half of 1.064 μm, or 0.532 μm. This laser, and the two-color (0.532/1.064 μm) AFTS system, is

Figure 8.19 Space-qualified laser communication laser. (Courtesy of McDonnell Douglas Aerospace.)

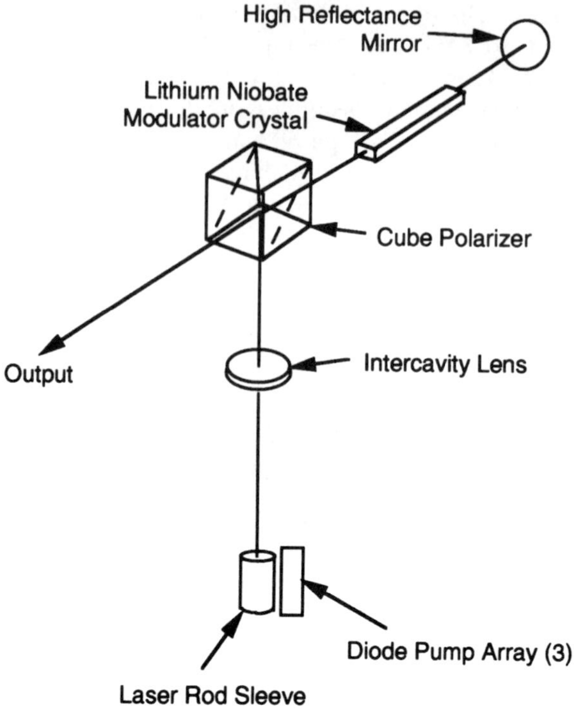

Figure 8.20 LCS laser optical layout.

described in detail in [18]. An external modulator was used at 0.532 μm to communicate from an aircraft to the ground at a 1-Gbps data rate.

The frequency-doubled Nd:YAG laser is a candidate for space since the transmit beam at 0.532 μm is narrower by a factor of two than a 1.064-μm beam for the same aperture, Avalanche photodiode detectors also are very sensitive at this wavelength, and the use of a two-color system simplifies the optical transmit to receive isolation in full duplex systems. Systems using FD Nd:YAG have recently been studied and offer a lot of merit. For example, their use has been postulated in a deep space data return from a distant spacecraft to Earth.

8.2.2.3 Semiconductor Diodes

The semiconductor diode made it appearance as a commercial component in the early eighties. Since then the quality, life, and power output of these devices has increased, primarily driven by commercial needs, as mentioned earlier. For laser communications, laser diodes offer a number of important advantages over other laser sources. A serious disadvantage is that not enough power has been available from a single diode to close synchronous long-range high-data-rate crosslinks. Another disadvantage is that no peak power operation is available from a diode. The devices are peak power limited and thus, the peak power is the same as the CW power. The following advantages are offered by laser diodes:

- *Very small*—the diode is about the size of a grain of sand and the entire package is thumbnail in size.
- *No modulator is required*—the diode is directly modulated by controlling the excitation current.
- *The diode is extremely efficient*—efficiencies up to 50% have been observed. This requires less power and less heat dissipation.
- *Wavelength is selectable*—a function of material composition, any wavelength over two separate ranges of wavelength is available for high-power diodes.
- *Very long life*—the diode is solid-state, diode life can far exceed spacecraft life requirements.
- *Voltage is low*—excitation is only a few volts, so no arcing problem.

The diode is a classic laser with a gain medium, two mirrors, and a pump source. This is illustrated in Figure 8.21. The elementary theory, operation, and types of diodes are covered in [18]. The single longitudinal-mode diode is the type required for crosslink applications since the output is diffraction limited and of narrow line width.

Diodes available in the early to mid-eighties had output powers less than 50 mW, CW. To close moderate data rate spaceborne direct-detection laser communications links with high reliability over a long life, it was and still is necessary to

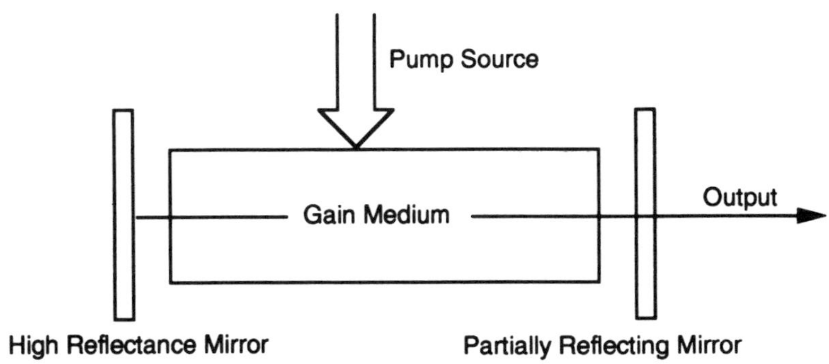

Figure 8.21 Classic laser diode configuration.

produce a power of at least 250 mW, CW. Three approaches may be followed with diode lasers to achieve this.

First, a 200 mW, CW diode, the model 5430 (which is currently available from *Spectra Diode Labs* (SDL)) or its predecessor, a single Model 5420 (which operates at 150 mW, CW), may be configured in a coherent system. With a 7 to 10-dB advantage over direct detection, adequate margin to close a one synchronous link at moderate data rate would be available. Second, a number of these high-power diodes (SDL 5420s or 5430s) can be aperture-shared to produce a power summation. This approach has been configured in test systems. And finally, a number of high-power diodes (SDL 5420s or 5430s) could be combined dichroically for direct detection to produce 1W or more of power using techniques developed in the early eighties. The latter two techniques, involving aperture sharing and dichroic combining, will be discussed in the following paragraphs.

Aperture Sharing. This approach was planned for an early Lincoln Lab laser communication experiment in the seventies. It was studied late in the seventies for application to the LCS during trades that included laser diodes as well as the Nd:YAG laser. The approach works as follows. The output of a number of individually collimated laser diode outputs are combined and inserted into a system pupil. The combined collimated beams then undergo an afocal magnification and are transmitted from the transmit telescope toward the cooperating terminal. Each transmitted beam produces a beam divergence in accordance with its transmit pupil aperture. As part of the pupil, it is apparent that the beam divergence of each is wider than that of a single beam transmitted from the telescope. Herein lies the rub with this approach. Not only is the combined beam divergence greater than a single beam, the irradiance of the beam is lower by the square of the divergence. As an example, if we aperture share, say, five diodes, each will produce a beam divergence of greater than twice that of a single diode transmitted from the full aperture. It may be calculated that the irradiance at the cooperating terminal is very close to that of the single diode using the full aperture, but the combined beam divergence will be more than twice the single diode's divergence. It is possible that for a given link this may be a preferred approach, particularly where pointing stability is a problem and achieving the highest link margin is not. Figure 8.22 illustrates the aperture-sharing approach.

Diode Combiner Lasers. Initially, the possibility of combining diodes was explored in the early eighties when single-mode diode outputs were 20 to 30 mW, CW. A means to combine the output of a large number of diodes was conceived [19]. Single-mode diodes were combined using dichroic combination to produce a high-power diffraction-limited beam that produces a minimum beam divergence using the full aperture of the telescope. Several other techniques for combining diodes

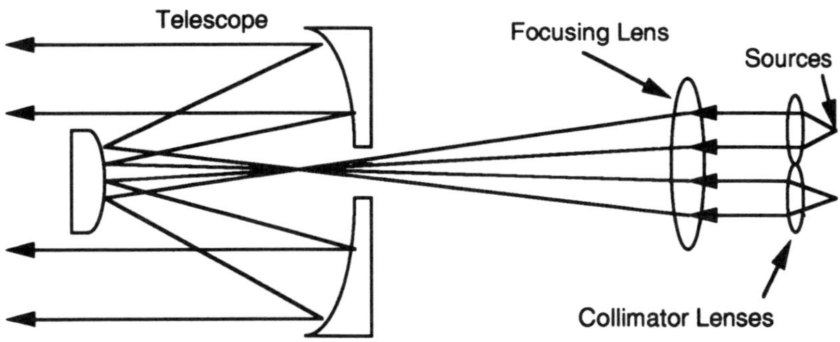

Figure 8.22 Aperture sharing approach.

have been explored in recent years. The one technique that has survived and been implemented in several iterations has been the dichroic approach.

Six generations of diode combiners have been built subsequently, the most recent being for a new-generation geosynchronous satellite program that, unfortunately was recently canceled. Dichroic combiner lasers have also been baselined for a number of other programs including NASA *Goddard Space Flight Center's* (GSFC) *Direct Detection Laser Transmitter* (DDLT) for the *Advanced Communication Technology Satellite* (ACTS), NASA GSFC's *Full Scale Development Demonstration* (FSDD), and the European Space Agency's SILEX program.

The generations of diode combiner lasers are described in [20]. An innovation during the development of the first diode combiner laser, the *high-data-rate laser transmitter* (HDRLT) for the Air Force assured the proper operation. The laser combines six diodes in a bandpass of 100Å, which amounts to a single-diode channel width of 20Å. Diodes produce a very consistent narrow CW linewidth, but when modulated at high-speed full on and off, mode hops produce a wide multiwavelength signal that can easily jump out of the assigned 20Å band for each laser. R. J. Smith, working on the HDRLT at the time, conceived of a way to stabilize each diode so that each diode would produce a signal that would stay in its assigned 20Å channel. The innovation was the creation of a three-cavity configuration for the laser. This was accomplished by the addition of a thin glass plate or etalon directly in front of the diode. This results in strong side-mode suppression and yields stable single-mode operation even in the presence of deep high-speed modulation.

Two other innovations as part of the HDRLT effort were a thermally controlled heat sink and an impedance-matching network at the diode, which permitted modulation from less than 50 megapulses per second up to one gigapulse per second. Since the optical paths of the diode beams differed considerably in length, each diode's coax line was trimmed so that the delays were compensated and the diode pulses were synchronized. During test it was determined that the approach worked well, the pulses were synchronized within 50 ps. The extinction ratio (max pulse height to min pulse height) was primarily limited to the relatively long fall time (330 ps) of the drive pulse. Still, it was found to range from 10:1 at 50 MPPS to 6:1 at 1 GPPS. Rise time was 100 ps. Use of an active pull-down circuit could be expected to improve the pulse symmetry by decreasing the fall time. The laser was designed to operate at 400 Mbps with a Manchester (biphase, level) modulation format.

Follow-on models of diode combiner lasers were considerably simpler in design. In the HDRLT, the collimating lenses for the diodes were of multiple-element design and were quite expensive. The last model uses a single molded asphere from Corning, which is exceedingly inexpensive although still diffraction-limited. The dichroic combining technique is illustrated in Figure 8.23. The combining technique is based on the fact that a narrowband filter transmits at its bandpass wavelength, but makes an excellent mirror for any wavelength on either side of its bandpass. The HDRLT used a glass block to mount the filters on. It was

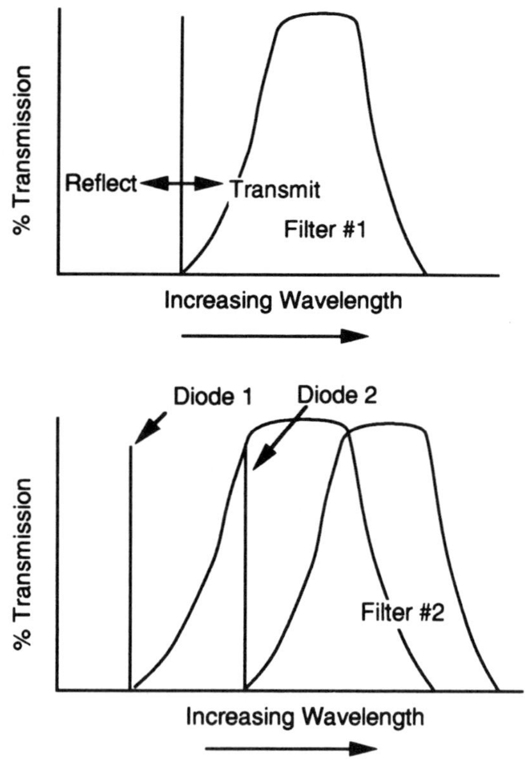

Figure 8.23 Dichroic combining technique.

necessary to temperature control the block to avoid temperature gradients that degrade wavefront quality. Individual diode steering for alignment was accomplished by pivoting the entire diode header in a ball and socket joint and subsequently locking it when the desired diode alignment was accomplished. The alignment proved difficult in practice. In the next generation, the approach was greatly simplified. The filters were not mounted on a glass block, which required temperature control, but rather were mounted ona precision metal frame that held the filters in precise alignment. The beams were combined with an air path in between the filters. The diode and headers were mounted in fixed position for better thermal control and the individual beam steering was done using Risley prisms.

A recent generation of diode combiner laser combined 16 diodes. Both polarization and dichroic combining were used. Two diodes were combined using polarization for each channel so that the bandpass is the same for the 16 lasers as it would be for an 8-laser dichroic-only laser. The polarization combining is accomplished by building two identical laser banks of 8 lasers each. There are 2 lasers at each wavelength. The first and second lasers are combined by taking the output of one of the banks and putting it through a half-wave polarizer, which rotates the plane of polarization 90 deg and then combining both lasers in a polarizing beam combiner cube.

The efficiency of the diode combiner is high considering the number of optical elements required. As the number of diodes increase, the efficiency falls and a point is reached where diodes are added and little is gained in increased power. This effect is plotted in Figure 8.24. It is seen that a 16-diode combiner is good but

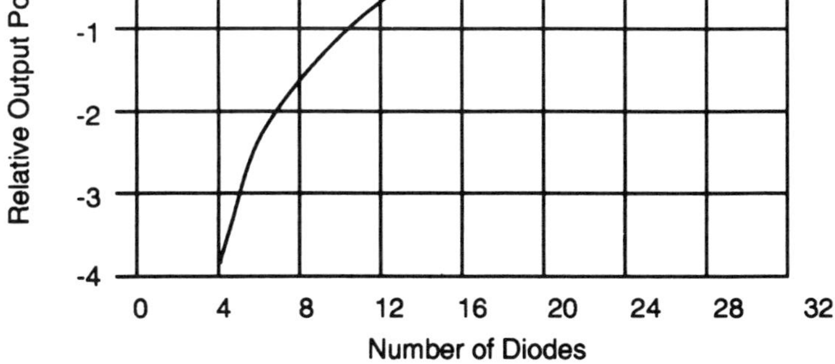

Figure 8.24 Diode combining efficiency.

diminishing returns accrue from the addition of more diodes. Actually, what with today's high-power diodes, the general trend has been to reduce the number of diodes in a system to cut down on weight, power, and complexity of the system.

An improvement of note in the past couple of years is the development of a hermetically sealed diode package specifically designed for diode combining. The *laser diode package* (LDP), dubbed the "Photon Bullet," contains the diode, the heater for the diode, some electronics, an air-spaced etalon, a diffraction-limited collimating lens, and a window that completes the seal.

Thermal tests have been run on this LDP and the results indicate less than a 0.2-dB loss due to mispointing of the diode beams over an 11°C temperature excursion. Structural analyses based on mount material stability indicate that the laser mount is stable over the proposed life of over ten years for the flight configuration.

8.2.2.4 Master Oscillator Power Amplifiers

The MOPA consists of a master oscillator laser diode followed by a broad-area amplifier section. The MOPA has been configured with separate oscillator and amplifier sections and as a single monolithic device. The master oscillator output is injected into the amplifier section and locks the amplifier emission to the oscillator frequency. The amplifier output is then an amplified, higher power version of the injected master oscillator.

The introduction of the SDL-5760 Series, 1W MOPA to the commercial market by SDL in the fall of 1992 heralded the beginning for a new era in semiconductor devices. Although one might imagine that high-power sources such as the MOPA would eliminate the need for combining techniques, benefits could accrue from combining the outputs of a number of MOPAs dichroically, including smaller aperture, reduced weight, data channelization, and lower power. MOPAs have been run at several watts and a 10W MOPA appears likely in a few years. A MOPA configuration is shown in Figure 8.25.

By combining MOPAs, it is possible to close just about any conceivable crosslink with modest aperture telescopes using direct detection. In addition, very high data rates are available by combining several lower data rate channels. Also, acquisition and tracking frequently demand separate channels for best performance as well as isolation. The addition of the MOPA to the laser stable also has the effect of relaxing and simplifying acquisition, tracking, pointing, point-ahead, alignment, and beamsteerer requirements. Further, the fact that fewer sources need to be combined greatly reduces the number of optical elements in a terminal. The MOPA is available in enough wavelengths to span a wide range of communication applications. The impact of the MOPA on terminal size, weight, and power is discussed and quantified in the trend projections of Chapter 11.

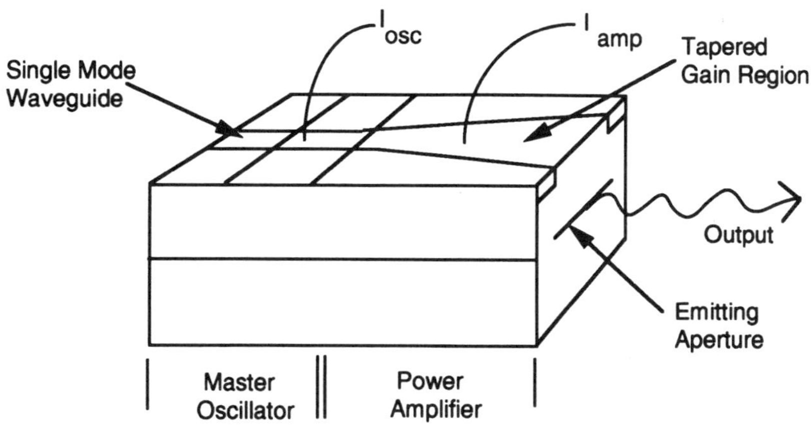

Figure 8.25 Diode MOPA configuration.

8.2.2.5 Erbium-Doped Fiber Amplifier

A fairly recent development in the telecommunications industry is the erbium-doped fiber amplifier (EDFA). For nearly thirty years it was known that lasing can be stimulated in an optical fiber. Developers of long-haul fiber systems need to increase the distance that signals can be sent between repeaters (signal boosters). Present repeaters detect the signal to be relayed, amplify it electronically, and then convert the amplified signal back to an optical signal for relaying down the fiber. A much more straightforward way is to develop a purely optical amplifier where the signal is never converted to an electronic signal. The EDFA fills this roll of pure optical amplification. The EDFA is modulation/bit rate scheme independent and a system can be easily configured.

The EDFA operates as shown in Figure 8.26. The incoming signal that is to be amplified, P_{in} at 1,550 nm, is directed to a *wavelength division multiplexer* (WDM) where it is combined with the pump signal from a pump diode at 980 nm, then enters an erbium-doped fiber, which has a laser transition in the 1,550-nm region. An optical isolator is used to avoid optical feedback and laser oscillation. The resultant small signal gain of the EDFA is about 30 dB with a resultant bandwidth of 35 nm. Very high speed is possible with an EDFA, with data rates of up to 100 Gbps attainable [21].

There are three envisioned uses for EDFAs. Figure 8.27 is a simplified illustration of the first use, the optical amplifier just discussed.

Figure 8.28 is the second use, an in-line amplifier for fiber-optic repeaters.

Figure 8.29 illustrates the third use, as an optical preamplifier. At the laser communications receiver terminal, the received signal would be input to the optical preamplifier, amplified, and then detected by a photodiode. The gain/noise charac-

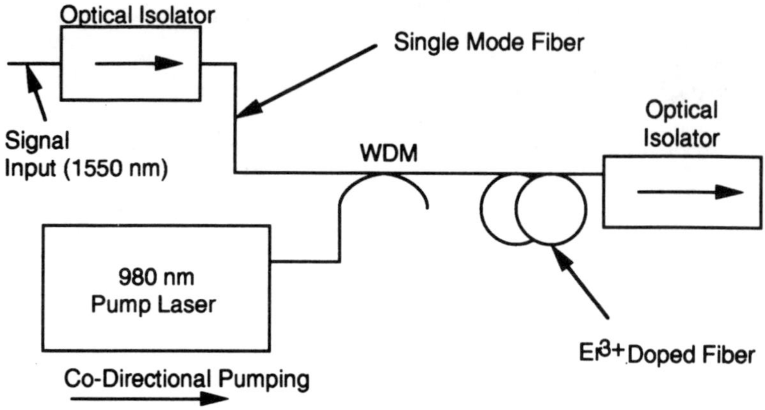

Figure 8.26 Erbium-doped fiber amplifier.

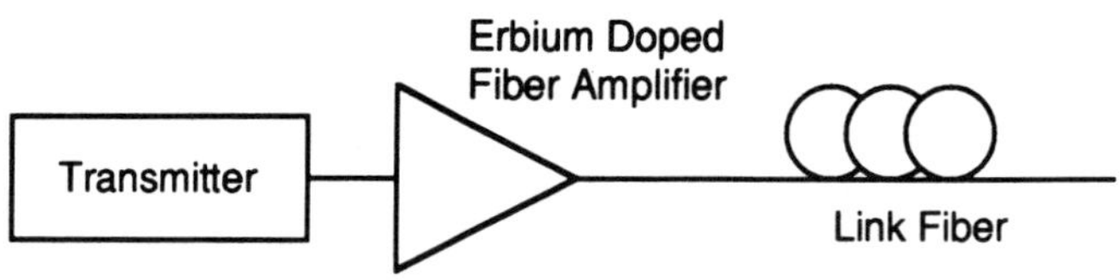

Figure 8.27 EDFA as an optical amplifier.

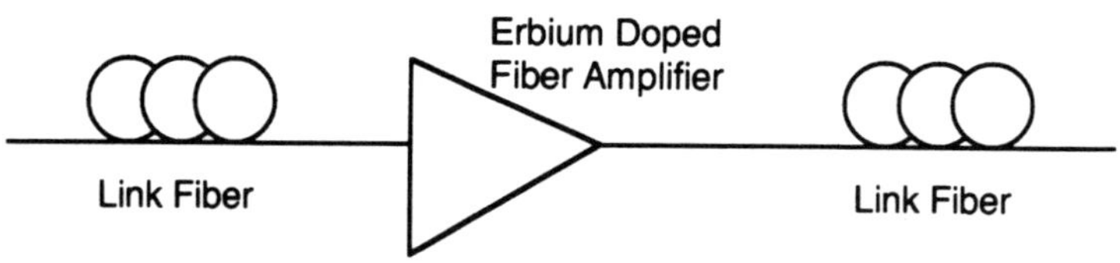

Figure 8.28 EDFA as an in-line amplifier.

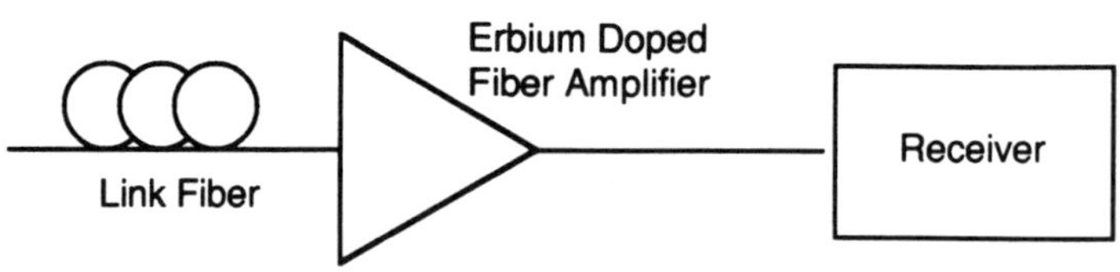

Figure 8.29 EDFA as an optical preamplifier.

teristics are such that this arrangement yields a sensitivity that comes close to that achieved by practical coherent-detection systems. The implementation of the EDFA direct-detection approach would be preferable to that of a coherent system. Sensitivity can be -45 dBm. We might envision an EDFA at both the transmit and receive locations in the link.

The current drawback for this approach is that a wavelength of 1,550 nm must be used and high-power diodes are not yet available at this wavelength. In the near future, higher power 1,550-nm diodes should be available and reliable, and as soon as this occurs we can realize the advantages offered by this approach.

8.3 DETECTOR TECHNOLOGY

Just as laser technology has evolved in recent years, detector technology has also evolved, although not as rapidly. The technology advancement focus has been directed towards lower noise and higher speed communications devices. Acquisition and tracking device technology advancement has emphasized the use of large-area array devices and high-efficiency quadrant dividing techniques. Research has also been performed on reducing the radiation susceptibility of detector devices, especially silicon-based detectors.

8.3.1 Photoconductive Detector Devices

Several types of detector device structures exist. The most commonly used devices for laser communication have been the photodiode (both avalanche and p-intrinsic) types. These devices are characterized as photoconductive where the output current or voltage is proportional to the input optical intensity. The temporal content of the incident optical signal is also preserved in photoconductive devices, which have sufficient detector gain bandwidth.

The most mature, low-noise photoconductive device structures have been fabricated using silicon technology. These devices provide excellent noise performance over the quantum response of silicon, namely 400 to 1,000 nm. Beyond 1,000 nm, the silicon responsivity falls off rapidly. Techniques do exist, however, to enhance the silicon device responsivity at long wavelengths. These include increasing the device temperature (at a cost of increased dark noise), increasing the device thickness (at a cost in device bandwidth), and employing a process patented by EG & G involving dimpling of the silicon surface. Using this dimpling process, enhancement of up to 3 dB in silicon responsivity at 1,000 nm has been demonstrated.

In addition to silicon, photoconductive detector devices have been fabricated from *indium gallium arsenide* (InGaAs), *indium phosphide* (InP), *indium gallium arsenide phosphide* (InGaAsP) and *germanium* (Ge). These devices are typically

Table 8.7
Typical Photodiode Detector Characteristics

	InGaAs PIN	*Ge APD*	*InGaAs APD*	*Si PIN*	*Si APD*
Quantum efficiency (maximum)	0.8	0.8	0.8	0.8	0.8
Response time	100 ps	500 ps	300 ps	300 ps	300 ps
Capacitance	0.5 pf	0.5 pf	0.5 pf	0.5 pf	0.5 pf
Dark current (unmultiplied)	10 nA	100 nA	10 nA	10 nA	10 nA
Dark current (multiplied)		100 nA	10 nA	1 pA	1 pA
Ionization coefficient		1	.5		.007*

ps = Picoseconds, pf = picofarads, nA = nanoamps, pA = picoamps.
* Best silicon value is 0.004.
After [24].

employed at long wavelengths where the responsivity of silicon is very poor. InGaAs is the detector of choice for the 1.3- and 1.5-μm fiber-optic systems. These longer wavelength devices are typically noisier than comparable silicon devices; however, their responsivities at long wavelengths outweigh the noise performance penalty.

Table 8.7 lists characteristics of typical laser communications detectors. These parameters will be discussed below in the context of communication receivers.

8.3.1.1 P-Intrinsic Photodiodes

The simplest photoconductive photodiode devices are the p-intrinsic (PIN) type. These PIN photodiodes exhibit noise that is characterized by dark current electron generation. No internal gain is available with these PIN devices. They simply convert the incident optical energy to output current.

PIN devices are commonly employed in high-background conditions, (where the lack of gain is not an issue), in coherent receivers (where local oscillator shot noise dominates), and in fiber amplifier receivers (where amplified signal-spontaneous beat noise dominates).

The critical parameters of interest for good PIN photodiodes performance include low dark current and high quantum efficiency (the light to current conversion effectiveness).

8.3.1.2 Avalanche Photodiodes

The *avalanche photodiode* (APD) detector offers the added advantage over the PIN photodiode of internal gain. The gain mechanism results from ionizing collisions of electrons or holes with valence electrons causing extra electron and hole pairs to be generated. This ionizing effect cascades until an avalanche of carriers has been created. This gain generation, however, does not come free. The ionization of carriers also creates excess noise in the photodiode. Webb et al. have characterized the performance of avalanche photodiodes in [22]. They have defined the critical parameters associated with the avalanche process and have identified the noise characteristics of the APD.

Critical parameters of interest for good APD performance include high gain (optimized with all other noise sources—background, preamplifier, and shot noise), high quantum efficiency, low dark noise (which is a function of gain and physical volume), and low excess noise (also a function of gain and ionization coefficient). The electrical bandwidth and capacitance of the device are also important parameters that must not be overlooked.

Gain values for silicon APD devices range in the 50 to 200 region. InGaAs APD devices exhibit much lower internal gain and thus, their dark currents are larger than silicon devices. Gains of 50 or less are typical.

Typical ionization coefficient k values for commercial silicon APDs are on the order of 0.02. Lower ionization coefficient devices have also been manufactured. Values on the order of 0.007 for silicon have been available for sometime for additional cost. Recently, a *special low "k"* (slik) device has been made available from EG & G [23]. This slik device exhibits the lowest *k* currently available—0.004. InGaAs devices exhibit much higher *k* values, on the order of 0.2 to 0.4.

The quantum efficiency (that is, the effectiveness of the detector to convert optical incident energy to electrical current), is a function of the incident wavelength and the thickness of the device. The thicker the device, the greater the quantum efficiency and better the device, up to a point. Recall that the dark current is also a function of device thickness. Trades must be performed to achieve the thickness for optimum detector performance. A typical silicon wavelength response is shown in Figure 8.30 [24]. Also shown in Figure 8.30 are typical response curves for InGaAs, Ge, and InGaAsP.

A typical communication APD will have a diameter on the order of 0.3 to 0.5 mm. These small devices offer excellent dark noise performance.

The detector electrical bandwidth is also a function of device thickness. The thicker the device, the slower the response. For high-speed devices, quantum efficiency must be sacrificed for bandwidth. Device capacitance must also be minimized. Excess capacitance increases the effective noise performance of the downstream amplifiers.

8.3.1.3 High-Gain Devices

In the past, *photomultiplier tubes* (PMTs) and other photoemissive type devices have also been considered for use in laser communication systems. These devices exhibit very large gain ($> 10^4$) and are very responsive in the visible spectrum. Drawbacks of most PMT-type devices reside in their lack of near-infrared responsivity and their large size and bulkiness. Additionally, permanent magnets were required to establish the fields in the photomultiplier detector.

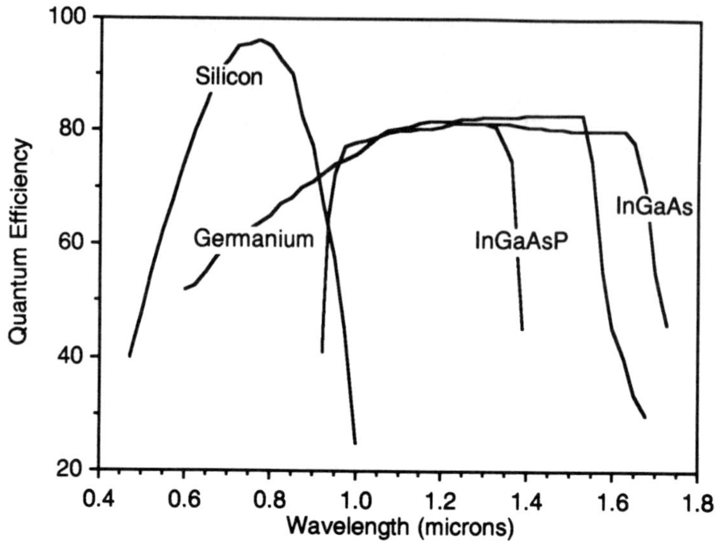

Figure 8.30 Typical detector responsivity. After [24].

Recently solid-state PMT-type devices have been introduced in small TO-type packages. These devices exhibit the same limited spectral performance and have moderate gain. While the devices are marketed as laboratory test devices, they may show promise as alternative high-gain receiver devices in the visible spectrum.

8.3.1.4 Quadrant Configurations

Both the PIN and APD can be configured as quadrant devices. Many of the characteristics are the same, but a few additional features must be identified here. When dealing with quadrant devices, care must be taken to understand the characteristics of the quadrant splitting technique. Several device types exist. Some employ total quadrant division with a dead zone (i.e., a nonactive region) and others use a sharing transition between quadrants. Quadrant crosstalk in the former is minimal, but rather large in the latter type devices.

Most quadrant devices to date have been fabricated out of silicon. The author's know of no reason why this cannot also be extended to InGaAs devices with proper investment.

Most current dead zone-type devices exhibit large dead areas, on the order of 50 to 100 μm between quadrants. Optical images from large-viewfield detectors yield spot sizes that are on the same order or smaller than the dead zone. This can result in appreciable losses in the detection process. To that end, development of image dividers, which reduce the dead zone, are very desirable. One such device had been developed by EG & G and was baselined on the *Follow-on Early Warning System* (FEWS) *Crosslink Set* (CLS) [25]. This device used an etched silicon divider with a back reflector that reduced the deadzone of a standard quadrant device to 2-4 μm. The etched silicon divider was fabricated and front-surface coated to reflect the incident optical energy away from the dead zone to the active areas of the quadrant device. The splitter was epoxied onto the surface of the quadrant device. Testing of the device yielded excellent performance.

Dark current is higher in a quadrant detector than in a communication device. Quadrant devices are typically large in diameter, on the order of 1.3 to 1.5 mm. The larger area yields larger bulk dark noise current.

8.3.1.5 Dual Signal Devices

Quadrant devices have also been fabricated that allow the signal incident on the device to be electrically filtered and split into two separate signals. One signal is low bandwidth, the other high bandwidth. The signals from the four quadrants are summed from the front side of the device, where this composite is used for communication. The low-bandwidth individual quadrant information is taken from the back junction of the device and used to develop the tracking signal. The concept for signal splitting from the quadrant detector was patented by Dr. Sam Green while working for McDonnell Douglas Corporation in the mid eighties [26].

8.3.1.6 Radiation Issues

APD and PIN photodiodes, especially in a silicon configuration, are susceptible to radiation effects, and to that end much work has been performed to minimize these effects. Guardring and guardband structures have been introduced into the design of the devices to reduce the effective silicon active volume, thereby minimizing the volume susceptible to radiation events. Other techniques, more passive than active, have also been used. Most radiation mitigation is designed into the system approach rather than the device itself. These approaches include peak clipping circuitry, high-z shielding and waveform selection for radiation effect minimization.

8.3.1.7 Detector Packaging

APD and PIN detectors are usually package in a hermetic can arrangement with a fused-silica or sapphire window. Custom packages, containing integral preamplifiers, thermal control circuitry, and gain bias circuitry have been produced for specific applications. Depending upon the application and volume of detector devices needed, custom packaging may be beneficial to the system design and life cycle costs.

8.3.2 Integrating Imaging Devices

The second type of detector commonly used in laser communication system designs has been the integrating imaging device. In this context, imaging identifies those devices that integrate incident optical charge for a period proportional to the readout time. With these devices, low signal levels are enhanced through integration; however, the temporal content is most often lost, except for very low frequency signals. Most integrating devices are large-array pixel-type devices. The large number of small pixels allows for operation in higher background conditions with a large viewfield when compared to the quadrant detectors mentioned above. This is easily seen as each pixel maps into an individual viewfield, which is a very small portion of the total viewfield.

Large-area array device development has piggybacked primarily off of the development of commercial imaging cameras. Most area array devices to date have been fabricated from silicon. Here, the responsivity is directly proportional to the silicon response. Thus, these devices are limited to 400 to 1,000 nm in wavelength range. Work has been done in developing devices fabricated out of platinum silicide for operation in the 3 to 5-μm spectrum. Similar type development may prove useful using either InGaAs alone or a combination of InGaAs and silicon to develop large-area array devices that are sensitive in the 1.3- to 1.5-μm regime.

Two of the most common large-area array devices, fabricated from silicon, are the *charge injection device* (CID) and the *charge coupled device* (CCD). Each is fabricated with a large number of active charge integrating pixels that store charge until saturation or readout occurs. The devices are similar in operation, but each has its unique characteristics. We will discuss each below.

8.3.2.1 Charge Injection Device

The CID is an area device whose capability resides in its ability to integrate incident optical signal photons on a large array of pixels. The CID has the capability of injecting the charge counts into storage registers without destroying the integrated charge level. In effect, the CID can sample the signal at different time intervals and output signal levels proportional to the total integrated charge at any time T. The fabrication techniques associated with the CID have limited the pixel densities to approximately 256 by 256 pixels.

Another unique feature of the CID is its ability to address individual pixel elements or combine the outputs of 4 or 16 adjacent elements (arranged in a square) into a single charge value. This superset pixel addressing can be advantageous in scanning the entire array during an acquisition search. The CID readout rate is limited by the speed of the electronics. Typical frame rates are on the order of 100 Hz. Subpixel addressing rates can achieve kilohertz readout rates.

The major noise contributors in the CID device are attributed to the dark noise (which can be large since the device size is large) and readout noise (which is a function of readout rate). Background is minimized through the physical pixel angular subtense and can also be further reduced through average background subtraction algorithms.

CID technology has been summarily superseded in the commercial world by the CCD and its large pixel densities. These large densities yield greater resolution for the imaging camera applications.

8.3.2.2 Charge Coupled Devices

Like the CID the CCD is a large-area pixelized device that integrates charge. CCDs have been fabricated with active areas of 2,048 by 2,048 pixels, offering unparalleled resolution. The CCD differs from the CID in that once it reads charge, the stored charge value is lost. Additionally, the CCD is a row readout device. An entire row (or in some cases, column) is shifted out serially to address an individual pixel. Post-processing identifies the location of the pixel and the value of the stored charge. Some CCDs are fabricated with an additional covered active area where charge is shifted and read out leisurely while the next frame of information is integrated.

Just like the CID, the CCD is dominated by bulk dark noise and by readout noise. Most commercial CCDs are addressed at video frame rates (i.e., 30 or 60 Hz). Devices have been fabricated that have readout rates in excess of 10 kHz for small arrays. Additionally, small arrays have been fabricated for tracking applications. One such device has been baselined in the European Space Agency SILEX demonstration system [27]. It is a 14 by 14 pixel device used for both acquisition and tracking of the beacon signal.

CCDs are very common today, but their application to laser communication systems must be optimized for the specific application and more ideally customized for the application at hand

8.3.3 Utility Detector Devices

Other optical detectors have found utility in laser communications system design. In addition to the detectors operating on the main received signal, detector devices have been used for alignment, power monitoring, and pointing feedback. Devices such as *lateral photodiodes* (LPDs), simple PINs, and quadrant cells have been used in these capacities.

The LPD is a bulk silicon device that has four voltage outputs that are proportional to the location of the received signal centroid incident upon its surface. Bandwidths of up to tens of kilohertz are possible and linearity is very good near the center two-thirds of the device. The LPD has been used in beamsteerer angular position feedback to allow off-axis angular positioning of a point-ahead device.

Simple quadrant cells are low-bandwidth versions of the silicon quadrant PIN photodiode. They allow position centroiding of a signal to achieve boresight alignment. Simple PIN detector devices have also been used internal to the optical system as waveform and power monitors.

8.4 SUMMARY

Technology choices for major elements of the laser communication system have been introduced in the previous sections. While these choices represent the past, current, and future state-of-the-art technology choices, they are by no means exhaustive. New technology advances are emerging every day. The designer is urged to keep abreast of development and direct the technologists' efforts towards the components that make the system approaches and designs simpler and more cost effective. It is imperative that technology be developed with a focus toward the systems design.

References

[1] Casey, W. L., and Phinney, D. D. "Representative Pointed Optics and Associated Gimbal Characteristics," *SPIE OE-LASE '88*, Vol. 887, Los Angeles, CA, Jan. 1988.

[2] "Zero-Backlash Helps Satellites Communicate," *Machine Design*, March 7, 1991.

[3] Lambert, S. G., et al. "Short-range multi-terminal satellite crosslink communications," *IEEE Military Communications Conf. MILCOM '92*, Vol. 3, 1170, San Diego, CA, 1992.

[4] Yoder, P. R. *Opto-Mechanical Systems Design*, New York, NY: Marcel Dekkor, Inc., 1985.

[5] Vukobratovich, D. "Introduction to Opto-Mechanical Design," Course Notes (unpublished), Optical Sciences Center, University of Arizona, Tucson, AZ, 1986.

[6] Hopkins, R. N. "Some Thoughts on Lens Mounting," Optical Engineering, Vol. 15, No. 5, Sept.–Oct. 1976.

[7] Bayer, M. "Lens Barrel Opto-Mechanical Design Principles," Optical Engineering, Vol. 20, No. 2, March–April 1981.

[8] Hopkins, R. E. "Optical Element Mounting and Alignment Technique," *Geometrical Optics*, SPIE Vol. 531, 1985.

[9] Seigman, A.E. *Lasers*, Mill Valley, CA: University Science Books. 1986.

[10] Melles Griot. *Optics Guide* 5, 1990.

[11] *The Photonics Design And Applications Handbook*, Book 3, Pittsfield, MA: Laurin Publishing Co., 1993

[12] McElroy, J. H., McAvoy, N., Johnson, E. H., Degnan, J. J., Goodwin, F. E., Henderson, D. M., Nussmeier, T. A., Stokes, L. S., Peyton, B. J., and Flattau, T. "CO_2 Laser Communications Systems for Near-Earth Space Applications" IRE Proc., 1977

[13] Bonek, E., and Lutz, H. "CO_2 Laser Communication Technology for Intersatellite Data Links," *ESA Journal*, 1981.

[14] Barry, J. D., Darien, A., Dawkins, B. T., Freedman, P., Heitman, J. M., Kennedy, C. J., Lyon, J. K., Mattassov, G., Matulka, D. D., Whited, C. E., and Zack, D. M. "1000-Mbits/s Intersatellite Laser Communication System Technology," *IEEE Trans. On Communications*, 1976.

[15] Ross, M., Freedman, P., Abernathy, J., Matassov, G., Wolf, J., and Barry, J. D. "Space Optical Communications With the Nd:YAG Laser," IRE Proc., 1978.

[16] Katzman, M., Ed. *Laser Satellite Communications*, Englewood Cliffs, NJ: Prentice-Hall, 1987.

[17] Kobylinski, R. A. "Coherent Nd:YAG Laser Communications," *SPIE OE-LASE '86*, Vol. 756, Los Angeles, CA, Jan. 1986.

[18] Begley, D. L. "Semiconductor Laser Diodes: Fundamentals and Use," *The Photonics Design And Applications Handbook*, Book 3, Pittsfield, MA: Laurin Publishing Co., 1993

[19] Casey, W. "A Diffraction Limited Dichroic Combiner Diode Laser," Patent Number:4,823,357, assigned to the United States as represented by the Secretary of the Air Force, April 18,1989.

[20] Jackson, J. E., Chenoweth, A., Dreisewerd, D., Lambert, S., and Casey, W. "The Evolution and Future of the Multiple Diode Laser," *SPIE OE-LASE 93'*, Vol.1866, Los Angeles, CA, Jan., 1993.

[21] Auge', J., Chesnot, J., Gabla, P. M., and Weygang, A. "Progress in Optical Amplification," *Microwave Journal*, June 1993.

[22] Webb, P. P., McIntyre, R. J., and Conradi, J. "Properties of Avalanche Photodiodes," *RCA Review*, June 1974.

[23] MacGregor, A. D., Dion, B, and McIntyre, R. J. "High Sensitivity, High Data Rate Receivers for ISL Using Low-Noise Silicon APD's," Spring 1989.

[24] Van Muoi, T. "Receiver Design For Optical Fiber Communication Systems," Course Notes, Plesscor, Inc., 1985.

[25] Green, S. I. "Optical Dividers for Quadrant Avalanche Photodiode Detectors," *SPIE Proc. on Free Space Laser Communication Technologies III*, Vol. 1417, Jan. 21–22, 1991.

[26] Green, S. I. "Tracking Atmospheric Optical Communication System," *SPIE Proc. on Free Space Laser Communication Technologies II*, Vol. 1218, Jan. 15–17, 1990.

[27] Mathur, R. P., Beard, C. I., and Purll, D. J. "Analysis of SILEX Tracking Sensor Performance" *SPIE Proc. on Free Space Laser Communication Technologies II*, Vol. 1218, Jan. 15–17, 1990.

Chapter 9

Configuring the System

9.1 OVERVIEW

With the tools established by the previous eight chapters, the system designer can now begin to configure systems based on the customer-established requirements. It is important as a system analyzer to understand the give and take associated with the design of a major satellite system. While the requirement for minimum size, weight, and power may exist on the task table of the satellite systems engineer, achieving this result may require various design iterations. Some iterations, while yielding optimum satellite performance, may not yield the lightest weight laser communication terminal design. It is, therefore, imperative that the laser communication system designer know the hardpoints (that is to say the boundaries) in these design configurations. While it would be beneficial to the designer to have these boundaries included in this text, unfortunately, they are learned only through experience. Even with experience, the designer must be adaptable since these boundaries are not fixed. They are constantly in motion and tied to technology advances and mission requirements.

This chapter will provide the laser communication system designer with insight into configuring a laser communication crosslink system. An example crosslink design will be undertaken from requirements to link definition. Along the way, insight into choices made and trades that should be considered are highlighted. Remember, the job of any good system designer is to understand the requirements established by the customer, understand his or her own system's limitations, and work with the customer to achieve the best results for the total system design. Included in this understanding should be the need to challenge system level requirements, which require the laser communication system to perform near or over its limits. Only through iteration will the system design achieve the desired optimization.

Table 9.1
System-Level Customer Requirements

Requirement	*Value*
Data rate	5 Mbps full duplex
Bit error rate	Less than one error in a million
Range	120 deg of orbital arc
Satellite altitude	Geosynchronous-stationary
Minimum acquisition time	Less than 5 min
Acquisition success probability	Greater than 99%
Weight	Less than 100 lb
Power	Less than 100 W
Installation	No obstruction of viewfield
Lifetime	Greater than 10 years
Reliability	Greater than 90%

9.2 THE REQUIREMENTS

We begin the discussion of the example system configuration with an understanding of the requirements. Table 9.1 outlines the assumed customer-required performance necessary from this crosslink system. As can be seen, the system is required to operate across a 120° orbital arc in geosynchronous orbit and transfer up to 5 Mb of data in a full duplex (both directions simultaneously) configuration. A bit error rate (BER) of less than one error in a million bits transferred is required (i.e., $P_{be} \leq 10^{-6}$). The initial link-up (or acquisition time) is not a driving concern and shall be required within 5 min of initial command with greater than a 99% probability of success. The system shall weigh less than 100 lb and consume no more than 100W. The crosslink installation shall also not obstruct any other satellite sensor viewfields. On-orbit life shall exceed a mean time to fail of 10 years with a probability of greater than 90%.

Sounds simple enough, right? Well, almost. The initial requirements, while important to the laser communication system designer, don't begin to yield the necessary information needed before a system can be configured. What's still needed? The environment, that is, the satellite and its surroundings, is required. Everything from potential installation locations to something as simple as interfaces with the satellite command bus must be understood to a first order, or at least estimated, before the design iterations can begin.

9.3 THE ENVIRONMENT

In establishing the environment for the laser communication payload, the system designer must understand not only the normal environmental issues (natural and manmade radiation, thermal, vibration, shock, acoustics, vacuum, contamination, etc.) for on-orbit, storage and launch conditions, but the designer must also understand the physical environment of the satellite itself. By physical environment, we mean those items, separate from environmental issues, which interact with the laser communications payload directly. Some of the normal satellite environmental issues with which the designer must be concerned are

Natural radiation;
Man-made radiation;
Thermal flux;
Vibration;
Pyroshock;
Acoustic;
Vacuum;
Contamination;
Aging.

Some of the physical interactions with which the crosslink and the satellite interact are

Satellite ephemeris accuracy;
Satellite attitude accuracy;
Potential mounting locations;
Command interface type and rate;
Viewfield limits;
Potential obstructions;
Power interface characteristics;
Data interface characteristics;
Solar angles to line of sight;
Other background sources.

The satellite ephemeris and attitude accuracies, and the transfer of these accuracies to the mounting location of the crosslink, contribute to the initial search uncertainty and can drive the acquisition system design if not properly managed. The mounting location can drive the gimbal design. The type and speed of the command interface with the satellite can have deleterious effects if not properly understood up front. The on-orbit vibration introduced into the crosslink optical bench can affect the pointing and tracking system design. These disturbances can come from solar array drives, payload gimbals, scanning mechanisms, and momentum wheels. Some discrete mechanism transients can also be of concern to the laser communication system designer.

Viewfield obstruction is another big issue for both the satellite and the laser communication system designers. Many satellites have restricted viewfield space. These result from solar panels viewing the Sun, radiators viewing deep space, *radio frequency* (RF) up and downlink antennas viewing Earth, and mission sensors viewing their respective fields of regard. With all these systems vying for the precious viewfield areas, minimizing the total viewfield surface volume is beneficial to an efficient satellite design. In addition, sensors and antennas that require boom placement or solar shielding sometimes obstruct the field of regard for the optical crosslink antenna (telescope). In these situations, workarounds must be established early in the satellite design. Either designing out or adding secondary subsystems sometimes solves these issues.

Other items, such as the data interface, the power interface, and software coordinate transformations, need to be understood as part of the overall system design. As the fidelity of these data becomes increasingly better, the system design is iterated to achieve a more robust and efficient crosslink design.

With the requirements established and the environment understood, the system ground-zero design iterations can begin. It is imperative at this stage of system design not to be driven to a point design. Some technology choices are very logical at this point. As we shall see in the next section, technology choices based on experience are made to simplify the trade space. The more open the trade space at the early stages of system design, the more flexible to satellite requirement iterations the design is, especially in data rate and installation. Driving to a point design too early can be detrimental to the overall laser communication system design.

9.4 MAKING SYSTEM CHOICES

Several top-level system trades have been exercised to reduce the trade space for the laser crosslink system design. While either direct or coherent detection could have been selected for this design, our example focuses on direct detection. The design employs classical direct detection. A classical direct-detection system provides a good and straightforward example that is easily understood by the first-time system designer. It is felt that extending this example to either amplified receivers or coherent detection is possible with the foundation laid in Chapters 1–8.

The laser technology selected for this example is direct drive, single-mode laser diodes. For efficient operation, the data rate requirement limits the laser technology selection to either single-mode laser diodes, externally modulated solid-state devices, or high-power master oscillator power amplifier devices. Selection of gallium arsenide laser diodes was made on the basis of efficiency, lifetime, simplicity of use, direct modulatability, detectability (as it relates to wavelength), and frequency availability (full duplex requirement).

9.5 DEVELOPING A POINTING BUDGET AND BEAM DIVERGENCE

Since the minimum beam divergence yields the minimum aperture-power factor, establishing a budget that estimates the error sources associated with pointing the laser beam is necessary. For the system example, the initial estimate of the closed-loop pointing budget is given in Figure 9.1. As is shown by the figure, three primary sources of error exist. The residual gimbal jitter is that amount of excess jitter not compensated by the tracking system. The tracking error is the *noise equivalent angle* (NEA) of the tracking receiver and is a function of the received spot size, the intensity distribution, and the *signal to noise ratio* (SNR) of the ratiometrically processed signal. The pointing error estimate is comprised of the transmit to receive alignment error (static and dynamic) and the accuracy of the pointing mechanism.

For our example system, we have employed a dual-loop pointing and tracking system as depicted in Figure 9.2. An outer loop is used to control gimbal motion,

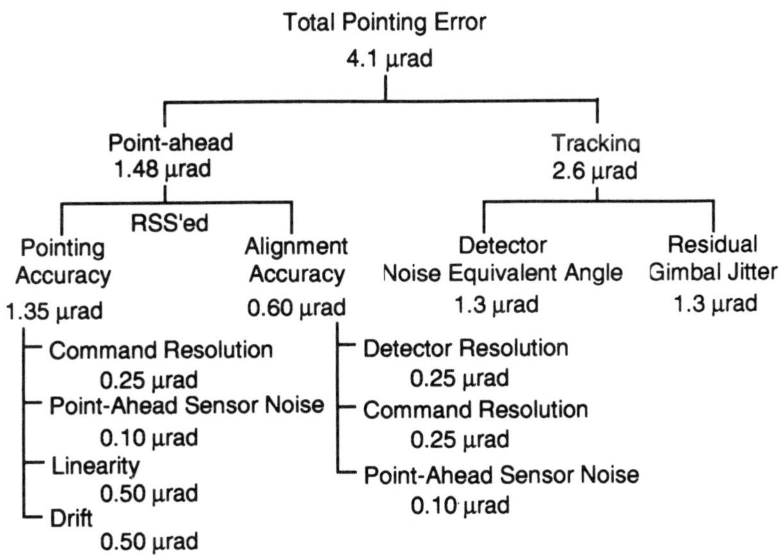

Figure 9.1 Closed-loop pointing budget.

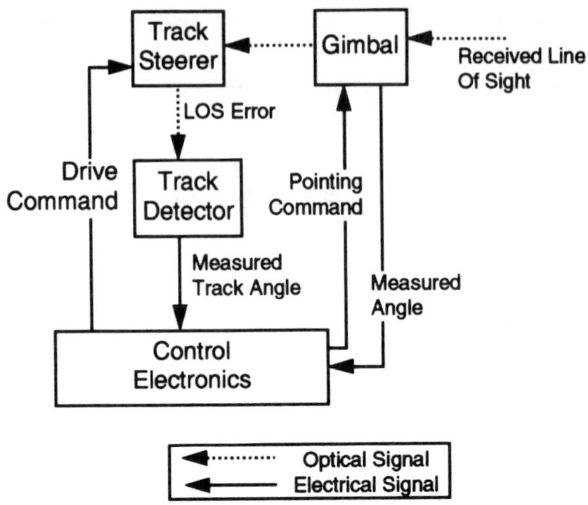

Figure 9.2 Dual-loop pointing and tracking system.

while the inner loop operates on the fine beamsteering mechanisms used for received signal boresight compensation. From this, the error contributions are collected and a total estimate of the pointing error (i.e., the pointing budget) has been developed. Our example illustrates a 3σ total error of 4.1 μrad. The error is characterized as 1.5 μrad of bias and 0.866 μrad of *root mean square* (rms) jitter.

Techniques do exist to make the pointing error even smaller by reducing either the track noise, the gimbal jitter, or the pointing error. The gimbal jitter can be reduced by increasing the rejection of the tracking system to eliminate the residual jitter. The pointing system noise can be reduced through use of increased accuracy angle feedback on the point-ahead mechanism. And the tracking error can be reduced by either increasing the received SNR on the tracking channel or by increasing the gain of the ratiometric transfer function. By far the easiest for the system designer is the latter; however, it may become necessary to tweak all three to achieve the desired compensation.

It is imperative that the system designer revisit the assumptions made in the development of the pointing budget as all links become formalized. A balance should exist between the communication and track links to approach optimization of the required laser power for the two links.

9.6 CONFIGURING THE COMMUNICATION LINK

Once the pointing budget has been established, the system designer's task now turns toward the communication link development. Top-level requirements are flowed down directly where applicable and derived inputs are also developed at this time.

For this example, the link requirements indicate a moderate data rate link. From experience we have learned that for links of approximately 50 Mbps or less, use of a common track and communication receiver can reduce the overall optical complexity and parts count. While separate track and communication detectors could also be used, our example has adopted a single-detector approach.

9.6.1 Data Rate and Modulation

The nominal data rate requirement is given as 5 Mbps; however, additional data bits need to be added for laser crosslink internal telemetry. An additional 5% of overhead has increased the instantaneous data rate to 5.25 Mbps.

The use of single-mode laser diodes lend themselves easily to 50% duty cycle modulation formats. The thermal properties of the diode, which are important for mode stability, are much more benign at this duty cycle than at lower duty cycles. Both *nonreturn to zero* (NRZ) and Manchester formats could be used to accomplish this. Our design example has employed Manchester modulation, selected for its ease of data clock recovery at the receiver.

9.6.2 Error Rate and Coding

With a required BER of $\leq 10^{-6}$, the use of link encoding is recommended to enhance performance and reduce the overall power required. When analyzing a laser communication link, one must be concerned with the effects of the burstiness of the pointing system. To that end, the BER requirement has been partitioned equally between the receiver and the pointing system. This has yielded a nominal derived error rate for the detection system of $\leq 0.5 \times 10^{-6}$ bit errors per second.

Surveying potential coding types has identified an industry standard rate one-half, constraint length seven convolutional encoder with Viterbi decoding available on a single chip. With this codec, the derived symbol error rate is 2.4×10^{-2} symbol errors per second. Of course, the number of raw bits transmitted has now been doubled from 5.25 to 10.5 Mbps.

From the established symbol error rate, the required SNR can be calculated. Using from Chapter 3,

$$P_E = \tfrac{1}{2}\,\mathrm{erfc}(z) \tag{9.1}$$

The SNR is calculated for a probability of symbol error of 0.024 to be 1.95. At this time it is beneficial to consider the receiver efficiency for end-of-life performance. What we have just calculated is the ideal performance that can be theoretically achieved. Degradation due to real hardware realization, aging and thermal drift can reduce this performance from theoretical values. For our example, we accommodate these effects by derating the ideal performance by approximately 2 dB. This yields a required SNR of 1.95/0.631 = 3.1.

9.6.3 Range Between Satellites

The 120° orbital arc between satellites in geostationary orbit equates to a range of 81,428 km. This is calculated using the geometry of Figure 9.3 and

$$R = \left[A^2 + B^2 + 2AB \cdot \cos(\theta) \right]^{1/2} \quad (9.2)$$

where A and B are the orbital altitudes of the satellites (19,323 nmi) plus the radius of Earth (3,441 nmi) and θ is the orbital arc angle (120°).

$$R = \sqrt{2 \cdot (19{,}323 + 3{,}441)^2 + 2 \cdot (19{,}323 + 3{,}441)^2 \cdot \cos(120°)}$$

$$R = 43{,}967 \text{ nmi (or } 81{,}428 \text{ km)}$$

Because of the stationary positioning of these satellites, range variation is minimal and not of concern in this link analysis. Should nonstationary orbits be part of the system design adequate consideration to dynamic range should be made.

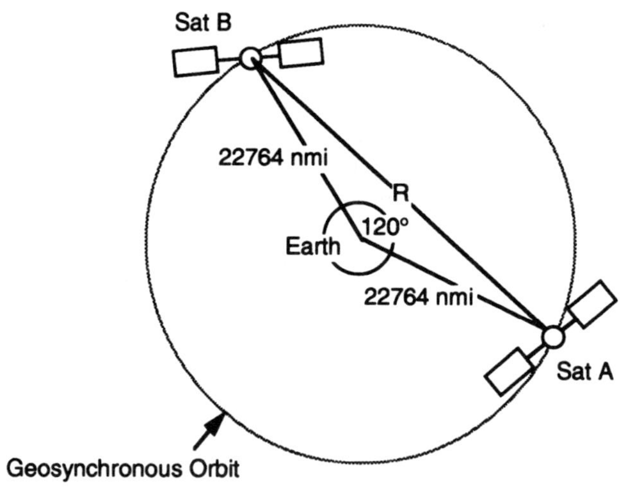

Figure 9.3 Range geometry.

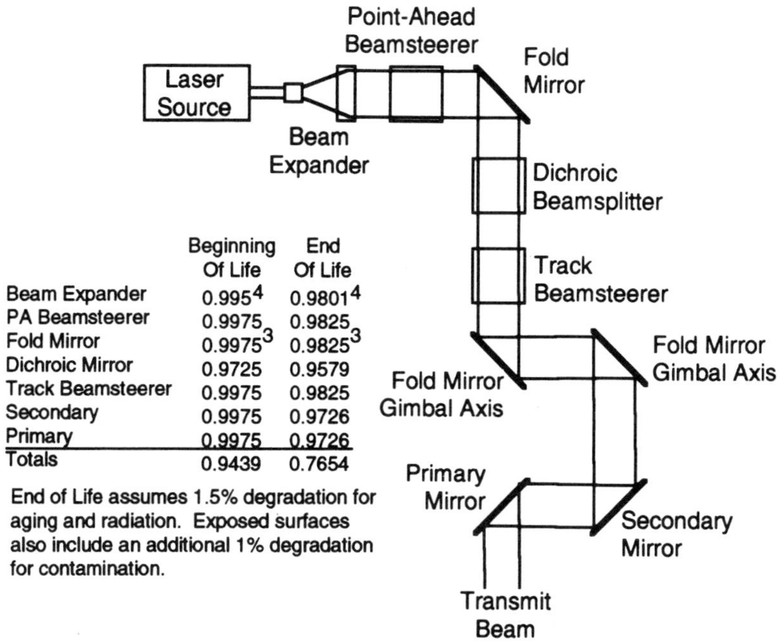

	Beginning Of Life	End Of Life
Beam Expander	0.995^4	0.9801^4
PA Beamsteerer	0.9975_3	0.9825_3
Fold Mirror	0.9975^3	0.9825^3
Dichroic Mirror	0.9725	0.9579
Track Beamsteerer	0.9975	0.9825
Secondary	0.9975	0.9726
Primary	0.9975	0.9726
Totals	0.9439	0.7654

End of Life assumes 1.5% degradation for aging and radiation. Exposed surfaces also include an additional 1% degradation for contamination.

Figure 9.4 Transmit path optical layout.

9.6.4 Transmission Budget

The transmission budgets for the transmit and receive paths are based upon the type and number of optical surfaces. For this example, the transmission has been established for an initial estimate based on a preliminary optical design. Figure 9.4 identifies the types and number of surfaces of the transmit path. A similar analysis has been performed for the receive path. The values are then derated for end-of-life contamination, aging, and radiation effects. The transmission numbers for the optical path are estimated from the preliminary optical layout and calculated following the analysis in Chapter 4. The transmissions are given as 76.5% for the transmit path and 55.4% for the receive path.

9.6.5 Aperture Sizing

Based on the established pointing budget, an optimum aperture size is computed from the wavelength and desired obscuration ratio. Optimizing the off-axis gain, based on the pointing statistics, yields an aperture size and transmit beam divergence. For our example, the transmit aperture size has been computed as 5.5 in.

This yields a beam divergence of 12.2 μrad when combined beam coalignment considerations are included. Chapter 3 outlines the details of the gain equation used to calculate the beam divergence and gain.

9.6.6 Receiver Parameters

Using a quadrant avalanche photodiode as a common receiver for communication and tracking requires a balance between both receivers. The noise sources present include background, detector dark, preamplifier, signal shot, and excess track signal noise.

9.6.6.1 Background Noise

Background energy is restricted through the use of narrowband filters and field stops. The source spectral width limits the passband width of the filter. The *field of view* (FOV) is restricted by the spot size and full angular track range of the receiver. In our example system, we begin with a viewfield of 100 μrad in angular diameter and a 60Å optical bandpass (estimated bandpass for diode combining up to four single-mode laser sources).

The background noise current density has been calculated for the Lunar disk on-axis as

$$P_B = H_B \cdot \frac{\pi}{4}(\text{FOV})^2 \cdot B_f \cdot \frac{\pi}{4}(D_{\text{aper}})^2 \cdot L_R$$

$$P_B = (0.00242) \cdot \left(\frac{\pi}{4}\right)(0.0001)^2 \cdot 60 \cdot \left(\frac{\pi}{4}\right)(0.0254 \cdot 5.5)^2(0.5535) \quad (9.3)$$

$$P_B = 9.68 \times 10^{-12} \text{ W}$$

Using this background value, the background noise current density is calculated as

$$\sigma_B^2 = 2qFP_BR_d$$

$$\sigma_B^2 = 2(1.602 \times 10^{-19})(3.177)(9.68 \times 10^{-12})(0.5429) \quad (9.4)$$

$$\sigma_B^2 = 5.35 \times 10^{-30} \text{ A}^2/\text{Hz}$$

Recalling from Chapter 5 that F is the detector noise figure and is calculated from the detector ionization coefficient, k, and avalanche gain, M.

$$F = Mk + (1 - k)(2 - 1/M)$$

$$F = (150)(.008) + (1 - .008)(2 - 1/150) \quad (9.5)$$

$$F = 3.177$$

Table 9.2
Characteristic Quadrant Avalanche Photodiode

Parameter	*Value*
Quantum efficiency	0.816
Ionization coefficient	0.008
Avalanche gain	150
Multiplied dark current	230 nA
Unmultiplied dark current	83 pA

All values derated for end of life.

R_d is the detector responsivity calculated from the quantum efficiency, η, electron charge q, the speed of light c, wavelength, λ, and Planck's constant h, and is calculated as

$$R_d = \frac{\eta q \lambda}{hc}$$
$$R_d = \frac{(.816)(1.602 \times 10^{-19})(825 \times 10^{-9})}{(6.626 \times 10^{-34})(2.998 \times 10^{8})} \tag{9.6}$$
$$R_d = 0.5429$$

9.6.6.2 Detector Dark Noise. Use of a QAPD detector is postulated and its parameters are characterized in Table 9.2. From these values the detector noise current density can be calculated as

$$\sigma_D^2 = 2qI_{\text{du}}/M^2 + 2qFI_{\text{dm}}$$
$$\sigma_D^2 = 2(1.602 \times 10^{-19})(2.3 \times 10^{-7})/(150)^2 + 2(1.602 \times 10^{-19})(3.177)(8.3 \times 10^{-11}) \tag{9.7}$$
$$\sigma_D^2 = 8.78 \times 10^{-29}\ \text{A}^2/\text{Hz}$$

9.6.6.3 Preamplifier Noise

In Chapter 5 we discussed the preamplifier noise effects. For our example, which has a communication bandwidth of 10.5 MHz, the secondary noise caused by the capacitance is negligible. Therefore, the preamplifier noise density can be calculated from

$$\sigma_{\text{pa}}^2 = (\text{NEI})^2/M^2 \tag{9.8}$$

where NEI is defined as the preamplifier noise equivalent current. In our example, the transimpedance amplifier chosen has an NEI of approximately 2.1×10^{-12} amps per root Hz. Using this value yields

$$\sigma_{pa}^2 = (2.1 \times 10^{-12})^2/(150)^2$$
$$\sigma_{pa}^2 = 1.96 \times 10^{-28}\,\mathrm{A^2/Hz}$$

9.6.6.4 Track Channel Shot Noise Contribution

Because of the choice of a single track and communication detector, and because of the associated track and communication channel splitting, the amount of track signal shot noise must be characterized and figured into the overall required signal calculation. For our example, the shot noise contributions can be calculated from

$$\sigma_{\text{ss-trk}}^2 = 2qFR_dP_{\text{rec-track, peak}} \tag{9.9}$$

The rationale behind using peak track received signal power is that this noise contribution is independent of communication signal strength and is usually temporally much longer than a communication bit or symbol period. Thus, the shot noise contributes to the communication noise budget in this manner.

At this stage of the link analysis, the designer usually can assume a value, based on experience for this track signal level. Under most conditions employing tone tracking, the received signal level will be on the order of 100 pW. We use 100 pW as an initial estimate for the track channel additive shot noise. Therefore, the signal shot noise contribution of the track channel to the communication channel is given for this example as

$$\sigma_{\text{ss-trk}}^2 = 2(1.602 \times 10^{-19})(3.177)(.5429)(100 \times 10^{-12})$$
$$\sigma_{\text{ss-trk}}^2 = 5.52 \times 10^{-29}\,\mathrm{A^2/Hz}$$

9.6.6.5 Total Noise Power

The total noise power density can then be calculated by summing the noise contributors, yielding

$$\sigma_{\text{total}}^2 = \sigma_B^2 + \sigma_D^2 + \sigma_{pa}^2 + \sigma_{\text{ss-trk}}^2$$
$$\sigma_{\text{total}}^2 = 5.35 \times 10^{-30} + 8.78 \times 10^{-29} + 1.96 \times 10^{-28} + 5.52 \times 10^{-29}\,\mathrm{A^2/Hz}$$
$$\sigma_{\text{total}}^2 = 3.44 \times 10^{-28}\,\mathrm{A^2/Hz} \tag{9.10}$$

Using equations developed in Chapter 5, the required signal power for this example is calculated as

$$I_{pk} = \frac{2qFB(1 + 1/N_{ext}) \cdot z}{(1 - 1/N_{ext})^2} + \sqrt{\left[\frac{2qFB(1 + 1/N_{ext}) \cdot z}{(1 - 1/N_{ext})^2}\right]^2 + \frac{4 \cdot \sigma_{total}^2 B \cdot z}{(1 - 1/N_{ext})^2}}$$

$$I_{pk} = 4.5 \times 10^{-11} + \sqrt{(4.5 \times 10^{-11})^2 + 5.53 \times 10^{-20}} \quad (9.11)$$

$$I_{pk} = 284 \text{ pA, peak}$$

and

$$P_{pk} = \frac{I_{pk}}{R_d}$$

$$P_{pk} = \frac{284 \times 10^{-12}}{.5429} = 523 \text{ pW} \quad (9.12)$$

where z is the SNR and N_{ext} is the extinction ratio and is given a value of 10.

9.6.7 Transmit Source Power

With all parameters established, we now turn to the link analysis to establish the necessary source power to close the link. It is normally necessary to establish an excess margin at end of life of greater than 3 dB to accommodate on-orbit degradation not planned, or designed for, in worst case assumptions. Thus, the required signal must be at least 3 dB less than the received signal level on the link.

Using the values established for this example, the link margin is illustrated in Table 9.3. All entries are shown for the link power margin. The required laser power is calculated to be 226 mW, peak, to achieve greater than 3 dB of margin. When laser power levels are considered moderate, trades can be performed at the system level to understand the effects of reducing the aperture size or increasing the beam divergence when operating at maximum available laser power. Recall from Chapter 3, the maximum available power can be a function of combined source energy. Power and weight penalties associated with increasing laser power in this way must also be understood to fully perform the trades necessary to achieve a minimized design.

In the event these power levels cannot be realistically reached given the technology available, iteration of the design choices must occur. Narrowing the beam divergence, increasing the receiver aperture size, or adding more complex coding schemes are just a few ways to modify the design to reduce the laser power needed.

Table 9.3
Communication Link Margin

Parameter	*Value*	*dB Value*
Transmit power	226 mW	−6.46
Transmitter optic transmission loss	76.5%	−1.16
Transmitter pointing loss	4.1 μrad (3σ)	−2.40
Transmitter Strehl loss	$\lambda/9.5$	−1.88
Transmitter gain	12.2 μrad	112.01
Range loss	81,428 km	−301.87
Receiver gain	5.5 in	114.52
Receiver transmission loss	55.3%	−2.57
Received signal level	1.04 nW	−89.81
Required signal level	523 pW	−92.81
Excess link margin		3.00

In the example, no additional power compensation is necessary at this time. Now that we have established the link closure for the communication system, we must examine the tracking link margin to establish a baseline link indicating closure on this link.

9.7 CONFIGURING THE TRACK LINK

Much of the hard work has already been accomplished by the time the system designer begins examining the track link closure. The beam divergence, aperture size, and range are already set. What remains is the computation of the track required signal and definition of the required track link source power.

For this example, we have chosen a simple yet very sensitive tracking system using a transmitted tone to provide the necessary temporal discrimination from the background and frequency discrimination from the communications channel. This tone tracking signal operates at a frequency of 10 kHz.

Ideally, this frequency, for greatest sensitivity, should have the smallest bandwidth available. For a tracking loop of approximately 150-Hz bandwidth, the tone frequency should be, as a rule of thumb, at least ten times the final tracking bandwidth, namely 1,500 Hz. We must also consider the mode stability of the laser diode source. Most single-mode diode sources, without spectral stability compensation, need to operate at modulation frequencies that are much higher, near hundreds of kilohertz to megahertz, to guarantee spectral mode stability. Even with spectral mode compensation, operating at modulation frequencies below 10 kHz is not recommended. The authors are aware of single-mode laser diode stability testing that has shown proven mode stability at 10 kHz. This stability was aided using etalon feedback to stabilize the diode frequency. It is this 10-kHz lower limit that sets the tone frequency for our example.

9.7.1 The Track-Required SNR

The required signal for the tracking system is directly related to the tracking noise established by the budget developed earlier. Based on the required noise equivalent angle and the angle slope factor, the required signal is calculated from the criteria set forth in Chapter 6. The tracking system parameters are outlined in Table 9.4. From Chapter 3, we recall that the track required SNR can be expressed as

$$\mathrm{SNR} = \frac{1}{(\mathrm{SF} \cdot \mathrm{NEA})^2} \tag{9.13}$$

where SF is defined as the angular slope factor and NEA is the rms noise equivalent angle of the tracking system.

9.7.1.1 Spot Size Determination

In examining this example, the 5.5-in receive aperture yields a diffraction-limited spot size on the tracking detector that is calculated as

$$\theta_{\mathrm{Spot}} = \frac{2.44\lambda}{D} = 15.85\ \mu\mathrm{rad} \tag{9.14}$$

Ideally, the smaller the spot, the larger the slope factor and thus the smaller the SNR required. However, detector deadzone and tracking linearity must also be considered.

Typically, the tracking control loop design tries to achieve a linear tracking region that is twice the 3σ tracking error. From the closed-loop budget, we see the tracking error and associated residual gimbal jitter equate to approximately

Table 9.4
Tracking System Parameters

Parameter	*Value*
Detector unmultiplied dark current	230 nA
Detector multiplied dark current	83 pA
Detector quantum efficiency	0.816
Detector ionization coefficient	.008
Background noise	Moon on-axis
Preamplifier noise	0.3 pA/Hz$^{1/2}$
rms Tracking error	0.433 μrad
Track loop bandwidth	150 Hz
Track tone frequency	10 kHz
Tracking slope factor	50,000/rad

2.6 μrad, 3σ. This yields a linear range of

$$\theta_{\text{Linear}} = \pm 2(2.6\ \mu\text{rad}) = \pm 5.2\ \mu\text{rad} \tag{9.15}$$

For an Airy distribution spot intensity profile, the angular linear range extends typically to about one-half of the spot size. Matching the $\pm 5.2\ \mu$rad to the spot half-size yields a spot size of

$$\theta_{\text{Spot}} = 2 \times 10.4\ \mu\text{rad} = 20.8\ \mu\text{rad, diameter} \tag{9.16}$$

We achieve this spot diameter by defocusing the spot in the focal plane. This enables the system to be somewhat desensitized to the high-gain diffraction-limited Airy spot. For the 20.8-μrad spot, the slope factor is calculated using the curves in Chapter 6. For our system, the dead zone gap half-width is given as $2\,\mu$m and the spot linear dimension is calculated as

$$D_{\text{Spot}} = \theta_{\text{Spot}} \times \frac{D_{\text{det}}}{\theta_{\text{fov}}} = \frac{(20.8 \times 10^{-6})(1.1 \times 10^{-3})}{(100 \times 10^{-6})} = 0.23\ \text{mm} \tag{9.17}$$

Thus, the ratio of spot radius to deadzone gap half-width is

$$\text{Ratio} = \frac{114\ \mu\text{m}}{2\,\mu\text{m}} = 57$$

From the curve in Chapter 6, we can see that the slope factor is approximately asymptotic to the normalized slope factor of 0.1. From this value and the gap half-width, the slope factor is calculated as

$$\text{SF} = \frac{0.1}{2 \times 10^{-6}} = 50{,}000 \text{ per radian}$$

Returning to (9.13), we can now calculate the required SNR as

$$\text{SNR} = \frac{1}{(\text{SF} \cdot \text{NEA})^2}$$

$$\text{SNR} = 2133.5 \text{ or } 33.3 \text{ dB}$$

where the SNR is typically calculated in the track loop 3-dB bandwidth or, for this example,

$$B_{\text{Trk-Loop}} = \frac{\pi}{2}(150) = 235\ \text{Hz} \tag{9.18}$$

Additional degradation to account for aging and design capability of the receiver results in an increase of the required SNR of approximately 1 dB. Thus, a value of 34.3 dB is used for this example.

9.7.2 Required Signal

The required signal is dependent upon the noise sources and the required SNR. For this tracking system, noise sources include the classical background, preamplifier, detector, dark bulk noises, and signal shot noise. In addition, the single communication-tracking detector implementation yields an additional noise source. Just as the track received signal level adds noise to the communication channel, the communication channel received signal adds additional shot noise to the tracking channel. In fact, the dominant noise source for this example is this additional cross-channel noise term.

Background and detector bulk noise currents are the same as those calculated for the communication channel. The preamplifier noise for the tracking channel is much less than the communication channel since the bandwidth requirements are a mere 10 kHz. Use of an audio type amplifier for each quadrant yields from (9.8),

$$\sigma_{\text{pa}}^2 = \frac{4(\text{NEI})^2}{M^2} = \frac{4(0.3 \times 10^{-12})}{(150)^2} = 1.6 \times 10^{-29}\ \text{A}^2/\text{Hz}$$

The track channel received signal shot noise contributed by the communication signal is calculated as

$$\sigma_{\text{com-ss}}^2 = qF\text{P}_{\text{rec-pk}}(1 + 1/N_{\text{ext}})R_d \tag{9.19}$$

where $\text{P}_{\text{rec-pk}}$ is the received communication peak signal level.

From the link of Table 9.3, the value of $\text{P}_{\text{rec-pk}}$ is given as 1.04 nW, peak, resulting in

$$\sigma_{\text{com-ss}}^2 = 3.18 \times 10^{-28}\ \text{A}^2/\text{Hz}$$

The total noise density for track is then calculated as

$$\sigma_{\text{tot-trk}}^2 = \sigma_{\text{b}}^2 + \sigma_{\text{d}}^2 + \sigma_{\text{pa}}^2 + \sigma_{\text{com-ss}}^2$$

$$\sigma_{\text{tot-trk}}^2 = 5.35 \times 10^{-30} + 8.78 \times 10^{-29} + 1.6 \times 10^{-29} + 3.18 \times 10^{-28}\ \text{A}^2/\text{Hz}$$

$$\sigma_{\text{tot-trk}}^2 = 4.27 \times 10^{-28}\ \text{A}^2/\text{Hz} \tag{9.20}$$

The required signal then follows from the Chapter 6 analysis and (9.11), and is given as

$$I_{pk} = \frac{2qFB(1 + 1/N_{ext}) \cdot z}{(1 - 1/N_{ext})^2} + \sqrt{\left[\frac{2qFB(1 + 1/N_{ext}) \cdot z}{(1 - 1/N_{ext})^2}\right]^2 + \frac{4 \cdot \sigma_{total}^2 B \cdot z}{(1 - 1/N_{ext})^2}}$$

where here B is defined as the bandwidth of the track loop (i.e., $B = B_{Trk\text{-}Loop}$) and z is the required SNR of the tracking system. The value I_{pk} is then calculated as

$$I_{pk} = 37.4 \text{ pA}$$

which can be expressed in terms of peak optical power by dividing by the responsivity, yielding

$$P_{pk} = 68.9 \text{ pW}$$

9.7.3 The Tracking Source Power

Using link analysis again, we established the source power, this time for the track link. Table 9.5 illustrates the tracking link analysis performed for this example. The power required can be made available in several ways. Two of the simplest involve either a separate track source diode, which is combined colinearly with the communication source, or adding the track signal directly to the communication waveform prior to modulating the diodes. The resultant waveform is given in Figure 9.5. For our example system, we have chosen to minimize the signal driver complexity and signal level matching complexity. This has led us to the use of a separate single diode for tracking.

Table 9.5
Track Link Margin

Parameter	*Value*	*dB Value*
Transmit power	30 mW	−15.27
Transmitter optic transmission loss	76.5%	−1.16
Transmitter pointing loss	4.1 μrad (3σ)	−2.40
Transmitter Strehl loss	$\lambda/9.6$	−1.88
Transmitter gain	12.2 μrad	112.01
Range loss	81,428 km	−301.87
Receiver gain	5.5 in	114.52
Receiver transmission loss	55.3%	−2.57
Received signal level	137 pW	−98.62
Required signal level	68.9 pW	−101.62
Excess link margin		3.00

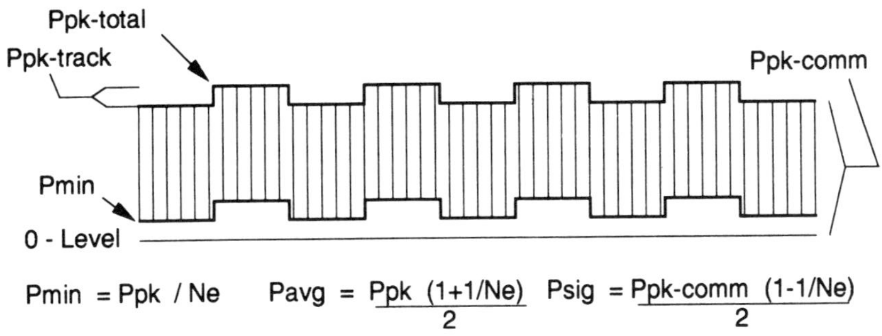

Figure 9.5 Track / communication waveform.

As is illustrated in the link table, the required laser power to achieve a greater than 3-dB link margin on the track link is given as 30 mW.

9.7.4 Communication-Track Link Iteration

Because of the unique interaction in our example between the communication and track links (recall these channels use the same detector), iteration of the links occur until the desired margin balance is reached. Notice the 30-mW track required power yields 137-pW received signal level on the track link. This is about 1.4 times larger than the initial estimate of 100 pW used in the communication link analysis that was previously performed. Updating the communication link noise density based on 137-pW excess track shot noise yields a required communication signal of 536 pW.

The updated communication link is shown in Table 9.6. This slight increase in noise power density requires 0.1-dB additional signal power on the communica-

Table 9.6
Updated Communication Link Margin

Parameter	*Value*	*dB Value*
Transmit power	232 mW	−6.35
Transmitter optic transmission loss	76.5%	−1.16
Transmitter pointing loss	4.1 μrad (3σ)	−2.40
Transmitter Strehl loss	$\lambda/9.5$	−1.88
Transmitter gain	12.2 μrad	112.01
Range loss	81,428 km	−301.87
Receiver gain	5.5 in	114.52
Receiver Transmission loss	55.3%	−2.57
Received signal level	1.07 nW	−89.70
Required signal level	537 pW	−92.70
Excess link margin		3.00

tion link (i.e., power transmitted is now required to be 232 mW). This excess noise increase results in a communication channel received signal level of 1.07 nW, which couples into the track link as excess communication shot noise. This slight increase in noise results in an increase in the required tracking signal of 0.026 dB. For this example, we have determined this is sufficient iteration. The margin increase is minimal and the additional signal required is also minimal. Additional iteration would be performed if the desired margins had not been established.

9.8 CONFIGURING THE ACQUISITION LINK

9.8.1 Uncertainty Area Establishment

Acquisition requires searching the uncertainty area to locate and establish the link between the satellites. For our example, we first must establish the uncertainty area size. Figure 9.6 shows the basic contributors to the uncertainty and the values associated with each. The largest contributor is the transfer of satellite knowledge to the crosslink terminals. With these values, an uncertainty area of 1 mrad has been established.

9.8.2 Acquisition Time and Beam Divergence

Link acquisition in a specified amount of time is also required. With the 1-mrad uncertainty, it is becomes obvious that scanning of the transmit source is necessary.

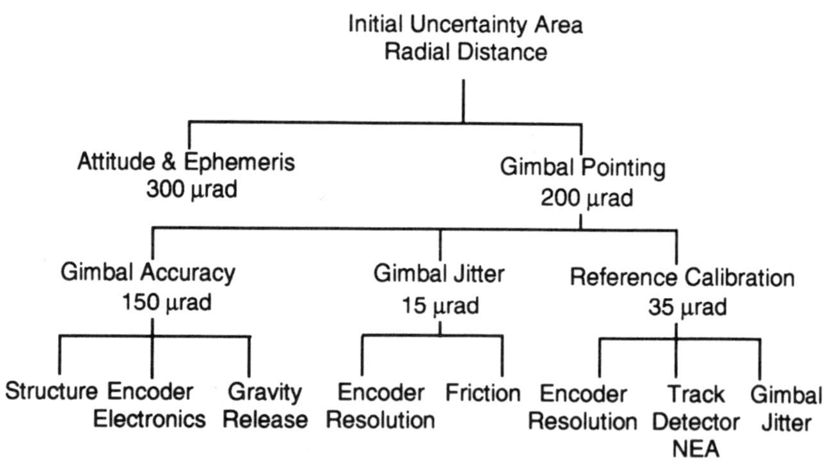

Figure 9.6 Initial uncertainty budget.

The added complexity and time associated with the scanning of the receiver weighs heavily against that method. The minimal starfield acquisition background allows the acquisition detector viewfield to be sized to the uncertainty area. With that, we have chosen a transmit scan and receiver stare mode of operation. Using a 1-mrad viewfield for the acquisition process drives the design to two detectors. The communication receiver cannot overcome the additional background noise level when using a 1-mrad viewfield.

Acquisition is a two-step process. First, we must have sufficient signal for initial detection and, second, sufficient received energy to allow closed-loop tracking to begin. If we choose to use a beam divergence that is larger than that required for communication, we must provide enough signal energy to support the transition to narrowbeam tracking.

The bound on beam divergence comes from the required acquisition time and transition power margin. We must accommodate the transition maximum beam divergence to track beam divergence power ratio by having additional power for acquisition. In addition, the beam divergence selected must accommodate the required acquisition time. If we are unable to accomplish this, the requirements must be re-examined and acquisition time or acquisition probability must be modified.

For our example, the tracking system uses the 12.2-μrad communication beam divergence. In order to expand this beam for acquisition, we must maintain the tracking link during the transition from wide acquisition beam to the narrow communication and tracking beam. Thus, any power density lost due to a larger beam must be overcome by additional source power.

In this example, we have chosen to use a separate track diode along with two communication diodes. For acquisition, it is easily seen that the use of all three laser diodes at full rated power can provide a combined source power for acquisition of approximately 450 mW. The combined power includes the effects of etalon control, collimating optics, and combining efficiency on standard 200-mW laser diode sources.

In addition, the convergence link is not dominated by the excess communication channel shot noise, as is the track channel. The difference in required signal when the communication channel excess noise is removed is on the order of 3 dB.

The square root of the ratio of the laser powers yields the divergence widening factor

$$\text{Divergence factor} = \sqrt{1.93 \cdot \left(\frac{450}{30}\right)} = 5.38 \tag{9.21}$$

A divergence of 65 μrad or less will allow transition to occur with sufficient power margin. We now must determine if acquisition can occur within a reasonable amount of time using a beam divergence of this size.

In Chapter 3, we established the time to complete a spiral scan of an uncertainty area using a beam divergence $\theta = 60$ μrad. For this example, given a reasonable overlap factor of 20% and a dwell time of 20 ms (sufficient to lock a phase lock loop and provide at least 10 ms of tracking signal) we calculate the time to complete one scan as

$$T_{\text{scan-clv}} = \frac{2\pi}{(1 - F_0)}\left[\left(\frac{R}{\theta_{1/e}}\right)^2 + \frac{1}{2}\left(\frac{R}{\theta_{1/e}}\right)\right]T_{\text{dwell}}$$

$$T_{\text{scan-clv}} = \frac{2\pi}{(1 - .20)}\left[\left(\frac{500 \times 10^{-6}}{35 \times 10^{-6}}\right)^2 + \frac{1}{2}\left(\frac{500 \times 10^{-6}}{35 \times 10^{-6}}\right)\right] \cdot (.02)$$

$$T_{\text{scan-clv}} = 22.7 \text{ sec} \tag{9.22}$$

This is well within our initial acquisition time requirement of 5 min. However, we must add additional time to account for gimbal slewing to the line-of-sight location (on the order of 18 sec (180° @ 10° per second) and any additional time required to scan the uncertainty multiple times if we need to enhance the acquisition detection probability. Thus, the use of a 60-μrad $(1/e^2)$ beam divergence becomes the value chosen for this example system.

9.8.3 The Acquisition-Required Signal Criteria

The required signal for the acquisition link depends upon the detection criteria set forth. For this example, we have chosen to provide sufficient loop SNR to lock a second-order phase lock loop. In Chapter 3 we established the loop lock criteria necessary for acquisition. The loop lock time is established using the following equation:

$$T_{\text{lock}} = \frac{4.2(\Delta f)^2}{B_L^3} \tag{9.23}$$

Given a frequency deviation Δf, of 10% of the acquisition tone frequency and a lock time of 5 ms yields

$$B_L = \left[\frac{4.2(0.1 \cdot 10{,}000)^2}{.005}\right]^{1/3} = 943.5 \text{ Hz}$$

For good loop performance, a 10-dB loop SNR is required. Thus,

$$\mathrm{SNR} = \frac{I_s^2}{N_o B_L} = 10$$

As before, an additional 1 dB is added to the required SNR to accommodate aging and build-to capabilities. Thus, the required signal is based upon an 11-dB SNR.

9.8.3.1 Acquisition Noise Density

Just like the tracking channel, the noise density for acquisition is determined from the preamplifier noise, the detector dark noise, and the background noise. For acquisition, the detector and preamplifier noise sources are identical to the tracking link calculated previously. The background noise, however, is determined from a starfield flux and a 1-mrad viewfield. Thus, from (9.3)

$$P_B = H_B \cdot \left(\frac{\pi}{4}\right)(\mathrm{FOV})^2 \cdot B_f \cdot \left(\frac{\pi}{4}\right)(D_{\mathrm{aper}})^2 \cdot L_R$$

$$P_B = (1.69 \times 10^{-10}) \cdot \frac{\pi}{4}(0.001)^2 \cdot 60 \cdot \frac{\pi}{4}(0.0254 \cdot 5.5)^2(0.5535) \quad (9.24)$$

$$P_B = 6.75 \times 10^{-17}\ \mathrm{W}$$

And as before, the background noise current density is calculated from (9.4) as

$$\sigma_B^2 = 2qFP_B R_d$$

$$\sigma_B^2 = 2(1.602 \times 10^{-19})(3.177)(6.75 \times 10^{-17})(0.5429)\,\mathrm{A}^2/\mathrm{Hz}$$

$$\sigma_B^2 = 3.73 \times 10^{-35}\,\mathrm{A}^2/\mathrm{Hz}$$

The total noise is then calculated from

$$\sigma_{\text{tot-acq}}^2 = \sigma_b^2 + \sigma_d^2 + \sigma_{\mathrm{pa}}^2$$

$$\sigma_{\text{tot-acq}}^2 = 3.73 \times 10^{-35} + 8.78 \times 10^{-29} + 1.6 \times 10^{-29}\,\mathrm{A}^2/\mathrm{Hz} \quad (9.25)$$

$$\sigma_{\text{tot-acq}}^2 = 1.04 \times 10^{-28}\,\mathrm{A}^2/\mathrm{Hz}$$

9.8.4 Acquisition-Required Signal

The acquisition-required signal power is calculated from the noise, bandwidth, and SNR values previously established. Using (9.11), the required signal is calculated as

$$I_{pk} = 2.5 \text{ pA}$$

Dividing by the responsivity yields

$$P_{pk} = 4.6 \text{ pW}$$

9.8.5 Acquisition-Required Laser Power

Table 9.7 establishes the required laser power at the source for the acquisition and transition links. For our design, this source power requires the combining of three laser diode sources to achieve the desired result. The required laser power for

Table 9.7
Acquisition and Transition Link Margins

Parameter	*Value*	*dB Value*
Acquisition Parameter		
Transmit power	55.6 mW	−12.55
Transmitter optic transmission loss	76.5%	−1.16
Transmitter pointing loss	1/e power level	−4.34
Transmitter Strehl loss	$\lambda/9.6$	−1.88
Transmitter gain	60 μrad $1/e^2$	99.49
Range loss	81,428 km	−301.87
Receiver gain	5.5 inches	114.52
Receiver transmission loss	55.3%	−2.57
Received signal level	9.2 pW	−110.36
Required signal level	4.6 pW	−113.37
Excess link margin		3.01
Transition Parameter		
Transmit power	450 mW	−3.47
Transmitter optic transmission loss	76.5%	−1.16
Transmitter pointing loss	1/e power level	−4.34
Transmitter Strehl loss	$\lambda/9.6$	−1.88
Transmitter gain	60 μrad $1/e^2$	99.49
Range loss	81,428 km	−301.87
Receiver gain	5.5 inches	114.52
Receiver transmission loss	55.3%	−2.57
Received signal level	75.4 pW	−101.28
Required signal level	35.7 pW	−104.47
Excess link margin		3.19

acquisition is given as 55.6 mW. The transition link requires the full 450 mW to achieve the desired greater than 3-dB margin. Once the link is established, trades can occur to optimize the design. Some flexibility still exists in this design. The additional acquisition time can be traded for a smaller beam divergence and less laser power for acquisition and transition.

9.9 SUMMARY

This chapter has provided a quick look at the trades and design issues required in the development of a laser communication system. Included has been an example design, which has evolved from requirements to link design. The link design is the first step towards development of the system-level partitioning and eventual system synthesis.

Iterations can, and should, occur to refine any design. The important features of any system design involve a complex interplay between the links that can only be optimized through an iterative process. Requirements should be understood and challenged where necessary.

When designing a system, be sure the delicate "balance of power" between the links is not misunderstood. While acquisition is important, in most crosslink design cases, acquisition time can be traded for beam divergence and hence power. The acquisition system design should not drive most link designs.

Laser crosslink tracking and communication go hand in hand. Without one, the other cannot exist. It is the mission to communicate, but without the ability to track, the laser crosslink is effectively nonoperative. A balanced design between these two links is imperative for good laser crosslink design.

Chapter 10

Trading RF and Laser Communications for Your System

10.1 TRADE STUDY CONSIDERATIONS

As demands on space communication systems become greater, it becomes more important that those who configure systems pay close attention to selection of the best communication technology for the job. This requires that a detailed trade be performed of the appropriate candidates. One factor that has hindered realistic trades in the past is the lack of understanding of laser communication systems. Communications system designers have historically specified *radio frequency* (RF) systems since this has been, until recently, the only mature technology available. Individuals in the traditionally RF communications community need to understand the differences between laser and RF communications in order to conduct realistic trade studies.

There are significant differences between RF and laser communication systems, and much of it results directly from the several orders of magnitude difference in wavelength. This difference is most noticeable in antenna size and beam divergence. *Extremely high frequency* (EHF) systems (e.g., 60 GHz, with over three orders of magnitude longer wavelength than the near infrared laser communication wavelengths), require antenna sizes measured in feet versus inches for laser communication. Similarly beam divergences are measured in fractions of degrees versus microradians for laser communications.

Of course, there are applications where laser communication does not compete with RF. For example, in broadcast applications, RF systems with broad-beam capability cover a much larger angular area than laser communication links could. Since cloud statistics enter into the availability equation for laser communication links, it is also not as well-suited as RF for single high-availability ground station applications. There are, however, even in the satellite to air and satellite to ground arenas, certain applications where laser communication is equal or superior to RF. Satellite to satellite crosslinks are one area where laser communications can compete successfully with RF systems.

To conduct a realistic trade study of RF versus laser communications, the important characteristics or factors must be identified and included in the trade. The trade study must be as quantitative as possible. There are factors, however, that are very difficult to quantify due to their inherently subjective nature and this, of course, can lead to much controversy. *The use of a consensus approach where several individuals establish the list of factors helps minimize the controversy.*

Important factors are identified and discussed in the following section. A simple methodology is used to structure an example trade based on assumed characteristics for both the RF and laser communication candidate systems. It must be pointed out that the characteristics assumed for candidate RF systems are based on a composite view of 60-GHz RF technology. The authors have attempted to assure that any uncertainty in assumed 60-GHz characteristics lies on the side of favoring the RF. The laser communication system characteristics for the example are based on current laser communication system technology [1], which is the current state of the art in direct-detection laser communication technology. A number of papers have been published recently describing various aspects of these systems. A few of these papers are referenced. The reader is referred to the *Proceedings of the SPIE on Laser Communications* for many excellent papers. The example trade study is given only to illustrate an approach that may prove useful in future trades. The reader should tailor the approach to his or her own needs since each trade study is unique and may differ in the importance of the factors and even in which factors are relevant.

10.2 FACTOR IDENTIFICATION

The factors that will be part of the trade study should be those that are most important to the crosslink and spacecraft designer. There will be instances where a requirement immediately leads to a selection of one over the other, such as the minimal dynamic reaction impact that favored laser communications over 60 GHz for the Air Force's *laser communication subsystem* (LCS) [2]. The factors to be considered for inclusion in the example trade study are discussed below.

10.2.1 Weight

Terminal weight is frequently the one most important factor in satellite-level trade studies. System weight is important in that it manifests itself in many ways. Additional weight will require additional fuel. It also adds directly to satellite structure. If the spacecraft weight addition is great enough, a larger size launch vehicle could be required, with a very significant increase in satellite system cost. This has happened in the past.

As an example, assume a launch is planned on a Delta launch vehicle with a capability of placing 2,000 lb in geosynchronous orbit. If the satellite weight, including station-keeping fuel, weighs 1,800 lb, a larger, heavier communication system of even 100 lb could wipe out the margin. The additional communication system weight increase is compounded with the fact that additional satellite weight may also be incurred in the form of additional structure, additional fuel, and, possibly, more power.

The satellite designer must then be forced to launch with a larger payload capacity booster, such as an Atlas II (which has a capacity of 2,600 lb). The launch costs have now increased by approximately $20 million ($60 million for Atlas II versus $40 million for Delta). Multiplying this by several launches over the program's life cycle yields a substantial increase in overall program cost.

10.2.2 Power

Power usage is also a prime factor since any power used by the terminal must be supplied by the satellite power system and requires more power system (i.e., more solar cell area, more batteries, a larger power regulator, and increased bus wiring), all of which increase weight of the satellite. Also, with increased power, more heat must be dissipated by the thermal control system, increasing its weight. The additional weight caused by an increase in power required will be added directly to the terminal weight increase. It has been estimated that the weight increase per watt of increased power will vary from 0.30 lb/W [3] to 0.50 lb/W [4].

10.2.3 Cost

This is obviously a very important factor, especially when the system *lifecycle cost* (LCC) is considered. There are two cost inputs to the LCC that are of interest. The first is the development or nonrecurring cost. The second is the recurring cost, which is the cost for each unit as produced. For a program of ten units the total, or LCC, would be the sum of the nonrecurring cost plus the cost for ten units (ten times the recurring unit cost). To accurately estimate the costs, the system designer should obtain these cost estimates through formal responses to *request for proposal* (RFP) or *rough order of magnitude* (ROM) estimates from system manufacturers.

10.2.4 Technical Risk

Always of serious concern in selecting an approach for communications system development is the maturity of the technology and/or hardware. The risk of not meeting the required performance, overrunning the cost, or slipping the schedule has many times overridden all other considerations, as well it should.

A number of aspects influence program managers regarding risk in selecting a communications system. For example, the shorter wavelengths of an optical system require a thorough understanding by the communication system builder regarding sensitivities to dimensional tolerances and thermal distortions, and very careful use of materials.

Optics also require end-to-end control over contamination, a basic nonproblem for RF systems. Elaborate clean-room facilities are required for not only the build and test of optical communication systems, but the satellite manufacturer must also follow the same clean-room practices as the communications systems builder. Coupled with cleanliness is the ability of the optical train to handle scattered light, whether related to "self blinding" effects or stray light rejection. Again, an understanding of these effects is necessary when specifying a laser communication system.

Other factors that must be addressed are those associated with parts availability and the level of parts qualification for space. Very nearly all optical parts for laser communication systems are considered nonstandard. Many standard parts exist for, and are used in, RF systems, although 60-GHz systems are not as "rich" in standard parts as are many of the more common EHF frequencies such as 20, 32, and 44 GHz, which have been common for many years.

System test for optical systems is a very exacting discipline where air-supported optical tables are necessary to mitigate the effects of vibration, a basic nonproblem with RF testing. Optical receiver sensitivity testing must also take place in a darkened enclosure to avoid unnecessary background signals generated by ambient light. It is necessary, when specifying an optical communications system, that a manufacturer be available that is thoroughly experienced in the design, build, and test of complex optical systems.

There have been advances in laser communication source technology and key components, but the same has also been true of RF systems where the use of microwave integrated circuit technology is reducing both risk and cost. A major risk item for laser communications has historically been the laser source. The LCS laser source, for example, was the major development challenge for the program. Recent advances in semiconductor laser diode technology have reduced the risk down to modest levels. Current designers can feel confident of sources, which are high power, long-life and reliable.

Risk will be viewed from two aspects. The first is the maturity of the transmitter, whether it be an RF or laser source. The second is the maturity of the overall system implementation. EHF *traveling wave tube amplifier* (TWTA) systems are certainly not new and their implementation should be considered rather routine. The same is true to a somewhat lesser extent for laser communications, with the first-generation LCS that was in production and the second-generation that was well on its way in development before program cancellation. The maturity of the transmitter and system should be scored against a maturity rating table similar to the *National Aeronautics and Space Administration* (NASA) criteria identified in Figure 10.1. A 9 is considered a maximum score and represents a system that has proven capability. The maturity is inversely related to the technical risk.

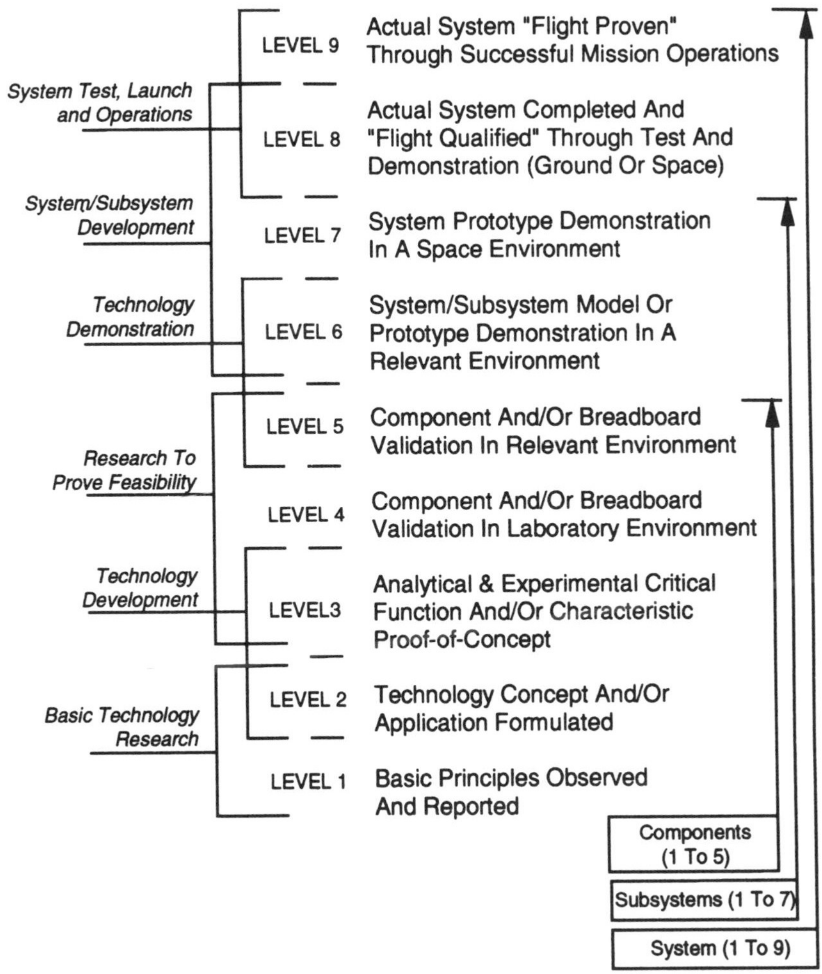

Figure 10.1 NASA maturity rating.

10.2.5 Immunity From Threats (Jamming)

Military systems must be configured to provide resistance to jamming. In the case of a national warfighting asset, such as a global communication system, vulnerability to jamming is exceedingly important. Even in today's relaxed strategic posture, pursuit of *anti-jam* (AJ) systems is still a reality. The speed with which the world changes should be evident, having observed the quickness with which the Cold War era ended. Laser communications has an inherently robust AJ capability, which arises from its narrow *field of view* (FOV), its very low sidelobes (high off-axis gain), and its design, which includes wavelength selective receiver optics. RF systems must respond to threat environments with additional equipment that makes the system hard to jam. This additional hardware must be factored into any trade study that requires AJ capability. This factor will be quantified in the example trade study.

10.2.6 Nuclear Scintillation Vulnerability

Because of the nature of ionizing radiation following high-altitude nuclear bursts, RF crosslink channels may experience severe fading and multipath effects due to the scintillation environment present. This is of major concern when the geometry is such that the links graze Earth or pass within several thousand kilometers of Earth's surface. This geometry can occur for crosslinks that cover twice-synchronous range (85,000 km). Conversely, laser communications, due to its much shorter wavelength, is orders of magnitude less susceptible to the fading channel environment.

The fading environment will put burdens on the RF systems that may include the use of additional power (larger apertures or higher power sources) and/or the use of data interleaving (additional requirements for more memory, data buffering, and data latency concerns). This is a factor that should be included in the trade if the application is military and the geometry of the link indicates applicability.

10.2.7 Data Rate Growth

Laser communications can be easily used at all data rates, from low to multigigabits per second. Data rate, per se, is a requirement and is not a factor. However, data rate growth should be considered for future system upgrades. Higher data rates force RF designers to larger antennas and higher power transmitter sources, with accompanying rapid growth in physical parameters. While laser communication systems can achieve the same, or greater, individual channel rates as RF systems, added capability exists through wavelength channelization. Multichannel systems are very easy to implement with laser communications.

10.2.8 Low Probability of Intercept

Low probability of intercept (LPI) is important where security or privacy is a requirement. Again, due to the short wavelength, laser communications systems are extremely difficult to intercept. Beam divergences are measured in microradians and sidelobes are low. Of course, this narrowness of the beams requires that laser communication systems track with a precision not required by RF systems. Fortunately, the precision-tracking technology to accomplish this is well in hand and has been demonstrated in the laboratory, in space, and in air to ground tests. The LPI characteristics of laser communications are so favorable that nonencryption of laser communication links has been postulated since it is practically impossible to intercept the transmissions.

10.2.9 Satellite Integration Impacts

Integration impacts include a collection of factors that denote the overall effect of integrating a communications system. One view is the "real estate" required to mount the terminals on the satellite. This factor is related closely to size. View fields are required for each antenna, whether it be RF or laser communications. The view must be unobstructed over the required view angles; otherwise a different arrangement, operational constraints, or an additional terminal may be needed, with significant weight impact. The requirement to provide a clear view throughout a range of angles is most stringent for RF systems, with their large antennas. It may be necessary to stow the antenna for launch and deploy it for on-orbit operation. This requires a deployment boom. These types of considerations are included in this factor. In most cases, a real understanding of the issue of "real estate" can only be achieved by performing a top-level layout so that the view fields and installation can be examined in three dimensions.

Due to the small size of laser communications terminals and antennas (telescopes), stowing, as a rule for launch, need not be considered. An additional factor, which is relevant to integration impact, is the dynamic reaction effect. This factor was dominant in the selection of the LCS over a 60-GHz system on its satellite. Most rotational movements of large structures such as an antenna must be countered by an opposite momentum correction of the spacecraft. For a spacecraft that requires precision stability while it performs its function, this is very disturbing and should be considered in the trade study. Attitude control fuel may also be used during the spacecraft momentum compensation correction.

10.2.10 Sun in the Field of View

Operation through the solar disk is a requirement for some systems where the continuous flow of strategic data is a requirement. Both RF and laser communications systems have been designed to operate through the disk with reduced capability. If the system design requires this type of operation, the constraints of each system in the area of data rate capability must be identified and traded.

It is generally accepted that RF crosslink systems operating at EHF frequencies can operate through the solar disk at reduced data rates, error rates, or both. However, as a note, the first-generation LCS tracks and communicates as the cooperating terminal transits the solar disk. This is accomplished by using the high peak power capability of the LCS's Nd:YAG laser to track in the exceedingly high Sun background. There are two optical techniques also used to ensure that the Sun background does not cause the LCS to lose track. First, a very narrow bandpass filter is used (20Å) and second, an adjustable field stop is closed down so that only the tracked spot plus a small amount of surrounding solar background is imaged on the detector. With the advent of even higher power single-mode diodes (200 mW), and now, *master oscillator power amplifiers* (MOPAs) at 1W, tracking through the solar disk may become a reality for even gallium arsenide–based laser communication systems [5]. As with the RF systems, the data rate or error rate (or sometimes both) is reduced for both laser and RF communication through the solar disk.

10.2.11 Reliability / Life

No evaluation or trade would be complete without scrutiny of the reliability and life aspects of the proposed designs. The major reliability issue or concern for both RF and laser communication systems has been the reliability of the RF or optical source technology. There are data on the reliability of TWTA sources, single-mode laser diodes, and *impact avalanche and transit time* (IMPATT) diodes. Life data on the newest solid-state devices for RF, the *power-high electron mobility transistor* (P-HEMTs), should be available shortly. No life data are yet available on 1W MOPAs for laser communications systems.

For consistency, the life/reliability of each system is rated. As an example, TWTAs have 5- to 7-year life. TWTA-based systems then should include at least a pair of TWTAs for a system design life of 10 years. The weight of the system reflects this. The degree of derating should be examined for transmitter components and included in the trade where available. The mode of failure should also be considered in the trade. As another example, in the multiple-diode combiners used in laser communication systems, the failure mode is "graceful" (i.e., as the diodes age, the current is turned up to compensate for output loss). If one or more diodes fail, the additional diodes pick up the slack with an increase in current and operation continues.

10.2.12 Mutual Interference

Where multiple antennas are required, antenna spacing must on occasion be made quite large to avoid mutual interference or self-jamming. The large sidelobes and high powers used for RF systems can cause serious problems when antennas are located too close together. Additionally, coupling of harmonics from and into additional payloads should also be a consideration when trading the system approaches. In one system, the authors are aware of coupling issues that drove the mechanical design of the satellite to mitigate the coupling of systems. Because of the low sidelobes and highly directional beams, there are no mutual interference effects with laser communications, unless of course the telescopes on both terminals are pointed directly at each other. An obvious advantage exists with laser communication when dealing with multiple onboard links.

10.3 FACTOR WEIGHTING

Before the characteristics of the systems are compared, each factor is weighted to reflect the importance of the characteristics. This weighting is application-driven and can incorporate whatever the designer wishes. As an example, the authors have conducted a trade incorporating factors used in a number of trade studies or discussed in a recent RF versus laser communications panel session at SPIE's OE-LASE'93 [6]. The following factors were used in previous trades and also mentioned at the above panel session, with the total indicated number of times in parentheses:

1. Weight (13);
2. Power (10);
3. Cost (8);
4. Technical risk (8);
5. AJ (7);
6. Nuclear (7);
7. Data Rate growth (6);
8. LPI (5);
9. Integration impact (5).

Additional attributes also added are

1. Sun in the field;
2. Reliability/life;
3. Mutual interference.

An examination of the above list shows that the weight and power are considered very important in any trade. Also, technical risk and reliability/life, by their very nature, are very important in making a selection. The weighting is set up to reflect the concern over reliability. In addition, integration impact is being emphasized because of its direct relationship to weight and satellite complexity. As with factor selection, weighting is best conducted using a consensus approach where concerned individuals provide the weighting prior to actually scoring the systems. The example then uses the weighting factors listed in Table 10.1.

Scoring of weight and power in terms of relative weighting will be handled using the trade technique outlined in the System Engineering Management Guide, 1983, published by The Defense Systems Management College, Fort Belvoir, Virginia. Whichever system is the best will form the base and achieve a perfect score (10). A heavier system will then be decremented in accordance with how much it exceeds the base by using a factor "utility curve" that permits scoring directly by reading the score off the curve. By controlling the shape of the curve to represent the range of acceptable values, the designer can structure the trade to accomplish whatever he wishes. The factor utility curve for weight and power will be shown in the next section. Cost will also be handled in a similar fashion.

A variance analysis should also be performed on the individual weightings to allow a measure of sensitivity to a particular parameter. In this way the designer can be cognizant of the sensitivity of his design to requirement and need changes. Care should also be taken when selecting a system design that has a high-category score variance. This could result from scoring very high in a particular category, which offsets a low score in another category. This may result in a design that is inflexible to change or one that is produced on the edge of the tolerance curve.

10.4 CANDIDATE SYSTEMS FOR THE EXAMPLE TRADE STUDY

The following system requirements, identified in Table 10.2, are assumed for the RF and laser communication candidate systems used in the example trade. These

Table 10.1
Example Weighting Factors

Factor	*Weighting*
1. Weight	15
2. Power	6
3. Cost nonrecurring	15
Recurring	15
4. Technical risk	15
5. AJ	2
6. Nuclear	2
7. Data rate growth	5
8. LPI	2
9. Integration impact	10
10. Sun in the field	2
11. Reliability/life	10
12. Mutual interference	1

Table 10.2
Assumed Requirements

Requirement	*Values*
Data rate	12 Mbps
Range	84,000 km
Sun in field	Yes
Life	10 years
Encrypted	Yes
Bit error rate	10E-6
Continuous data	Yes
Satellite stabilization	3-axis
Communication:	Full duplex
Terminals required per satellite	2

do not represent specific requirements for current or future systems, but are included as examples only.

The following postulated systems are compared in the example trade study:

RF System 1

- 60-GHz traveling wave tube amplifier (TWTA);
- Two TWTAs per terminal to meet 10-year life;
- Transmit power 10W;
- 36-in antenna;
- Deployment boom;
- Beam waveguide gimbals;
- Encrypted;
- Weight 140 lb/terminal, which includes boom (20 lb);
- Power 150 W/terminal.

RF System 2

- 60-GHz solid-state power amplifier (IMPATTS)*;
- Transmit power 2W;
- 48-in antenna;
- Deployment boom;
- Encrypted;
- Weight 145 lb/terminal, which includes boom (20 lb);
- Power 140 W/terminal.

*A variation on this system is to use P-HEMTs in lieu of IMPATT diodes. Currently, P-HEMTs are being developed and life data should be available shortly.

Laser Communication System

- GaAlAs multiple diode;
- Ten 150-mW laser diodes derated to 82 mW*;
- Diode mean time to fail much greater than 20 years.
- 5.5-in aperture gimbaled telescope;
- Quadrant photodiode detector for communication and track;
- Encryption not required due to LPI capability;
- Modulation biphase, level;
- 100-μrad track FOV;
- Rate 1/2 convolutional encoding;
- Electronics radiation hardened CMOS, ASIC integration;
- Weight 106 lb/terminal*;
- Power 126 W/terminal*.

*Current technology—incorporation of the 1W MOPA will drive terminal weight down in the future as predicted by trend projections [1].

10.5 THE EXAMPLE TRADE

As previously discussed, an example trade of RF versus laser communications for the systems defined in Section 10.4 is conducted to allow the readers to familiarize themselves with the process. The authors believe that only with this type of rigorous trade can a fair choice of technologies for the design requirements be made. The trade starts with the weighting factors. Some additional explanation is required and follows.

1. *Weight*: The lightest system is Lasercom, so it is the base. The RF systems will be scored based on the following factor utility curve shown in Figure 10.2.
2. *Power:* The system using the least power, laser communications, is the base. The RF systems will be scored based on the weight factor utility curve in Figure 10.2. The .42 lb/W weight burden for additional power is reflected in the ratio of the weightings of weight and power in the matrix.
3. *Cost:* To avoid specific costs in dollars, a comparative technique that does not require use of direct dollar cost is used. Cost units, in place of dollars, are used to illustrate the cost portion of the trade.

   The nonrecurring cost can be compared with a curve similar to the weight/power curve (Figure 10.2). If it is assumed that the cost of the development of a laser communications system is the same as that for the development of a comparable RF system, comparison of their nonrecur-

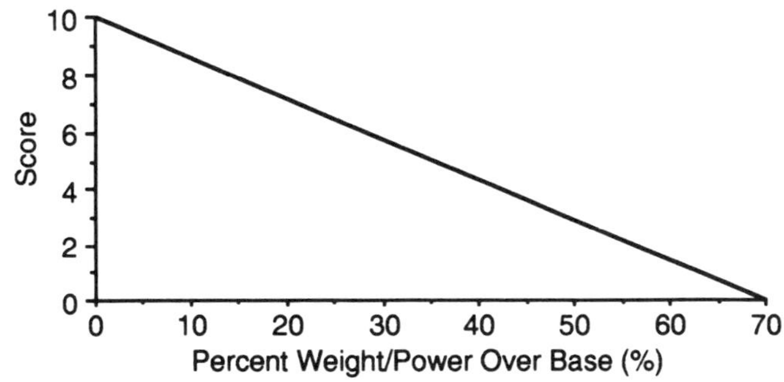

Figure 10.2 Weight / power factor utility curve.

ring cost is made by a percent of new development cost curve as illustrated in Figure 10.3.

A similar approach can be taken with the recurring cost by the use of a cost unit ratio curve such as that illustrated in Figure 10.4.

4. *Technical risk:* Two parts are considered here. One rating for the transmitter and another for the system. Most of the risk in systems of this type lies in the transmitter, so it is the most heavily weighted. The NASA maturity rating table is used to rate the systems. TWTAs have long been mature, so they rate the highest in maturity (and score), followed by laser communications and the SSPA 60-GHz system.
5. *Anti-jamming:* Without a very lengthy analysis, quantitative values here are hard to obtain. EHF 60-GHz systems can be configured for mitigating jamming threats, but this involves more sophisticated electronics than envi-

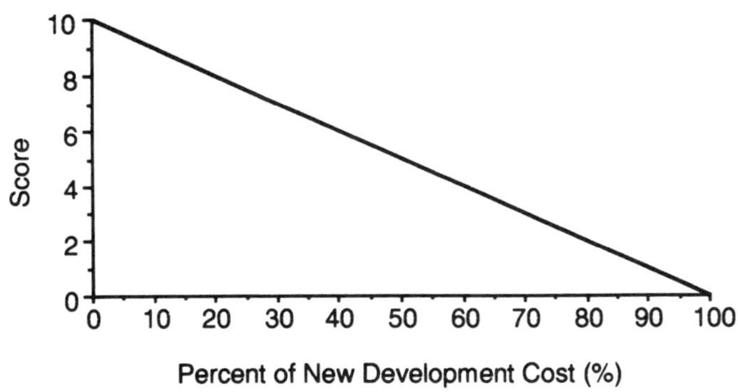

Figure 10.3 Development cost utility curve.

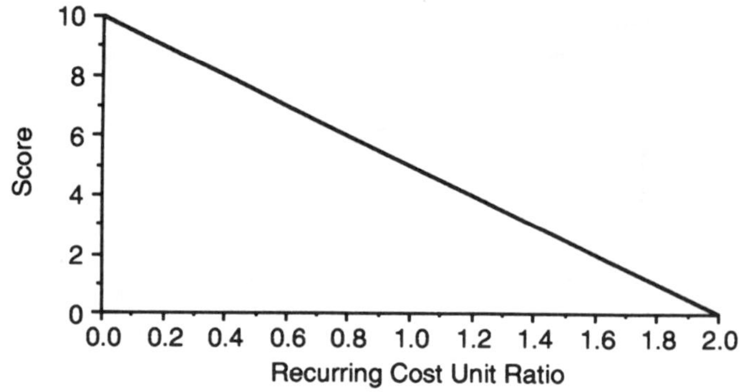

Figure 10.4 Recurring cost unit ratio utility curve.

sioned for the candidate systems shown here. There would also be a weight and power penalty for providing additional jam resistance to the RF systems. With laser communication's inherently high jam resistance, it is scored slightly higher than the RF systems.

6. *Scintillation:* Nuclear scintillation events can cause severe fades with RF when the signals are grazing Earth. Laser communications is inherently protected from this vulnerability by its short wavelength. As a result, and without a lengthy analysis (these have been performed), laser communications is again scored slightly higher than the RF systems since we are considering a synchronous range system.
7. *Growth:* Laser communications data rates are easy to increase, in this case to perhaps 25 Mbps, by adding diodes to the multiple diode laser. For the RF systems, which are assumed to be power-limited, the antenna size would have to be increased and there would be a significant penalty in size and weight. Based on these considerations, the RF systems are scored slightly lower than laser communications.
8. *LPI:* Again Lasercom is scored slightly higher than the RF systems. The beams are exceedingly narrow and the sidelobes very low. Laser communication's LPI characteristic is so effective that encryption of the signal is normally not required.
9. *Integration:* This consists of three parts.
   a. Real estate—area is required to mount the system—Lasercom is scored slightly higher.
   b. Deployable boom—RF for most installations requires a boom. This adds complexity to the satellite design.
   c. Dynamic reaction—the larger the antenna, the greater the reaction that must be countered when the antenna is accelerated/slewed. Laser communications scores higher on these attributes than the RF systems.

10. *Solar Operation:* Each of the three systems can be configured to track while transiting the Sun and communicate at reduced rates; however, the laser communication system envisioned here will require additional hardware to accomplish this. As a result the RF systems score higher here.
11. *Reliability / life:* Two items will be scored for this factor, namely, life and failure mode. As mentioned previously, TWTAs have a known life of 5 to 7 years and the failure mode can be considered catastrophic. For the SSPA, the life is very long provided the temperature can be controlled. With laser communications using the multiple diode laser [7], the diodes are maintained at a discrete temperature, the life is in excess of 20 years operating at full rating, and the failure mode is "graceful" (i.e., as one or more diodes fail the current can be turned up and operation continued).
12. *Interference:* This is completely an RF phenomenon; laser communications is not susceptible to self-jamming, and as a result, it will be scored higher than either RF system.

With these criteria established, the trade matrix can now be completed. Table 10.3 represents the example trade performed.

The weight and power scores were obtained using the weight/power factor utility curve. The recurring and nonrecurring cost scores were based on the respective utility curves for each factor. All other scores were obtained as discussed earlier in this section. It may be seen from the above total scores that laser communications scores higher in this example than either of the RF systems. Weight and power are strong drivers for the differences in score; however, laser communications does have other inherent advantages over RF systems, including a lower recurring cost. The situation could easily change if, for example, backward compatibility were to be considered under integration impact. In this instance, the incumbent TWTA RF system would have a higher total score. This illustrates that the trade setup is highly specific to each program's needs and the outcome is not known in advance. An important point to be made. The trade is most objective if the individuals who score the factors don't know the factor weightings.

As has been illustrated, it is possible to set up a realistic trade study between competing RF and laser communication systems. The realism is achieved by carefully gathering data on each system, by consulting with those who are knowledgeable in the technology, and soliciting technical proposals from designers and builders of the equipment. Ideally, the satellite designer would be included in the trade and would provide the sensitivities of integration and limits on weight, power, and so forth. The crosslink designer can then use the input data in a quantitative fashion to perform the trades. The *inclusion of the right factors, the proper weighting, and scoring through consensus* are of the utmost importance if the trade is to provide a realistic assessment of the competing system approaches and is to be judged objectively.

Table 10.3
Example Trade Matrix

Factor	*Weight*	*Lasercom Score /*	*Weighted Score*	*60 GHz TWTA Score /*	*Weighted Score*	*60 GHz SSPA Score /*	*Weighted Score*	*Comments*
1. Weight	15	10	150	6	90	5	75	Lasercom lighter than RF
2. Power	6	10	60	7	42	8	48	Lasercom lower power
3. Cost								TWTA nonrecurring zero
Nonrecurring	15	5	75	10	150	7	105	Lasercom & SSPA need
Recurring	15	10	150	7	105	8	120	develop. Lasercom lower recurring
4. Technical risk	15	5	75	10	150	6	90	TWTA systems built
5. Anti-jam	2	10	20	8	16	8	16	Lower sidelobes for Lasercom
6. Nuclear	2	10	20	5	10	5	10	No effect on Lasercom
7. Data rate								Lowest growth impact on
Growth	5	10	50	5	25	4	20	Lasercom
8. LPI	2	10	20	8	16	8	16	Lasercom no encryption
9. Integration								Lasercom small, no booms,
impact	10	8	80	6	60	6	60	minimal momentum impact
10. Sun in field								Reduced data rate for both RF
	2	5	10	8	16	8	16	and Lasercom
11. Reliability								SSPA Life-Redundant TWTAs-
Life	10	8	80	8	80	8	80	Lasercom graceful degradation
12. Mutual								RF has problem
Interference	1	10	10	8	8	6	6	
Totals: Lasercom		800						
60 GHz TWTA		768						
60 GHz SSPA		662						

References

[1] Dreimeyer, D., Chenoweth, A., Dreisewerd, D., Lambert, S., Morris, T., and Casey, W. "Size, Weight, and Power Trends in Laser Crosslinks," *SPIE OE-LASE '93*, Vol. 1866, Los Angeles, CA, Jan. 20–21, 1993.

[2] Deadrick, R., and Deckelman, W. "Laser Crosslink Subsystem-an Overview," *SPIE OE-LASE '92*, Los Angeles, CA, Vol. 1635, Jan. 23–24, 1992.

[3] Wertz, J., and Larson, W., Editors. *Space Mission Analysis And Design*, Chapter 11.4 Power, Space Technology Library, Kluwer Academic Publishers, ISBN 0-7923-0971, 1991.

[4] Internal communication with McDonnell Douglas Aerospace's Space Electrical Power System [*SEPS*] Program, Jan. 6, 1993.

[5] Lambert, S., Dreimeyer, D., and Casey, W. "Operation of a Diode-Based Direct Detection Laser Crosslink Through Solar Crossing," *MILCOM '93*, Boston, MA, Oct. 11–14, 1993.

[6] Mecherle, S., et al., "RF versus Laser Communications," Panel Discussion, *SPIE OE-LASE '93*, Volume 1866, Los Angeles, CA, Jan. 20–21, 1993.

[7] Jackson, J., Chenoweth, A., Dreisewerd, D., Lambert, S., and Casey, W. "The Evolution and Future of the Multiple Diode Laser," *SPIE OE-LASE '93*, Los Angeles, CA, Volume 1866, Jan. 20–21 1993.

Chapter 11

Size, Weight, and Power Trends in Laser Crosslinks

11.1 BASELINE RATIONALE

With ever-increasing improvements in the technologies that support direct-detection laser crosslinks, projections for even smaller, lighter, and lower power crosslinks are realistic. The rapid increase in diffraction-limited laser power is one of the main drivers for the trend toward smaller, more efficient terminals. Most particularly, the power summing of multiple high-power laser diodes and the parallel trend toward higher power single-diode sources is yielding projected powers in excess of several watts. This rapid increase in diode power coupled with long life enables a reduction in crosslink aperture with resultant relaxation in acquisition, pointing, and tracking requirements as well as a reduction in terminal size, weight, and power. Based on projected improvements in laser power and efficiency, detector sensitivity, electronics size and efficiency, and terminal simplicity, the physical characteristics of advanced terminals can be projected.

For higher data rates of 100 Mbps and above, laser crosslinks are generally considered to be lighter and lower power than 60-GHz crosslinks [1]. This is particularly true for operation at twice-synchronous range. What is not generally understood is that *current direct-detection laser crosslinks retain this weight and power advantage at much lower data rates than 100 Mbps (i.e., even at 1 Mb).* The reason for stating this is that two recent trade studies at 1 to 10 Mbps have established laser communications as the system of choice. The trades considered not only the size, weight, and power of the competitive *extremely high frequency* (EHF) 60-GHz and laser communication crosslinks, but also the more subtle but definite effect of the impact of the crosslink on the spacecraft design. Future laser crosslinks will be even lighter and smaller, and require significantly less power than current design terminals. Starting from an already low present base, weight and power trends will be projected so that crosslink designers can plan future laser crosslinks with realism. Since data rate and range are also parameters of interest, they are included in the projections. The baseline weights and powers given in this book are based on existing technology and experience with comparable direct-detection hardware systems.

The scaling for each system is based on work performed by McDonnell Douglas [2]. Each system uses the McDonnell Douglas laser communication link analysis program, LZRLNK, as a base. All pertinent link parameters have been included. There is, therefore, link closure realism in the scaling and projections that result.

11.2 CURRENT TECHNOLOGY

The current technologies that form the basis for scaling are derived from second-generation hardware designs. The first-generation laser crosslink, a design which was in production, consisted of early-eighties technology. The *laser crosslink subsystem* (LCS) weighed 317 lb and required approximately 300W of power. Detailed features of the LCS can be found in [3].

Major improvements have been made in laser communication terminal design over the past few years that make laser communications much more attractive to crosslink designers. These improvements are reflected in the second-generation crosslink designs. To appreciate the improvements gained from technology and design advances, it may be noted that a current second-generation optical crosslink design consisting of a suite of three optics heads with a common electronics weighs 240 lb and consumes 200W (two terminals operating). Although there is considerable redundancy in the individual terminal design, major redundancy is inherent in the use of three terminals. Direct detection is mature [4], low risk, and well understood. It forms the basis for the baseline and projections of this chapter.

The second-generation crosslinks considered here as baseline encompass the following technologies:

- High-power single-mode laser diodes with long life;
- Simplified optics—number of elements greatly reduced;
- Straightforward power-summed diode lasers—six generations of development.;
- Modern *avalanche photodiodes* (APDs) with improved sensitivity;
- Improved *quadrant APDs* (QAPDs);
- Two-axis high-speed beamsteerers;
- Gimbaled Gregorian Cassegrain telescopes—beryllium for light weight;
- Optical bandpass filters with proven stability over time;
- Optics designs, which are rugged;
- Reduced numbers of mechanisms.

Trends are projected for five systems which represent real near-term applications. Four of the systems were described at MILCOM'92 [5–8]. The fifth system is a twice-synchronous-range high-data-rate crosslink system.

Table 11.1
System Requirements Summary

Parameter	*System 1*	*System 2*	*System 3*	*System 4*	*System 5*
Data rate	12 Mbps data 100 Kbps command	650 Mbps	1–6 Gbps forward 500 Kbps from GEO 20–40 Kbps from ground	1 Mbps	0.5–4 Gbps forward 100 Kbps return
Number of channels	1	2	1–6	1	1–4
Range	84,000 km	69,450 km	40,000 km	9,000 km	84,000 km
Orbital altitude	GEO-GEO	GEO-GEO	LEO-GEO	LEO-LEO	GEO-GEO
Bit error rate	10E-6	10E-6	10E-6	10E-6	10E-6
Navigation/pointing			GPS Knowledge		
Modulation format	Biphase level	Biphase Level	Biphase level	Biphase level	Biphase level
Operation near Sun	within 2 deg	within 2 deg	within 2 deg		within 2 deg
Lifetime, 96% Prob.	10 years	10 years	10 years	3 years	10 years
Encryption/decryption	Yes	None	None	Yes	None
Fwd. error correction	Rate 1/2	None	None	Rate 1/2	None
Terminal data proc.	Yes	Yes	None	Yes	None
End-of-life margin	3 dB	3 dB	3 dB	3 dB	3 dB

Courtesy of McDonnell Douglas Aerospace.

11.3 SYSTEM REQUIREMENTS

The requirements listed in Table 11.1 are assumed for the terminal designs that will be projected. Hardware demonstrations have been performed on the second-generation terminal components so that the baseline configurations tabulated in the next section represent realistic system appraisals using mature laser crosslink technology. Electronics for the high-data-rate systems require development, using high-bandwidth specialty electronics including discrete *gallium arsenide* (GaAs) components where needed.

11.4 BASELINE SYSTEMS

The five systems considered for parametric scaling share a common second-generation technology base:

- High-power single-mode laser diodes in hermetically sealed packages;
- Power-summed diode combiners using dichroic/polarization combining;
- Projected diode lifetimes in excess of 20 years based on accelerated life tests;
- Two-axis gimbaled beryllium telescope;
- Optical encoders for gimbal position;
- Two-axis high-speed beamsteerers for track and point-ahead;

Table 11.2
Baseline Sizing Parameter Summary

Item	*System 1*	*System 2*	*System 3*	*System 4*	*System 5*
General					
Data rate forward, Gbps	0.012	0.65	1 2 4 6	0.001	1 2 4
Data rate return, Mbps	Same	Same	0.5	Same	0.1
Number channels	1	2	1 2 4 6	1	1 2 4
Fine pointing	Yes	Yes	Yes	No	Yes
Transmitter					
Aperture, in	5.5	12	6	1 in	7.5
Output power, W	0.5	1.2	0.5 1 2 3	3	1 2 4
Laser type	MD	MD	MD	MD	MD
Channel power, W	0.6	0.6	0.5 GEO 0.1 GEO	3.0	1.0 LEO 0.1 GEO
No. diodes per channel	10	8	6 2	20	10 2
Diode output power, mW	82	103	114 63	150	137 63
Combining efficiency, %	73	73	73 80	100	73 80
Receiver					
Aperture, in	5.5	12	6 LEO 24 GEO	3	7.5 LEO 24 GEO
Diffraction limited	Yes	Yes	Yes No	No	Yes No
Detector type(comm)	1 QAPD	2 APD	1 QAPD 1–6 APD	1 QAPD	1 QAPD 1–4 APD
Filter bandwidth, Å	160	80	60 60	80	100
Background	Moon	Moon	Moon Earth	Moon	Moon Earth
Electronics					
Bandwidth, MHz	20	650	1 1000	2	1 1000
Logic family	CMOS	GaAs	CMOS GaAs	CMOS	CMOS GaAs
Integration	ASIC	ASIC	ASIC	ASIC	ASIC
Packaging	SEM-E	SEM-E	SEM-E	SEM-E	SEM-E
Overall system	1 OU	1 OU	2 OUs	3 OUs	1 OU
Size per unit, ft^3	1.0 EU 1.8 OU	1.1 EU 5.5 OU	2.3 EU 1.9 OU	0.3 EU 0.3 OU	1.0 EU 2.5 OU
Weight, lbs	106	198	202 230 286 34239		130 151 193
Power, W	126	192	254 305 407 51060		161 202 290

SEM-E = standard electronic module E-size; OU = optical unit; EU = electronic unit; ASIC = application specific integrated circuit; CMOS = complementary metal oxide semiconductor; GaAs = gallium arsenide; GEO = geosyncronous Earth orbit; LEO = low Earth orbit; APD = avalanche photodiode; QAPD = quadrant APD; MD = multiple diode. Courtesy of McDonnell Douglas Aerospace.

- High-speed single and quadrant avalanche photodiode detectors;
- *Application-specific integrated circuit* (ASIC) level electronics;

Table 11.2 represents the baseline sizing for each of the five systems to be projected. The baseline will be the point of departure for the trend projections. The weight and power are consistent with the performance requirements of Table 11.1.

11.5 TECHNOLOGY TRENDS FOR PROJECTION

Increasing laser power offers the potential for a reduction in aperture, with a resultant decrease in terminal weight and power. This reduction in aperture also reduces the imaging optics size, with a further reduction in weight and volume. There are advances in electronics also, with reductions in size and power even as speed continues to increase. Also, there are gains in detector technology although their effect on weight and power will not be as dramatic as that due to increased laser power.

Single-mode diode power has gone from 50 to 70 mW a few years ago, to 100, 150, and recently to 200 mW with the commercial introduction of the Spectra Diode Labs Model 5430 rated at 200 mW, continuous. Combining of single-diode lasers, now in its sixth generation of development, forms the baseline source for the systems to be projected. The diodes that form the baseline lasers have projected lifetimes in excess of twenty years based on accelerated life tests. New laser sources being considered for space laser communications are the phased array and the broad area *master oscillator*, *power amplifier* (MOPA). A commercial 1W MOPA is currently available, also from Spectra Diode Labs. Higher power devices are in work. With the increased powers, confidence exists in predicting the effect an increase in device power will have on laser communication system size, weight, and power. Actually, it is likely that the predictions of power are conservative and power will increase faster than predicted in Figure 11.1. Note that the MOPA curves cover both a single MOPA and a MOPA combiner where dichroic/polarization combining techniques may be used to further increase power. In Figure 11.1, the design start year is used to identify the beginning of detailed design. In this context, performance of the devices has been proven. Space packaging and life testing will be required as a part of the design program so that systems will be expected to enter the production phase four to six years following the design start date.

Increased laser power leads to a reduction in telescope aperture, in accordance with the relationship indicated in Figure 11.2. The duplex communication curve assumes that both the transmit and receive apertures vary identically in size while the second curve varies only the transmit aperture. All other parameters remain fixed. The increased source power is traded for a smaller aperture with a larger

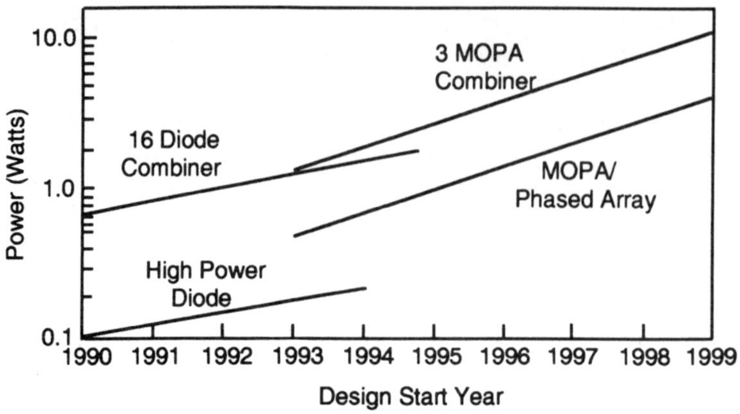

Figure 11.1 Transmitter power projections. (Courtesy of McDonnell Douglas Aerospace.)

beam divergence. The indicated curves are normalized to a 5.5-in aperture and a 600-mW transmit power, both current second-generation characteristics. This is the design configuration for System 1 where range is twice-synchronous and the data rate is 12 Mbps. The duplex communication curve is characteristic of Systems 1, 2, and 4, while Systems 3 and 5 have the stronger aperture scaling. As the aperture changes, the gimbaled telescope assembly weight and the imaging optics weight also decrease in accordance with the curves of Figure 11.3. This weight scaling is correlated with time using the transmit source power and the aperture scaling of Figures 11.2 and 11.3, respectively.

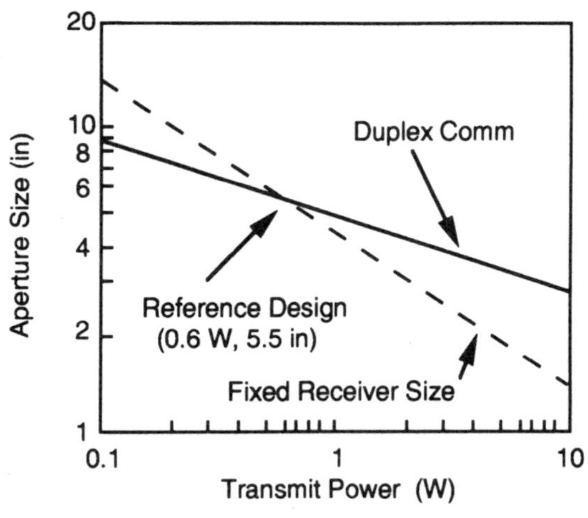

Figure 11.2 Aperture scaling with power. (Courtesy of McDonnell Douglas Aerospace.)

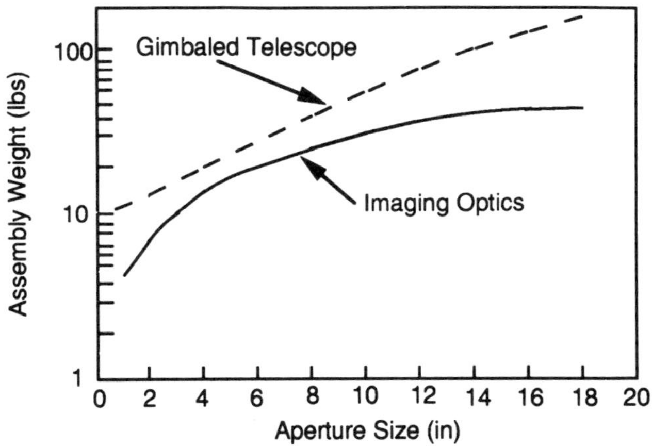

Figure 11.3 Gimbaled telescope / imaging optics weight scaling with aperture. (Courtesy of McDonnell Douglas Aerospace.)

A corresponding decrease in electronics and structure weight is expected with time due to systematic design improvements. This scaling, along with projected source efficiency improvement, is summarized in Table 11.3 based on historical design experience from the LCS through the second-generation system. These scaling rules are applied to each of the system projections in turn to arrive at the weight and power trend curves.

Terminal power consumption is also expected to decrease over time even though laser power is increasing. Gimbal power will drop due to the smaller mass of the telescope and gimbals. Further, the power efficiency of electronic devices has historically improved with time. As power decreases, less heat input is required to maintain optical alignment in combiner source lasers. The source efficiency increase is reflected in the overall power consumption projections. At some point, added laser power causes an increase in overall power, as will be shown in the trend

Table 11.3
Electronics, Structural Weight and Laser Source Efficiency Trends

Year	*Electronics*	*Structure*	*Source Efficiency*
1992	1.0	1.0	0.08
1993	0.9	0.9	0.20
1995	0.729	0.81	0.25
1997	0.590	0.729	0.25
1999	0.478	0.656	0.25

Courtesy of McDonnell Douglas Aerospace.

curves. Terminal design then becomes a tradeoff between prime power and terminal weight. Weight is most frequently a driver. Added power is obtained in a satellite at the cost of about 0.4 lb/W of power. Future power systems will obtain power at even less weight per watt.

Power consumption also increases with data rate, the amount depending on the electronics technology used. Eventually a crossover occurs, causing an alternate technology to be preferred. The projections incorporate the optimum electronics technology for the given data rate of the system. The indicated power consumption is expected to be conservative.

11.6 TRENDS

The results of the scaling are shown in the figures that follow. These figures project system weight and power for the five system configurations over the rest of this decade. A 1-dB improvement in detector performance is anticipated over the time frame but is not factored into the analyses. The later years may use integrated optics as described in [9] to achieve the projected weight reduction.

11.6.1 System 1: A 12-Mbps Full Duplex Twice-Synchronous-Range Crosslink

Figure 11.4 shows the power and weight trends for System 1. The baseline 5.5-in aperture beam divergence is 11.2 μrad and the pointing error (including jitter and point-ahead) is 4.1 μrad. The receiver field of view (FOV) is 100 μrad and a

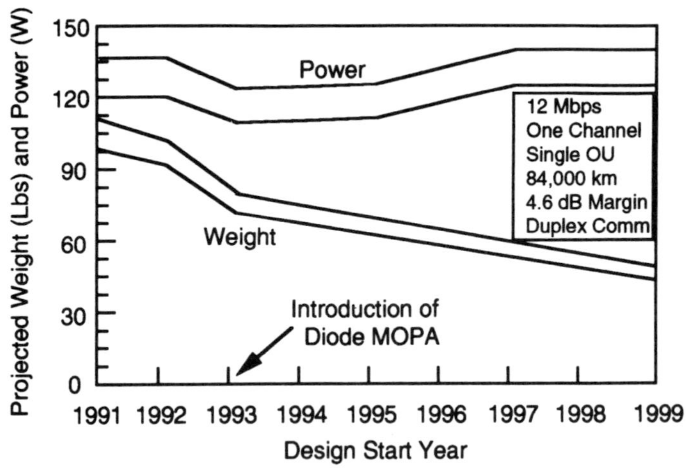

Figure 11.4 System 1: A 12-Mbps GEO to GEO terminal projection. (Courtesy of McDonnell Douglas Aerospace.)

quadrant APD is used for both communication and track. Rate one-half convolutional encoding is used to improve performance. The electronics are radiation-hardened *complementary metal oxide semiconductor* (CMOS) with a bandwidth of 24 MHz. The figure shows that satellite designs initiated in 1999 can anticipate a terminal operating at 12 Mbps over twice-synchronous range, which will weigh 50 lb or less and consume no more than 140W. The 1999 configuration assumes 6W (two 3.3W MOPAs) of transmit power with a 2.6-in aperture.

11.6.2 System 2: A 650-Mbps Geosynchronous to Geosynchronous Crosslink

Figure 11.5 shows the projected weight and power trends for System 2, a 650-Mbps design operating at 1.5 times synchronous range. The baseline configuration for this system has two 325-Mbps channels with 0.6W/channel of optical power and a 12-in transmit/receive aperture. The transmit beam divergence is 8.2 μrad and the pointing error is 2.8 μrad. The receiver FOV is 100 μrad, but the communication detector is a small area APD (low capacitance) required to achieve the high data rate. Tracking is accomplished with a QAPD. The high-bandwidth electronics are GaAs with a bandwidth of 325 MHz. Coding is not used. The figure shows that the 1999 satellite designs can anticipate a 650-Mbps terminal operating at 1.5 synchronous range, which will weigh 80 lb or less and consume no more than 205W. The 1999 configuration assumes 4.8W (three 1.8W MOPAs) of transmit power per channel with a 5.6-in aperture.

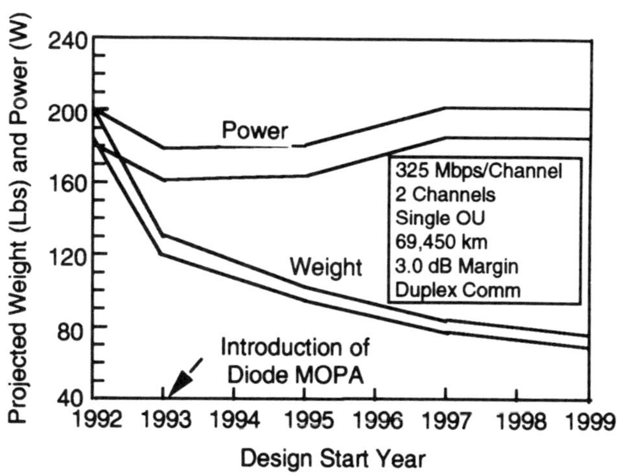

Figure 11.5 System 2: A 650-Mbps GEO to GEO terminal projection. (Courtesy of McDonnell Douglas Aerospace.)

11.6.3 System 3: A 1- to 6-Gbps Low Earth to Geostationary Crosslink

Figure 11.6 shows the projected weight and power trends for System 3, a 1- to 6-Gbps design operating at a one synchronous range. Two identical optical units with a common electronics unit comprise the high-data-rate transceiver suite, the transmit aperture being 6 in. Although the data flow is two-way, the high data rate flows only one way, from the 6-in aperture to a 24-in receiver. The data rate in the opposite direction is low (500 Kb). This data flow is characteristic of a data-gathering satellite mission where the data is relayed through a crosslink to the user on the ground. Each optical unit on the data gathering satellite consists of six 1-Gbps channels each with 0.5W transmit power. The 24-in aperture receiver, being of low optical quality (i.e., a "photon bucket"), is lightweight. The weight and power projections are for the dual optical unit configuration (two 6-in aperture optical units with a common electronics).

The baseline channel power is achieved by six diodes running at 114 mW, each with a combining efficiency of 73%. The transmit beam divergence is 11.3 μrad and the pointing error is 4.1 μrad. Again, the FOV is 100 μrad with a small, fast APD communication detector. The channelization is accomplished by dichroic combining at high efficiency. Coding is not employed. The electronics have a bandwidth of 1 GHz. The projection shows that the 1999 configuration would be a one-way, fully redundant, high-data-rate terminal operating a 6 Gbps over a one synchronous range and would weigh no more than 110 lb and consume 330W. The 1999 configuration assumes a 9.6W (three 3.5W MOPAs) transmit channel with a

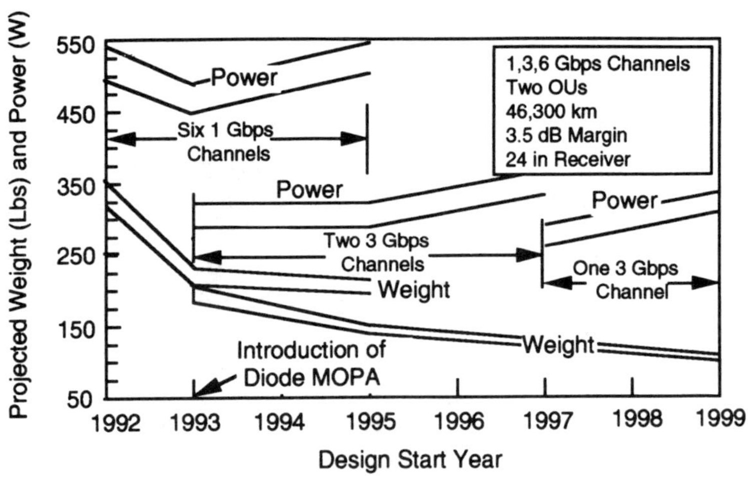

Figure 11.6 System 3: A 6-Gbps LEO to GEO terminal projection. (Courtesy of McDonnell Douglas Aerospace.)

2.6-in aperture. Power consumption of 40W can be traded for a slight increase in aperture size if laser power is fixed at the 4.8W 1997 level. Furthermore, a comparable power consumption, 3 Gbps channel, near term design is possible using 0.42W MOPAs. This system weighs only 200 lb as compared to the 350-lb 6-channel baseline.

11.6.4 System 4: A Short-Range 1-Mbps Crosslink

Figure 11.7 projects the weight and power for System 4. The baseline configuration for this system is a suite of three optical units with a common electronics unit. Each optical unit is a gimbaled package using spatially isolated transmit/receive apertures. The 3W 500-μrad transmit beam is provided by 20 aperture-shared diodes, each running at 150 mW. The receiver is a Cassegrain telescope with a quadrant APD detector for track and communication. Gimbal-only tracking is used with a pointing error of 150 μrad . The electronics are radiation-hardened CMOS with a bandwidth of 2 MHz. Rate one-half convolutional encoding is employed to improve performance. It should be noted that transmit power is held constant for this projection so that the performance trend is entirely due to the anticipated reduction in electronics and structure weight, a portion of which would be assigned to the gimbals. New, lightweight gimbal designs indicate that the projections should be conservative. A modest improvement in electronics efficiency and lower gimbal power accounts for the decrease in power. By 1999, we can anticipate that a short-range three-terminal suite can be built that weighs less than 25 lb and consumes no more than 52W.

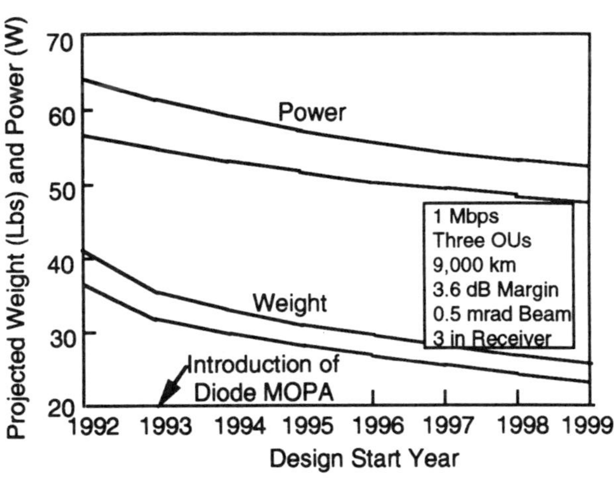

Figure 11.7 System 4: A 1-Mbps short-range LEO to LEO terminal projection. (Courtesy of McDonnell Douglas Aerospace.)

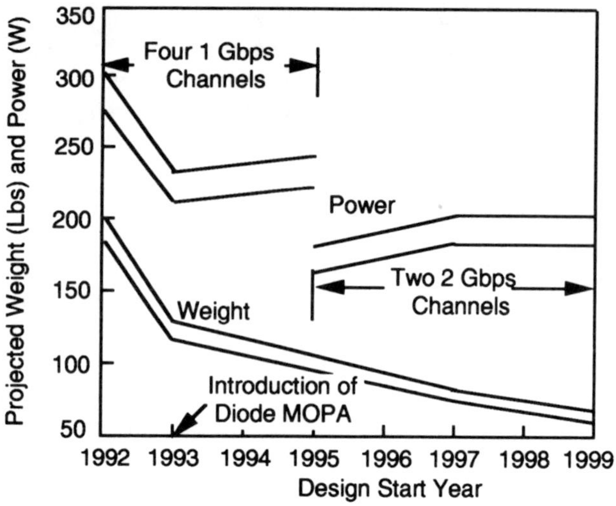

Figure 11.8 System 5: A 1-Gbps, twice-synchronous-range LEO to GEO terminal projection. (Courtesy of McDonnell Douglas Aerospace.)

11.6.5 System 5: A 1- to 4-Gbps Geosynchronous to Geosynchronous Link

Figure 11.8 shows the projected weight and power for System 5. The baseline consists of a single optical unit supported by an electronics unit. The optical unit has up to four 1 Gbps channels with 1W of optical power per channel. A 4.3-in transmit aperture is coupled with a 24-in aperture "photon bucket" receiver. The channels are again dichroically combined/separated with high efficiency. The data return rate is 100 Kb for command and control.

The baseline channel power is provided by ten diodes operating at 140 mW with a combining efficiency of 73%. The transmit beam divergence is 14.2 μrad and the pointing error 2.8 μrad. The receiver FOV is 100 μrad and the communication detector is a small APD. Coding is not used in this design. The electronics have a bandwidth of 1 GHz. The figure shows that a design initiated in 1999 can be expected to weigh approximately 65 lb and draw no more than 200W. The 1999 configuration assumes two 4.8W (three 1.6W MOPAs) 2-Gbps transmit channels with a 4.3-in aperture.

11.7 SIZE PROJECTIONS

The terminals will not only decrease in weight and power, but also in size. The baseline terminal size is given in Table 11.2 for the five systems considered here.

The size is shown for the optical and electronic units. To give an idea as to the size scaling, a projection of the 1999 version of System 1 (an aperture size of 2.6 in compared with the baseline 5.5 in) yields an electronics unit volume of 0.5 cubic feet and an optical unit volume of 0.6 cubic feet. This is a 60% reduction in total volume. The volume of the other systems will also decrease proportionally with the exception of System 4, which is already quite compact.

11.8 SUMMARY

Although laser communication terminals are already light in weight and low in power consumption, future designs are projected to be even smaller and lighter by a significant amount [9]. This is illustrated in Figure 11.9, which shows the projected weight trend for full duplex twice-synchronous-range terminals. The figure highlights the significant weight reduction between the first-generation LCS using early 1980s technology and the current second-generation system. The original 310-lb terminal weight has been reduced to 80 lb for the second-generation terminal given an equivalent 1 Mbps system. A second-generation system operating at 100 times the LCS data rate with today's laser technology can be designed for one-half the

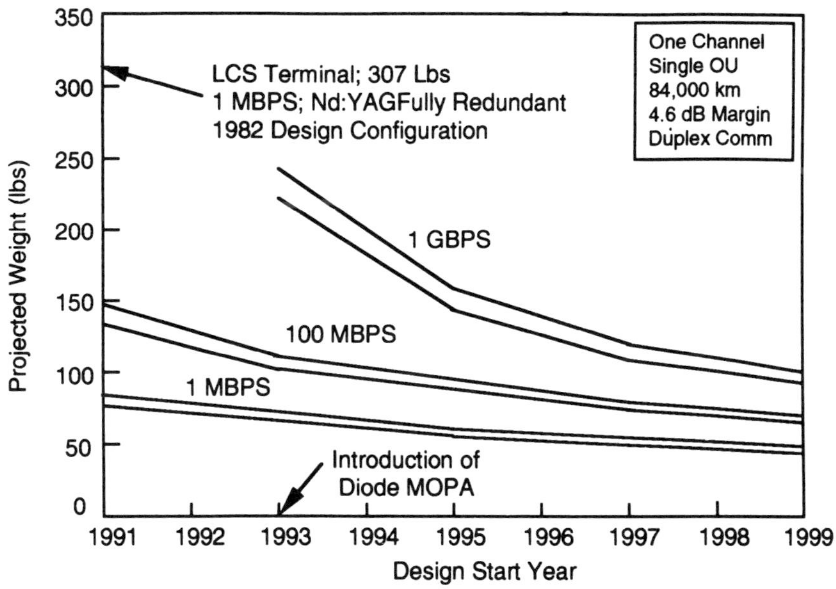

Figure 11.9 Twice-synchronous-range full duplex GEO to GEO weight projections. (Courtesy of McDonnell Douglas Aerospace.)

original LCS weight. The evolution of MOPA sources is expected to further reduce the weight for this system to less than 70 lb by the end of this decade. The MOPA also opens the window for full duplex twice-synchronous-range gigabit crosslinks. Projected future weights for these gigabit-class crosslinks are less than 100 lb based on the MOPA power trends shown in Figure 11.1.

References

[1] Chan, V.W.S. "Intersatellite Optical Heterodyne Communication Systems," *Proc. of the SPIE*, Vol. 1131, Los Angeles, CA, 1989.

[2] Driemeyer, D., Chenoweth, A., Dreisewerd, D., Lambert, S., Morris, T., and Casey, W. "Size, Weight and Power Trends in Laser Crosslinks" *SPIE OE-LASE '93*, Vol. 1866, Los Angeles, CA, Jan. 1993.

[3] Deadrick, R. B., and Deckelman, W. F. "Laser Crosslink Subsystem-an Overview," *SPIE OE-LASE '92*, Vol. 1635, Los Angeles, CA, 1992.

[4] Chenoweth, A. J., et al. "Maturity of Direct Detection Laser Crosslink Technology," *SPIE OE-LASE '93*, Vol. 1866, Los Angeles, CA, 1993.

[5] Dreisewerd, D. W., et al. "A Twice Synchronous Range GEO to GEO Optical Intersatellite Link," *IEEE Military Communications Conf. MILCOM '92*, Vol. 1, 177, San Diego, CA, October 1992.

[6] Dreisewerd, D. W., et al. "A GEO to GEO High Data Rate Optical Crosslink Approach," *IEEE Military Communications Conf. MILCOM '92*, Vol. 1, 183, San Diego, CA, October 1992.

[7] Dreisewerd, D. W., et al. "A High Data Rate LEO to GEO Optical Crosslink Design," *IEEE Military Communications Conf. MILCOM '92*, Vol. 3, 1163, San Diego, CA, October 1992.

[8] Lambert, S. G., et al., "Short-Range Multi-Terminal Satellite Crosslink Communications," *IEEE Military Communications Conf. MILCOM '92*, Vol. 3, 1170, San Diego, CA, October 1992.

[9] Carlson, R. T. "Technologies and Techniques for Lasercom Terminal Size, Weight, and Cost Reduction," *SPIE OE-LASE '90*, Vol. 1218, 6, Los Angeles, CA, 1990.

Chapter 12

Laser Communications Applications

Application of laser communication technology was discussed briefly in Chapter 2. In this chapter, the discussion is expanded and included are results of several top-level sizing studies on technology application to the required link application needs. This chapter organizes applications into four distinct laser communication link categories: satellite crosslinks, satellite to air or ground terminals, submarine laser communication, and deep space links.

12.1 SATELLITE CROSSLINKS

A crosslink, or communication link between two satellites, may be needed to solve certain requirements of a satellite communications architecture. One example is political; the desire to avoid reliance on a ground station located on foreign soil. Another is the need to transmit data from an information-gathering platform that is located on the other side of the Earth from the user. The data may be perishable, that is, it is needed in nearly real time for maximum usefulness, or may be so voluminous that placing sufficient data storage capability onboard the satellite is not practical. Yet still another reason may be a need to avoid multiple up and down links, which add complexity, potential unavailability, and communication data quality loss to the overall data transfer. Additionally, needs may arise for short-range crosslinks internetting small terminals in large constellations. As indicated previously, laser communications offers the user a number of unique advantages over *radio frequency* (RF) systems, including size, weight, power, and integration ease on the spacecraft. Integration ease issues include compactness of terminals, elimination of complex frequency planning and authorization, and RF interference issues.

Three of the main design drivers of a crosslink are data rate, range, and link symmetry. Once these three requirements are established, the design can proceed with some degree of certainty. For convenience in describing the missions and technologies, this section is divided into categories that include lower data rates at long range, high data rates at long range, and short-range links. Although all crosslinks are duplex, symmetric and asymmetric links may be implemented using slightly different hardware implementations. An example of an asymmetric link is the gathering of large quantities of data by a platform and the transfer of that data at high data rates to a user. The return link direction sends data at much lower data rates for command and control of the data gathering platform. Figure 12.1 illustrates the possible data link options for laser communication crosslinks.

12.1.1 Lower Data Rate, Geosynchronous to Geosynchronous Crosslinks

Inclusive of this category reside a number of applications. Satellites tasked to perform strategic and tactical early warning, survivable *command, control, and communications* (C^3), interoperable satellite command and control, and worldwide navigation and timing missions, are all candidates for crosslinks. Crosslinks are currently considered one of the main technology investments for military space. Military space has highlighted the lack of crosslinks as a deficiency in meeting the operational needs of their military satellite command development. Key benefits of the crosslinks are seen as manpower savings (from the elimination of ground stations) and improvements to communications security. It is easier to jam uplinks than downlinks, for example.

Additional benefits derived from using laser crosslinks are in the area of reduced weight, power, secure communications, and ease of integration with the satellite. For navigation satellites, the primary benefits of crosslinks are considered also to be savings in manpower from elimination of ground stations and improved satellite station-keeping autonomy. Of course, the benefits of crosslinks can be realized by either RF or laser crosslinks.

The first-generation *laser crosslink subsystem* (LCS) and a second-generation satellite laser crosslink design are examples of laser communication application to crosslink needs. Laser crosslinks were selected competitively for both systems over RF. Both the first and second-generation crosslinks were designed to operate at below 10 Mbps but at twice-synchronous range (i.e., ~ 85,000 km). They were also symmetric links in design, which means that only one terminal design was needed since identical capabilities can be selectively configured for bidirectional operation.

Unfortunately, from an affordability viewpoint, spacecraft are becoming increasingly more expensive. The U.S. military space mission is changing due to the decline in the threat from the former Soviet Union and the resulting reduction in U.S. defense investment. Desert Storm demonstrated the value of space assets in modern combat; however, the military space command will continue to take budget cuts as well as realign its emphasis from strategic deterrence to the conduct of theater operations. As a result of the increasing cost of spacecraft, there is a strong

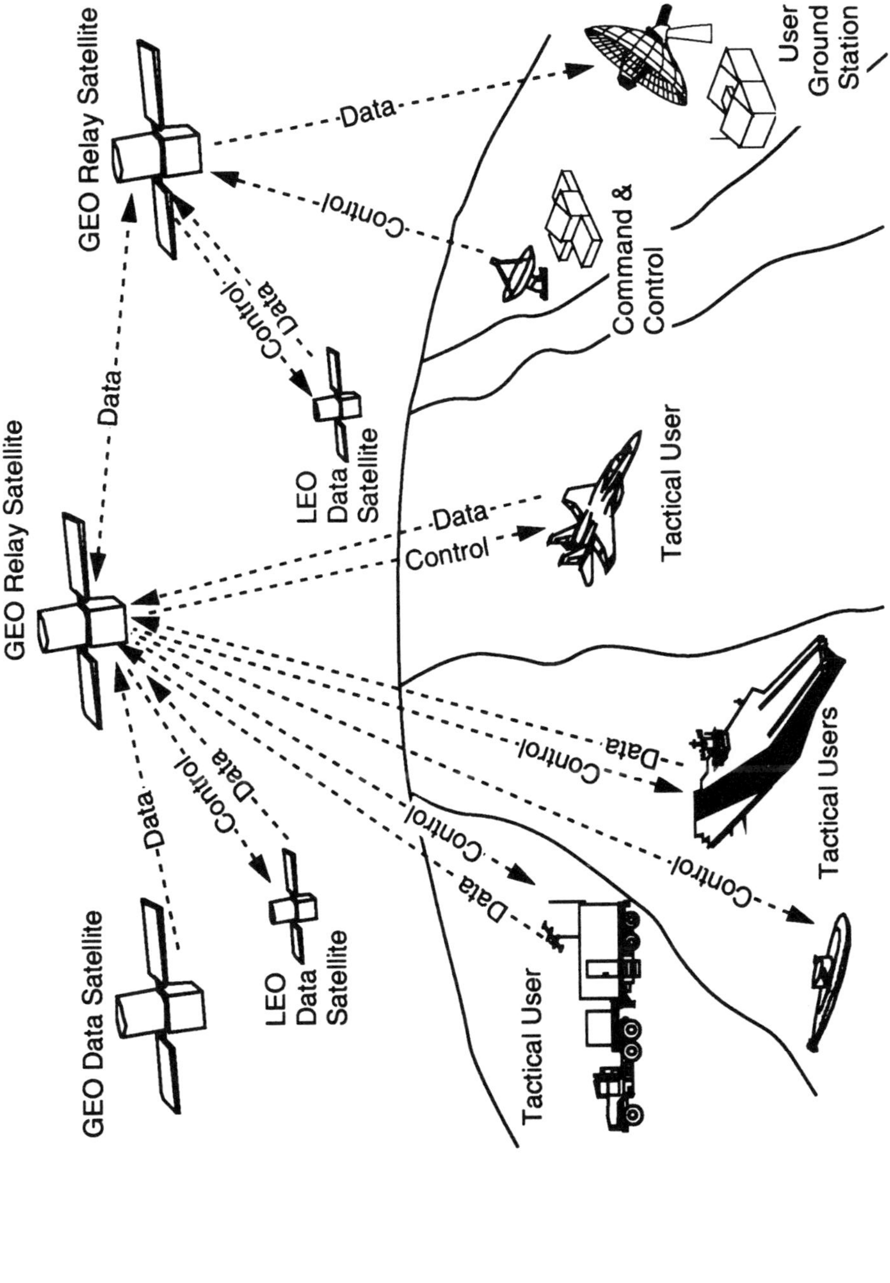

Figure 12.1 A military satellite communications architecture.

trend to put up smaller, much less costly satellites that perform a single function and don't combine a number of functions as do many larger, more expensive spacecraft. In summary, it may be said that crosslinks are needed in military satellite network architecture and that satellite networking is essential for military communications in the foreseeable future. Preparation for battle is currently being done on the way over to the theater. There is a need for video conferencing and high-data-rate transfer of maps and other intelligence data to the aircraft while they are on their way from the *continental United States* (CONUS) or other fixed bases to the theater of operation. Suitable aircraft for laser communications terminals would be those subsonic, wide-body aircraft that carry command functions.

The desired data rate for the military combat operation is about 50 Mbps where video data would be sent from battle field sensor platforms back to command and control centers. There is also a need to send maps and other video data from CONUS to command and control areas and forward areas. Long-range (twice-synchronous range) relay satellites obviously fit into this view of future operations. Long-haul trunk communication via satellite relay is a necessity for deployed military forces. The military can make use of land lines or terrestrial microwave communications, but cannot rely on having it available because not all countries have the infrastructure for it and because they can be suddenly made unavailable through hostilities or political acts. Trunk communications via satellite relay makes sense for warfighters from another standpoint as well—their communications are peak-load driven in that they need to operate in an area for a relatively short time and be able to start operations quickly. An optimum solution is to be able to move one or both ends of the communications trunk around where it is needed without having to engage a whole new trunk when major location changes are made. The satellite relay topology is much closer to the optimum solution than the fixed network topology, which may not even provide trunks to the desired locations. High-data-rate trunk communication through satellite relay has already enjoyed some successes, but has also encountered some major obstacles associated with the physics of the technical approach used. Technology has only recently become available to make long-haul trunk communications via satellite economically practical for the military.

Two general classes of satellite communication technology are now available for high-data-rate implementation. It is possible to provide long-haul communications using a combination of optical and RF communications relays through existing satellite and aircraft assets. Fixed or mobile ground terminals can also tie into this relay network with antennas as small as 6-in in diameter. Transmitted data can be simultaneously transmitted in either direction at rates from 50 Mbps up to 1 Gbps.

High-data-rate RF up/down links require large apertures and/or high-power transmitters at both the satellite and the user terminals. High-data-rate optical up/down links to the ground offer limited availability at any one location due to low cloud penetration ability. Using optical up/down links to a relay aircraft flying above the cloud deck (typically 20,000 to 35,000 ft) allows for aperture sizes of less than 6 in while using long life, low-power solid-state lasers. RF links between the aircraft and ground or sea-based terminals are then able to use relatively small apertures and low-power solid-state transmitters due to the greatly reduced range. This RF operation is conventional, easily implemented, and well understood. This type of operational scenario is most cost-effective when the aircraft used is a type of command and control aircraft that is normally present in the area of conflict or when the aircraft has other high-data-rate generating or receiving missions. The data-rate range mentioned previously is the same as that supported by the North American *synchronous optical network* (SONET) standard. Obviously, existing military communication units will not be capable of the higher data rates, but by using the *asynchronous transfer mode* (ATM) capability of the SONET architecture, lower data rate communications can be multiplexed into and out of this high-speed communications trunk. This multiplexing/demultiplexing process would occur at either the relay aircraft, the mobile ground terminal, or the fixed ground terminals These then would act as gateways in a global architectural plan.

A GEO to GEO Laser Crosslink Solution

A laser solution to the requirements just stated (i.e., a 50-Mbps full duplex symmetric crosslink at twice-synchronous range at today's low-risk direct-detection technology) is examined. An inherent danger in embracing a new technology is the likelihood of selecting the most advanced elements of the new technology. This has been done in the past and lessons should be learned from it; otherwise, earlier mistakes are destined to be repeated. For this reason, this book addresses solutions that are primarily tied to direct-detection technology as opposed to coherent detection. Direct-detection technology is mature at this point in time [1], a condition that was not true even five years ago. Coherent development is proceeding and strides have been made, but the greater complexity makes for increased risk that is not necessary to assume since direct detection can currently do the job with acceptable risk. The reasons for this are the maturing of the single-mode diode source and its long life, the high powers attainable (growth from 50 to 200 mW in the past five years), diode-combining to achieve even higher power, the advent of the *master oscillator power amplifier* (MOPA) laser source, and the maturity of acquisition/tracking. Also, the remaining key components of a laser communications system all assure low risk. Recent risk assessments of a second-generation laser communications system place it on an equal footing with the risk of developing a comparable performance 60-GHz crosslink.

A top-level laser communications terminal design to meet the requirements for a 50-Mbps symmetric full duplex link operating at twice-synchronous range would consist of a single optical unit and an electronics unit. The terminal would be designed for a reliability of 98% over a mission life of ten years. This would be accomplished by providing component redundancy for all critical components, including detectors. The design could be a derivative of the second-generation laser crosslink, which was in the demonstration and validation phase of the contract when the total satellite program was terminated.

A communication link summary for this 50-Mbps link is shown in Figure 12.2. The link design would provide for a 4-dB end-of-life link margin at the twice-synchronous range. Optical losses would be estimated and include degradation estimates due to energetic radiation. A diode combiner laser would combine the output of 16 diodes to achieve the power output of 1.2W. This would be accomplished with an optical combining efficiency of 73%, which is easily achievable based on the design of six generations of this laser type, including a 16-diode version. The diodes would be operated well below their continuous power output rating of 150 mW, which projects to a life far in excess of a 10-year mission life. A weight and power breakdown for the 50-Mbps terminal design is shown in Table 12.1. The weight and power are scaled from the second-generation design which, because of its technical maturity, is well understood.

Table 12.2 is presented for four cases that explore this design at data rates from 5 to 50 Mbps. It should be noted that the same telescope, a 5.5-in aperture off-axis unobscured design similar to the second-generation design was used in estimating the system capabilities. A diffraction-limited 5.5-in telescope was built and performed in test as designed for the second-generation design. It is interesting to note the relative flatness of the terminal weight and power values as the data rate

Table 12.1
50-Mbps Terminal Weight and Power Breakdown

Component	*Weight (lb)*	*Power (W)*
Optical unit		
Gimballed telescope assembly	25.0	11.0
Laser transmitter	19.2	22.9
Detectors/preamps	16.0	1.8
Optics/mechanisms	11.0	4.0
Electronics	2.0	13.4
Cables/connectors	7.0	
Structure/thermal control	16.7	18.8
Electronics unit	1 0 U	1 0 U
Processor/control module	6.1	22.4
Data handling/OU interface	3.6	9.7
Power regulation module	2.1	34.0
Structure/thermal control	16.5	
Totals	122.0	136.0

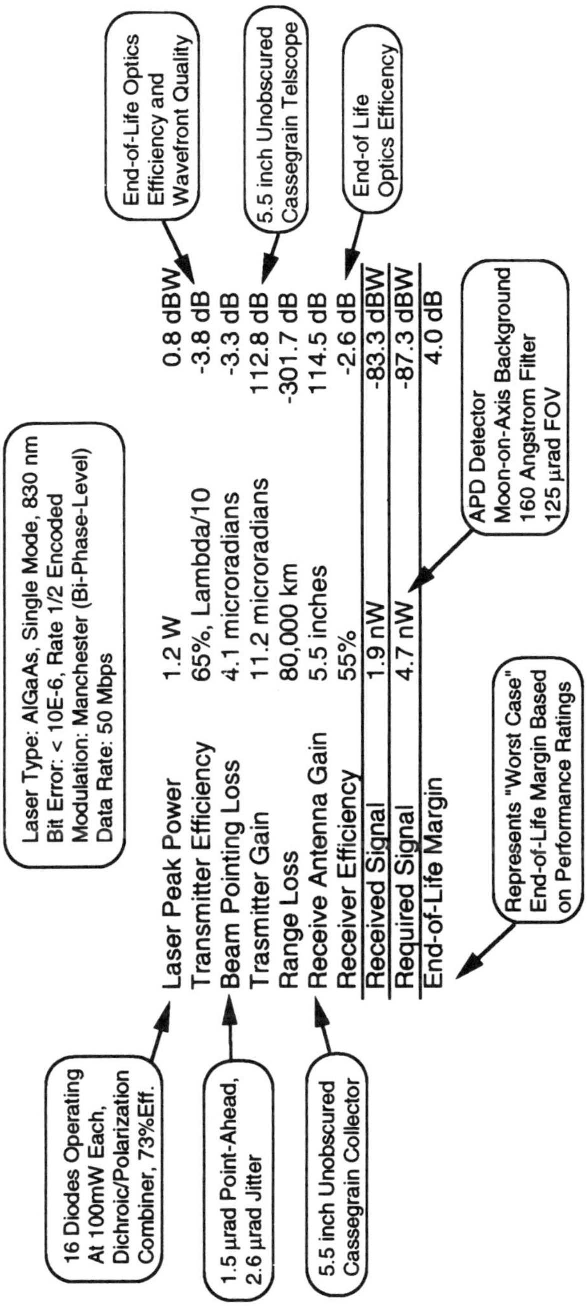

Figure 12.2 50-Mbps GEO to LEO link summary. (Courtesy of McDonnell Douglas Aerospace.)

Table 12.2
Characteristics Of Four Twice Synchronous Range Crosslinks

	System			
Parameter	*5 Mbps*	*20 Mbps*	*30 Mbps*	*50 Mbps*
Transmitter output power, (w)	0.4	0.8	0.8	1.2
Aperture size, (inches)	5.5	5.5	5.5	5.5
Beam divergence, (μrad)	11.2	11.2	11.2	11.2
Number diodes, comm/trk	8/0	8/0	8/1	16/1
Separate source for duplex comm	Yes	Yes	Yes	Yes
Range, (km)	80,000	80,000	80,000	80,000
Background	Moon	Moon	Moon	Moon
Bandwidth, (Å)	80	80	100	180
Detector type (comm)	QAPD	QAPD	APD	APD
Configuration	1 OU	1 OU	1 OU	1 OU
Weight, (lb)	105	105	109	122
Power, (W)	117	124	127	136

increases from 5 to 50 Mbps. The primary increase is due to the increase in the number of diodes. A telescope size increase would provide a much steeper rise in weight.

12.1.2 High-Data-Rate Long-Range Crosslinks

Both military and commercial satellite links are handling increasingly higher data rates. Imagery for commercial use, intelligence, and war fighting is on the climb. Serving as evidence of this in the commercial sector is the apparent need for hundreds of channels of video for viewer selection, which, coupled with the advent of *high definition television* (HDTV) is creating a requirement for very high data rates. The adoption of digital TV is also having an impact on the data rate.

Recently, private industry was given the go-ahead by the government for using surveillance satellite technology to produce imagery for the commercial market. Several firms are currently working on fielding high-resolution imagery for commercial use. Resolutions as good as 1m, requiring data rates on the order of 300 Mbps, are being planned. Although these early systems will use onboard recorders, the time may come, as the operations expand, when crosslinks will be necessary. In the military area of satellite data gathering, higher and higher data rates are required to transfer the imagery, both for intelligence and for war fighting. It is estimated that future systems will demand data rates in excess of 1 Gbps. These data rates will need to be crosslinked since recorders are unable to handle this quantity of data and further, near real-time access may be desired.

The increase in data rate also affects the *National Aeronautics and Space Administration* (NASA) programs. Increasingly, higher data rates approaching 1

Gbps or so appear to be required in future systems to handle multiple mission data simultaneously. Again, crosslinks are required for relaying the data from data gathering platforms to user collection site(s).

12.1.2.1 High-Data-Rate Low Earth Orbit (LEO) to GEO Crosslinks

The military data gathering mission is discussed. This mission requires very high data rates and crosslinking for data access. Early needs may be in the low gigabits per second range with future systems requiring even higher rates. The data flow is highly asymmetric in that the high-data-rate transfer occurs in only one direction, toward the user, while in the other direction, command and control data is relayed at low data rates on the order of hundreds of kilobits per second.

Figure 12.1 shows the scenario for this application. High-bandwidth data gathered by mission spacecraft or aircraft is relayed to a GEO satellite via a crosslink. From here the data would be relayed via a crosslink to one or more GEO platforms for downlinking to the user on the ground. Note the directional high-data-rate data flow. RF systems have difficulty with these data rates, placing laser communications in a favorable position for adoption as the crosslink of choice. Laser optical communications is looked on as not being suitable for up/down links due to cloud cover impenetrability. However, the up/down link is also an area where RF is currently having the most trouble and is creating "choke points." Very high RF powers and very large RF antennas are required at these high data rates for both the satellite and the ground terminal to close the links. With the desire to move to smaller, cheaper, single-purpose spacecraft, the adoption of very large RF antennas and high-power RF transmitters for crosslinking and up/downlinking would be a step in the wrong direction.

A good solution for the military application would be the use of lasers for crosslinks and portions of the up/down links. This approach blends the best of RF and laser communications technology. The approach is to crosslink with optical communications and up/down link with laser systems to subsonic aircraft flying above/around the clouds. These wide-body types of aircraft are not new acquisitions, but rather are already in the inventory. Once the data is received by the aircraft, it is downlinked to the user via RF, a very low risk technology that can accommodate clouds and weather. Uplinking of command and control data to the mission satellite could be accomplished via the aircraft or directly with RF.

An alternative approach to cloud cover mitigation is the use of several ground stations where the probability is high (98%) that one of the ground terminals will have a cloud-free line of sight and the data can be transferred. Long-haul fiber-optical linking between ground stations and the user provides instant access to the data. The fiber-optic linking is capable of handing the data rate. Actually, the commercial fiber-optics market is moving rapidly toward the same high data rates needed by the military.

This mission has been studied to a considerable extent [2]. A 4-Gbps data rate for a direct-detection optical crosslink from a LEO to a GEO is currently technically achievable. With a link range of 25,000 nmi and channelized asymmetric data transfer at 4 Gbps, the link can be closed into a 24-in collection aperture at GEO. A command and control forward link of up to 100 Kbps complements the mission data.

The channelization is very simple to accomplish and the approach is shown in Figure 12.3. At the GEO receiver, the channels are separated dichroically and each channel is detected by an APD. The acquisition and track channel can be configured to be an additional wavelength and would be detected by a quadrant APD to produce pointing angle information for update of the GEO aperture pointing system. A block diagram of the GEO system is shown in Figure 12.4 An abbreviated link margin for a LEO to GEO communication link is tabulated in Figure 12.5. Chapter 11 outlines weight and power trends for such a system.

12.1.2.2 High-Data-Rate GEO to GEO Crosslink

For a data gathering mission GEO to GEO crosslink, the range is twice-synchronous. The link would be sized based on the single direction of data flow (asymmetric). For the transmission of high data rates with a symmetric system, where the data must flow at a high data rate in both directions simultaneously, a symmetric design would be necessary. Also, since each terminal would be identical, cost savings would result over a link design using two terminal designs. A scenario

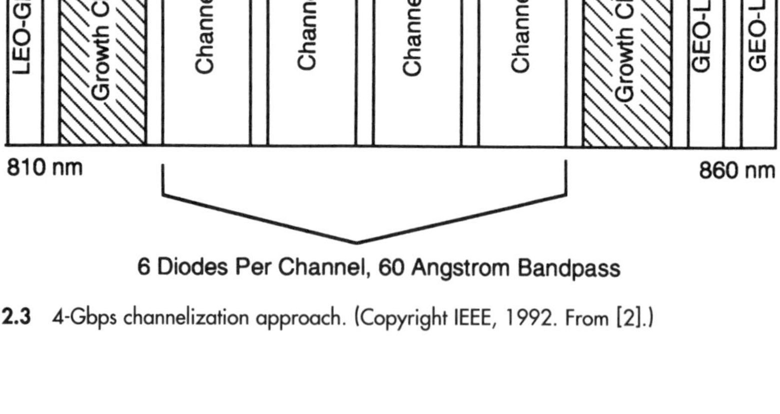

Figure 12.3 4-Gbps channelization approach. (Copyright IEEE, 1992. From [2].)

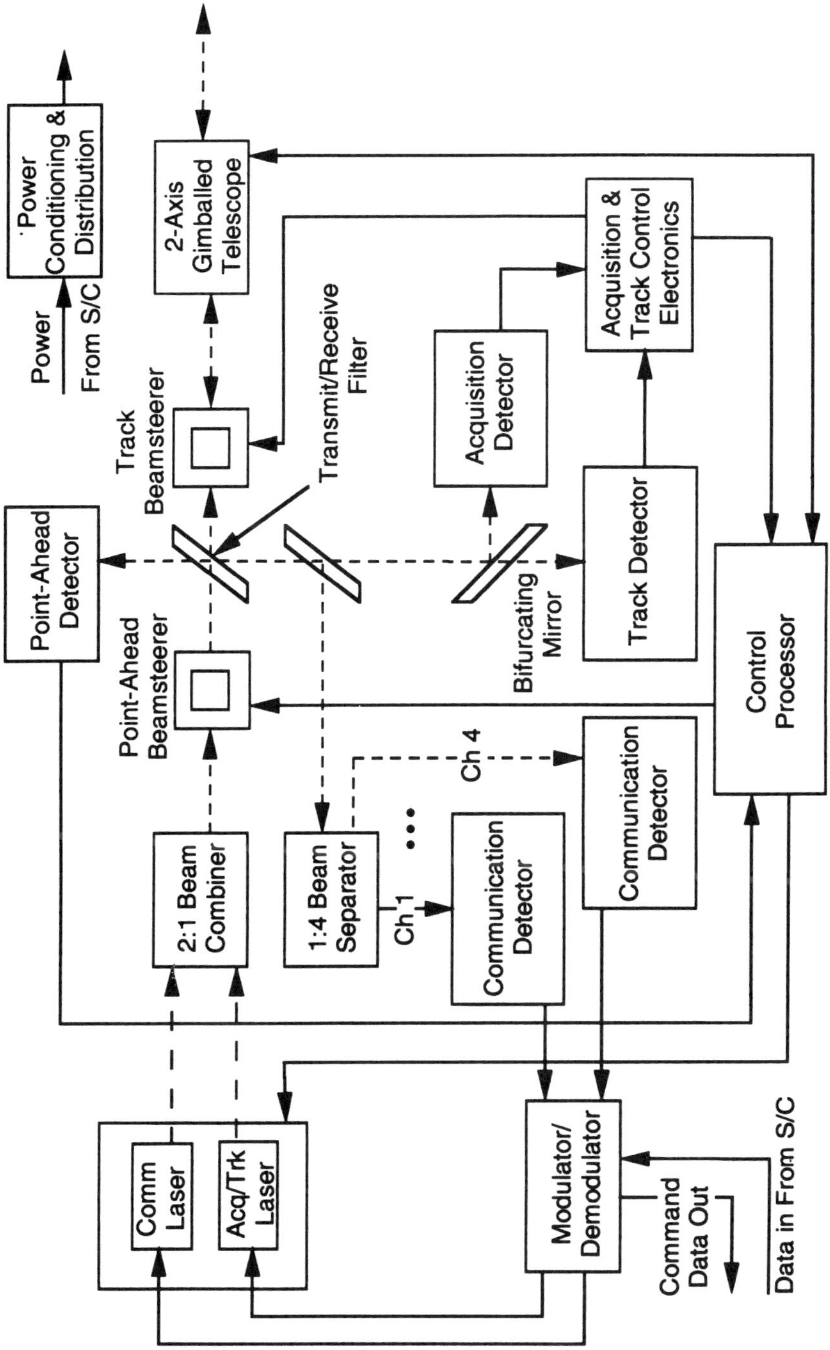

Figure 12.4 4-Gbps GEO terminal block diagram. (Copyright IEEE, 1992. From [2].)

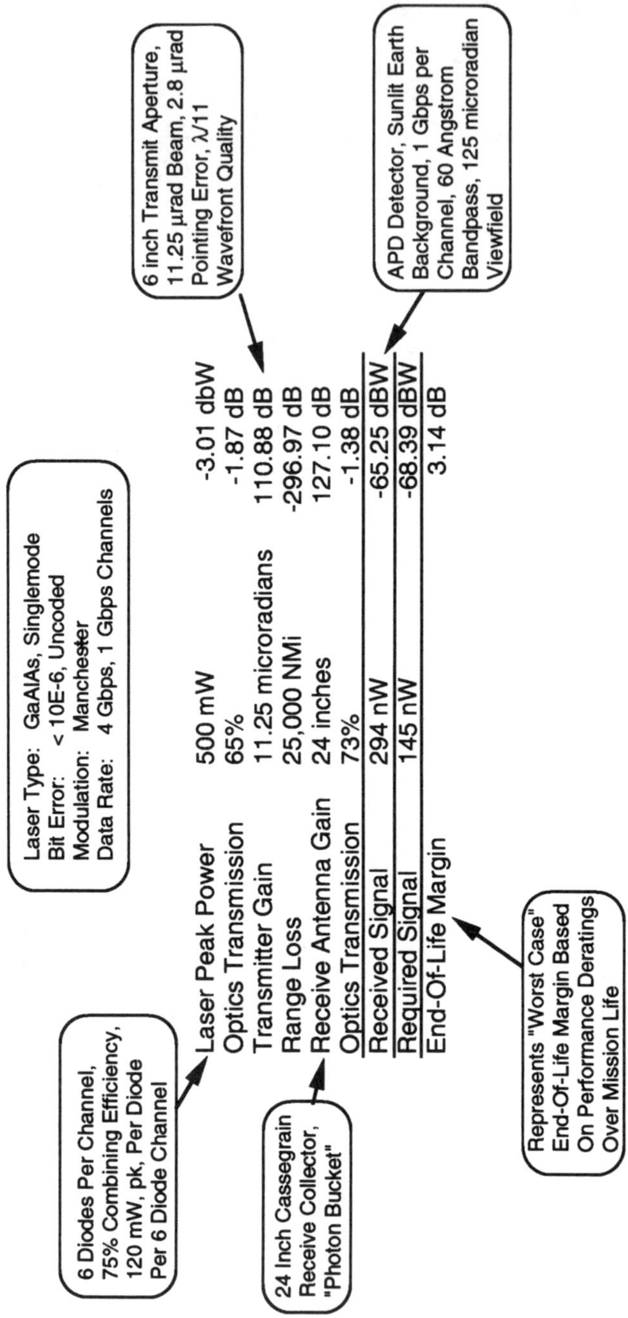

Figure 12.5 4-Gbps LEO to GEO link summary. (Copyright IEEE, 1992. From [2].)

defining this mission could be based on NASA needs for gathering data from a number of user satellites in LEO and relaying via GEO to a central site in CONUS for distribution to scientific users. This is similar to the current NASA *tracking and data relay satellite system* (TDRSS) [3]. These satellites are required to be interchangeable, so there would only be one crosslink design. For the NASA mission these crosslink relays would operate at a range of 1.5 times synchronous range. The range is sometimes implicitly expressed as a GEO arc angle, which is the angle in the equatorial plane between GEO satellites measured from the center of the earth. The GEO arc range is plotted as a function of angle in Figure 12.6.

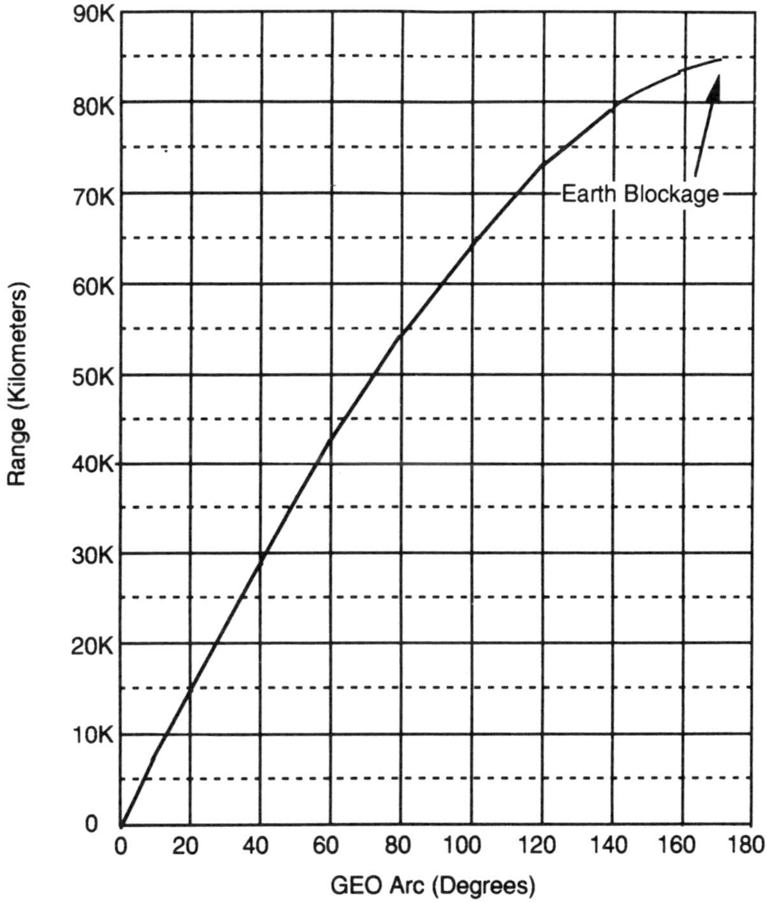

Figure 12.6 GEO to GEO arc range.

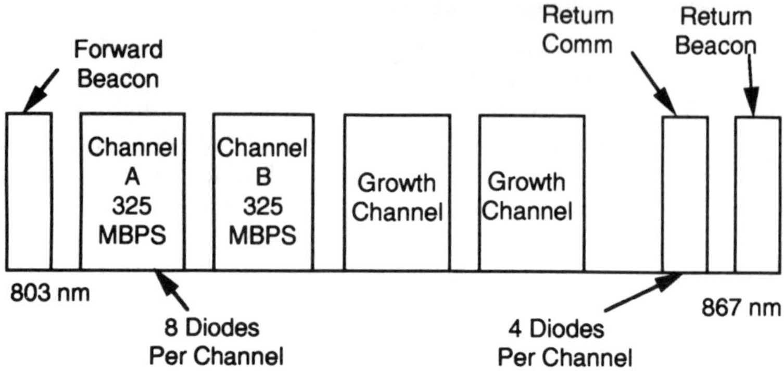

Figure 12.7 Two 325-Mbps channelization approach. (Copyright IEEE, 1992. From [3].)

Again, channelization could be used to reduce overall aperture size. The channelization for such a system could be as shown in Figure 12.7. Note that because of the large width of the available gallium arsenide laser spectrum, there is room for growth with at least two additional channels.

12.1.3 Short-Range Crosslinks

Recently a number of short-range applications have emerged that could use laser communication's small size to an advantage. Military applications such as short-range laser crosslinks for Brilliant Eyes (BE) for the Strategic Defense Initiative (SDI) have been studied in some depth [4]. The BE spacecraft envisioned was small and the crosslink terminals were expected to weigh little. How little will be indicated as the discussion proceeds.

Commercial applications are also appearing. Recently, the use of laser communications for global personal communication system satellite crosslinks has been studied with promising results. The large amount of internetting required in a large satellite constellation suits itself extremely well to the point-to-point nature of laser crosslinks. In addition, the use of multiple terminals for satellite internetting would be facilitated by a system very insensitive to frequency contention, such as is laser communications.

12.1.3.1 A 1-Mbps Short-Range Crosslink

An application for short-range crosslinks having a 1-Mbps data transfer rate over 9,000 km has been studied [4]. Multiple links onboard a single satellite were envisioned. A large number of satellites comprised the network for global connectivity. Since the constellation required a great number of satellites, cost was a major driver and was directly related to weight. Satellites must be small and lightweight to achieve the cost goals. The size launch vehicle required and how many small satellites could be included per launch were important. To achieve the global coverage desired by these systems, multiple orbital planes were envisioned. One concept is illustrated in Figure 12.8 for three planes. Full duplex communication connectivity between satellites was to be established fore and aft to form a LEO ring network in-plane. Cross-plane connectivity was to be established via selected nodes to facilitate data transfer between orbital planes. In these situations, a minimum of three terminals (fore, aft, and across) were required per satellite.

Because of the high frequencies associated with optical links, narrow beam divergences, small fields of view, and small apertures could be used. These advantages worked directly to reduce spacecraft design complexity. The terminal was simple, using a single frequency for transmit and receive. Separate apertures sufficed to provide transmit and receive signal isolation in space. Complex frequency management was not required to mitigate self-interference. High-order network management functions were not necessary to allocate frequency or time slots to the crosslink function, as would be necessary with conventional RF approaches.

The small antenna size coupled with the relatively large beam divergence enabled use of a lightweight gimbaled optical package (optical head), illustrated in Figure 12.9. Fine, high-speed tracking with internal beamsteerers was not needed, hence, gimbal-only tracking was used. No-point-ahead was required since the beam divergence was large enough to encompass the point-ahead angle. Separate transmit and receive apertures were used for isolation, as shown in Figure 12.9. The receive aperture was a 3-in Cassegrain telescope and the transmit aperture was

Figure 12.8 Internetting LEO constellation.

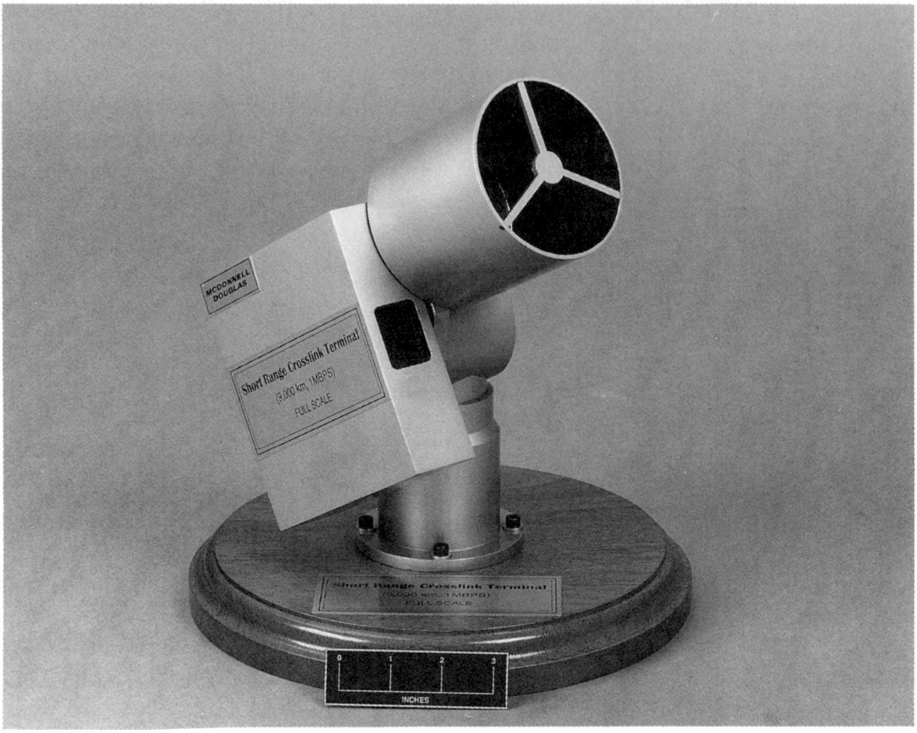

Figure 12.9 1-Mbps short range terminal. (Copyright IEEE, 1992. From [4].)

the rectangular opening slightly down and to the side of the receiver aperture. The basic material for the optics and structure was beryllium for light weight.

The transmitter is illustrated in Figure 12.10 and consisted of 20 hermetically sealed 150-mW laser diode packages capable of outputting power levels of nearly 3W. The elliptical beams were circularized by an optical element to produce a 500-μrad divergence beam. New technology advances in device power output will reduce this to a single emitter.

The receive optics were simple, as illustrated in Figure 12.11. Seven optical elements were used. The primary and secondary mirrors of the Cassegrain telescope relay the light through a bandpass filter, used to reject unwanted background signals. From the filter, the light is focused on a bifurcating mirror (mirror with a hole in it). Light within the communication detector field of view (FOV) is passed through the hole in the mirror and relayed to the communication/track detector. Optical energy at off-axis angles greater than the FOV of the communication/track detector, as during acquisition, is relayed to the wide FOV acquisition detector.

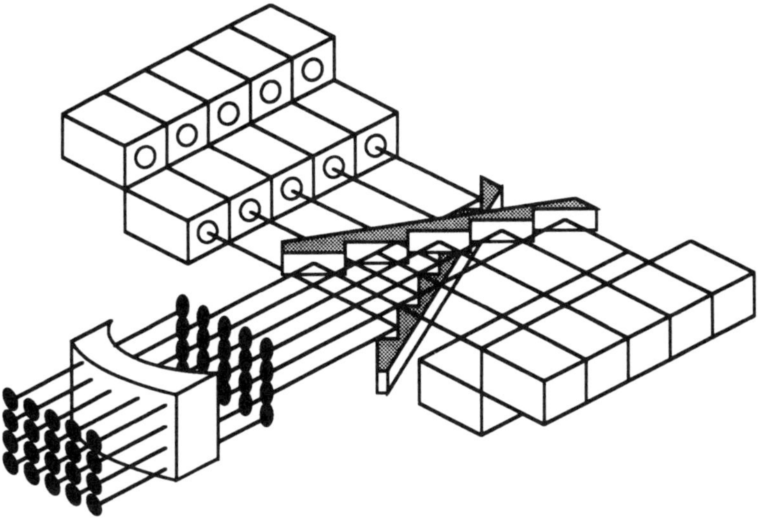

Figure 12.10 1-Mbps short-range crosslink transmitter concept. (Copyright IEEE, 1992. From [4].)

Both detectors are *quadrant avalanche photodiodes* (QAPD). An end-of-life link margin summary is shown in Table 12.3.

A three-terminal suite required for the satellite was estimated to weigh 28 lb and consume no more than 58W. A single terminal could be considered to weigh about 9 lb, although the weight would be slightly more since the suite of three optical heads shared a common electronics.

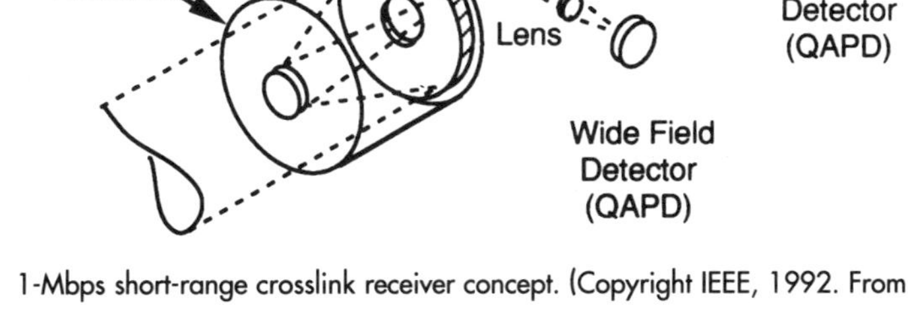

Figure 12.11 1-Mbps short-range crosslink receiver concept. (Copyright IEEE, 1992. From [4].)

Table 12.3
1-Mbps Short-Range Crosslink Link Margin Summary

Item	*Parameter*	*Value*
Laser peak transmit power	3 W	4.77 dbW
Optics efficiency	94%	−0.26 dB
Transmitter gain	500 μrad	78.42 dB
Free space loss	9000 km	−282.74 dB
Receive antenna gain	3 in	109.25 dB
Optics efficiency	78%	−1.05 dB
Received signal level	692 pW	−91.60 dBW
Required signal level	192 pW, 1 Mbps, $R = 1/2$, $k = 7$ convolutional Viterbi soft decision, Manchester, 1E-6 BER	−97.17 dBW
Margin		5.57 dB

12.1.3.2 A Commercial Short-Range Crosslink

A short-range crosslink for commercial internetting of a constellation providing global personal communications has also been conceptually configured. The system consisted of a constellation of satellites in several orbital planes with a number of satellites in each plane. The LEO satellites were required to communicate only with their nearest neighbor. Four crosslink terminals were to be mounted on each satellite to facilitate data transfer in each of the four directions. The maximum range required was 4,400 km. Separate transmit and receive apertures were again selected for isolation. The entire terminal was estimated to weigh less than 8 lb. Thus, four units would weigh 24 lb and consume about 50W. Each terminal was small, the dimensions being less than 6-in in diameter and 9-in high. (Figure 12.12).

12.1.3.3 Other Short-Range Applications

NASA has studied applications for small, lightweight laser communications terminals. NASA is seeking a solution to a class of user's small LEO satellites that do not possess the power and weight required to implement an RF link to a GEO satellite relay unless the science weight and power budgets are significantly reduced, an undesirable situation.

To implement a lightweight, low-power approach satisfactory to small satellite users, optical communications could be baselined. The system must be simple, small, and of low power. The data rate required would be approximately 1 Mbps and the range being from a LEO to a GEO. The GEO satellite in this application would be a small low-cost relay satellite. The scenario envisioned would be as shown in Figure 12.13. The optical link from the LEO to the GEO is 1 Mbps with a

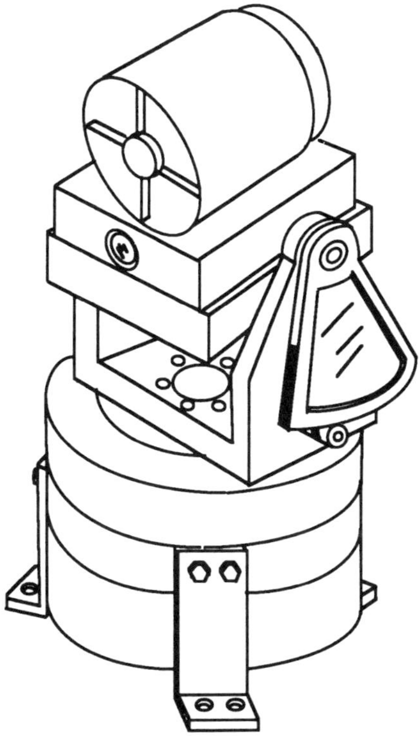

Figure 12.12 Commercial crosslink terminal concept.

return data rate of 1 Kbps. The downlink from the GEO to the ground station is K_a-band RF with the uplink at 1-Kbps S-band RF, which can penetrate clouds. Here again was a case of the apportionment of communication link tasks to both RF and optical, applying each to the task for which they were best suited.

As satellite communications evolve and become more sophisticated, increasing emphasis will be placed on use of optical crosslinks in the future.

12.2 SATELLITE TO AIR / GROUND

These applications, for the most part, are extensions of crosslinks since they represent the beginning or end of data flow that involves a crosslink. An example of this is where a data gathering LEO transmits the data to a GEO, which in turn transmits it to another GEO. The data must then be returned to the user on the ground for analysis. In addition, the ground has control of the spacecraft and issues commands via uplink. It is the uplink/downlink portions of the overall communications link shown in Figure 12.1 that are discussed in this section.

The uplink and downlink from a satellite to the ground or an aircraft require the passage of the signal through the atmosphere, so its many effects on the signal must be understood in some detail. Clouds, of course, can prevent laser communications with the ground or an aircraft. Mitigating approaches to cloud cover blockage do exist. The application of uplink/downlink segments operated in conjunction with crosslinks will be discussed.

12.2.1 Atmospheric Effects on Laser Signals

The atmosphere is an imperfect medium for transmission of laser communications signals. It is not homogeneous and further, it is dynamic in its behavior. The atmosphere can be considered as producing four effects on the passage of light. The first two are scintillation and absorption/scattering. In this atmospheric channel, minute variations in the index of refraction of air cause the beam to scintillate (or vary in amplitude and phase over time) and the beam to wander. In addition, atmospheric gas molecules and aerosols scatter and absorb the light in its passage. The effect is small in many instances, but at shallow elevation angles the losses increase rapidly and become extremely large. Third, clouds reflect, absorb, and temporally and spatially spread the incident energy. The losses due to certain clouds are very large, while for other cloud types the effects are small and easily overcome. Cloud statistics are a function of location and season, which are important factors in ground-site selection. Through multiple-site diversity, cloud cover

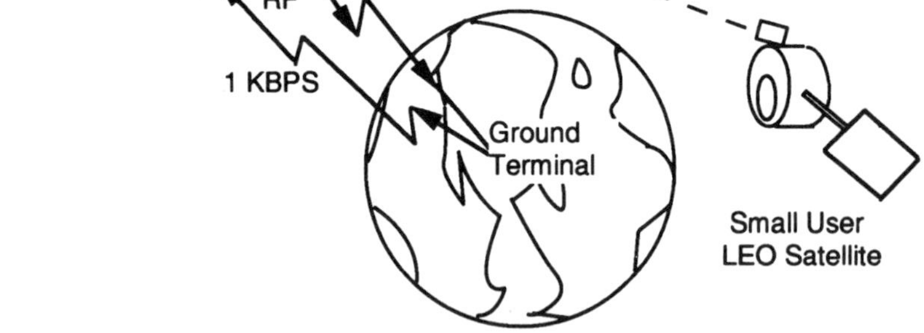

Figure 12.13 Lightweight, low-cost LEO to GEO 1-Mbps crosslink scenario.

effects can be mitigated. Finally, there are aircraft boundary layer effects due to the platform speed, which needs to be understood. The degradation in transmission due to turbulent flow over the aircraft window is discussed and the losses are described for subsonic as well as supersonic flow.

Assessing atmospheric effects is complex and the mathematics are far too lengthy to be included here. References are included that can be used by the reader to delve deeper into the physics of the atmospheric channel. The engineering of atmospheric laser links can be performed using [5, 6], which contain engineering approaches to quantifying the effects. Parametric analysis curves of laser communication links for satellite to aircraft and satellite to ground are included to give the reader an understanding of the effects the atmosphere has on the links.

12.2.1.1 Scintillation

Scintillation can be described as the destructive and constructive interference of optical waves caused by changes in index of refraction along the optical path. It is readily observed by the twinkling of stars. The changes are caused by minute variations in the air temperature, which in turn alter the localized index of refraction. These small localized variations in refraction, or blobs, act as lenses of various sizes that refract the light as shown in Figure 12.14. It is noted that the wavefront at the receive aperture can be characterized by either plane or spherical wavefronts. Spherical waves is the generic case and is the focus of the following discussion.

The fluctuation in the index of refraction that produces scintillation is termed the structure constant, C_n^2. The structure constant is a measure of the difference in the index of refraction "here," $n(r_1)$ and "there," $n(r_2)$. The structure constant has been determined by a number of different techniques including temperature measurements and radar [7–9]. A broad descriptor is required to characterize propagation effects for laser communication links.

The index of refraction varies not only between blobs, but also with time. Models of the structure constant should therefore be probabilistic in nature. A model was proposed for the structure constant, but it requires knowledge of the local wind speed at the various altitudes. A more general model that accounts for

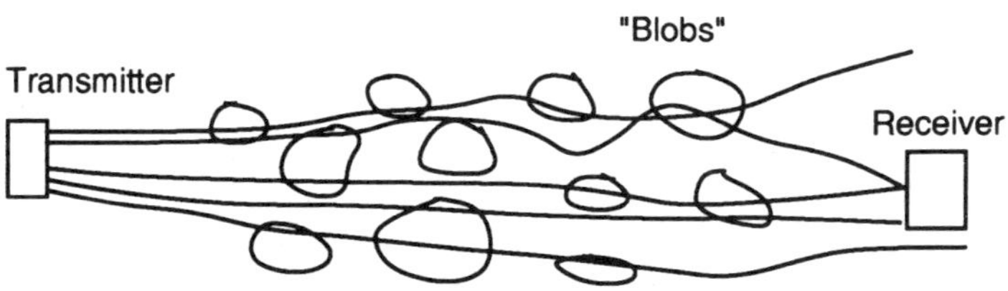

Figure 12.14 Atmospheric effects due to "blobs."

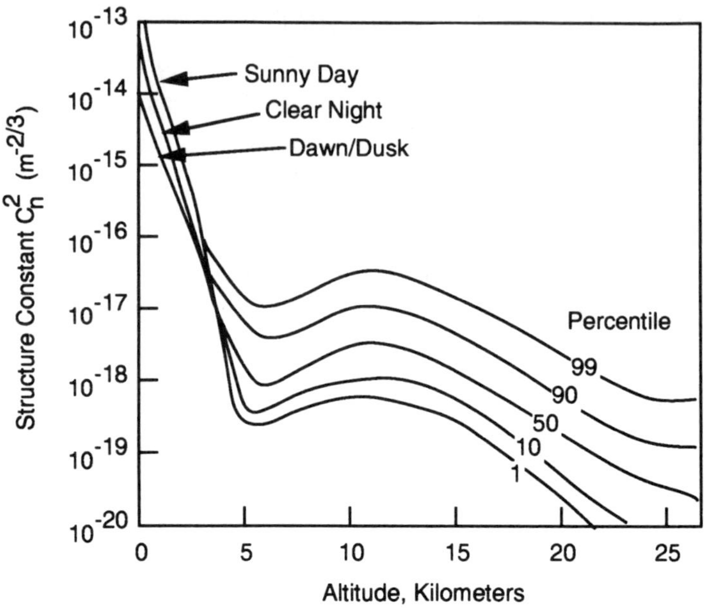

Figure 12.15 A Probabilistic structure constant model as a function of altitude. (After [10].)

the probabilistic nature of the wind was developed [10] and the probability curves of the structure constant as a function of altitude are plotted in Figure 12.15.

12.2.1.2 Absorption and Scattering

The propagation of a laser beam is affected by the atoms and molecules of the atmosphere. The first effect is absorption of photons by the atmosphere and the second is the scattering of the photons. Unlike scintillation, the losses due to absorption and scattering are a time-invariant process. The atmospheric transmission due to absorption and scattering is accurately modeled by the programs HITRAN, FASCODE, and LOWTRAN. These programs are standard references but are difficult to use in making parametric trades. The absorption coefficients of McClatchey [11] have been approximated [5]. Figure 12.16 is the approximation of the absorption coefficient versus altitude. Another important variable is wavelength. There are certain wavelengths where the absorption is extremely high. When picking a wavelength, it is important to determine the absorption for the specific wavelength chosen. For laser communications wavelengths of less than 1.1 μm, there is only one significant absorption band at about 0.9 μm where the absorption coefficient is slightly over 2 dB/km. The rest are below 0.5 dB/km.

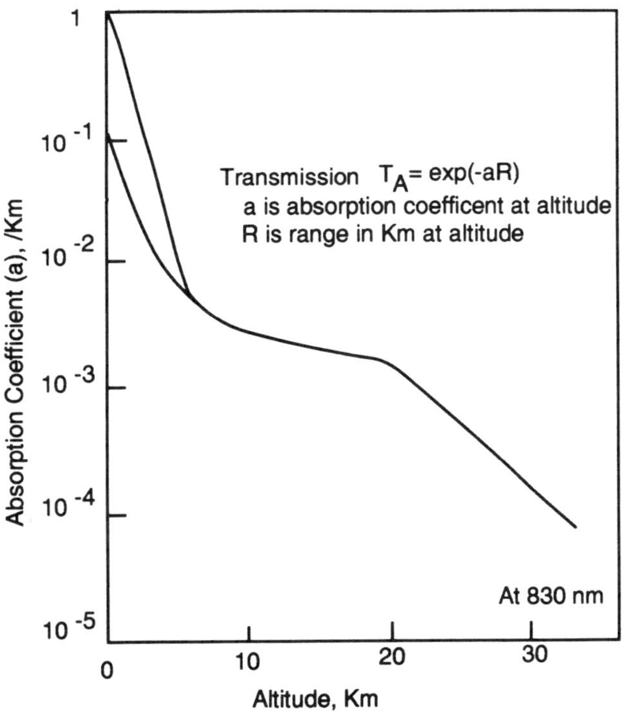

Figure 12.16 Absorption coefficient with altitude form 830 nm. (After [10].)

12.2.1.3 Atmospheric Channel Effects

The atmospheric channel characteristics, as noted in the previous discussion, fall into two categories: (1) fluctuations in the signal amplitude and phase, and (2) attenuation. Although the effect of attenuation can be overpowering in a link design, the fluctuations in amplitude and phase due to scintillation give rise to a number of important effects. Figure 12.17 illustrates the time variation of the signal and the relationship of the effects of scintillation. Statistically, it is important to define the variance of the signal fluctuations. It is also of interest to define the fade time below a specified fade depth and the associated fade frequency.

For acquisition and tracking, it is important to know the mean positive zero crossing rate, also called the apparent frequency or the quasi-frequency. For linear systems, the dynamic range is of great importance in the circuit design. Reference [5] gives the expressions required to calculate scintillation characteristics. There are four: (1) the log-normal amplitude distribution, (2) the variance of the log amplitude for a point receiver and a spherical wave, (3) the variance for log amplitude for

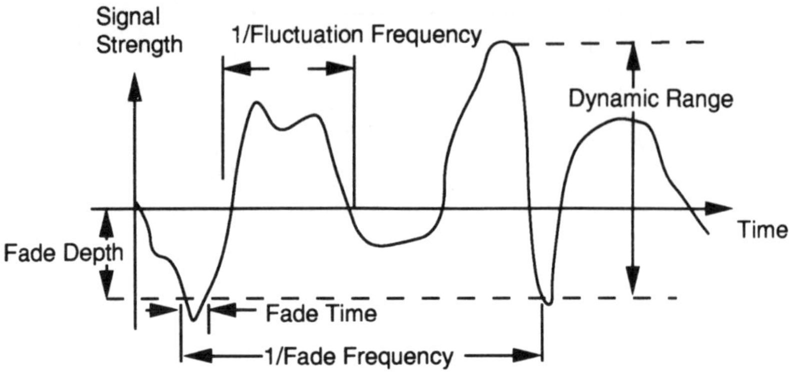

Figure 12.17 Atmospheric temporal fading channel characteristics. (After [10].)

a finite aperture, and (4) the probability of a fade. Reference [5] also includes the five equations for calculating the temporal characteristics of the scintillation: (1) the spectrum of the amplitude scintillation, (2) The quasi-frequency, (3) the fade and the surge frequencies, (4) the duration of a fade, and (5) the predominate frequency components of the angle of arrival fluctuations.

Beam spread and wander are additional atmospheric channel effects that can be quantified for engineering a link through the atmosphere. Reference [5] provides equations that can be used to quantify (1) the spherical coherence of the atmosphere (spatial coherence length), (2) long-term beamwidth, (3) short-term beamwidth, and (4) the angle of arrival fluctuations. If the beam spreading is considered, the effect at the transmitter is the greatest. Beam spreading consists of two components, a short-term effect (snapshot), and long-term effect (a time exposure). If the phase at the receiver is a primary concern, the dynamics immediately in front of the receiver are most important. A parameter of interest is the spatial coherence length. As defined in [6], the spatial coherence length, r_0, is related to the "aperture size of the atmosphere." An important consequence of this effect is that the signal to noise ratio (SNR) of a coherent system does not increase [6] as the aperture size becomes much larger than r_0.

The design of an air-path laser communications system generally includes a margin to overcome the fading environment. There are two approaches: (1) an exceedance approach, and (2) an expectation approach. These approaches differ fundamentally in that in the exceedance approach, the margin is assumed to be sufficient to overcome a fade depth of a given probability. The goal in the expectation approach is to design a system that, on the average, meets a certain performance level. One would then expect the expectation approach to be less demanding than the exceedance approach, and it is.

The particular approach used in designing the air link depends on the required signal criteria. The required signal for an acquisition link is based on the probability of detection and the false alarm rate, which is the single-pulse statistic. Therefore, the exceedance approach is used.

One the other hand, the required signal for a communication link is based on the bit error rate (BER), which is the average probability of error, so an expectation approach is used here. The reader is referred to [5, 6] for the equations for calculating these parameters.

Clouds effect a propagating beam in four different ways: absorption, reflectance, beam coherence, and pulse coherence. As with the atmospheric channel calculations, [6] contains equations useful for calculating these parameters given certain cloud characteristics. Figure 12.18 shows the cloud types and the approximate ranges of their occurrence. Note that above the 35,000-ft altitude, very little in the way of clouds exist. The impact of this in mitigating cloud cover effects on laser communications up/downlinks will be discussed subsequently.

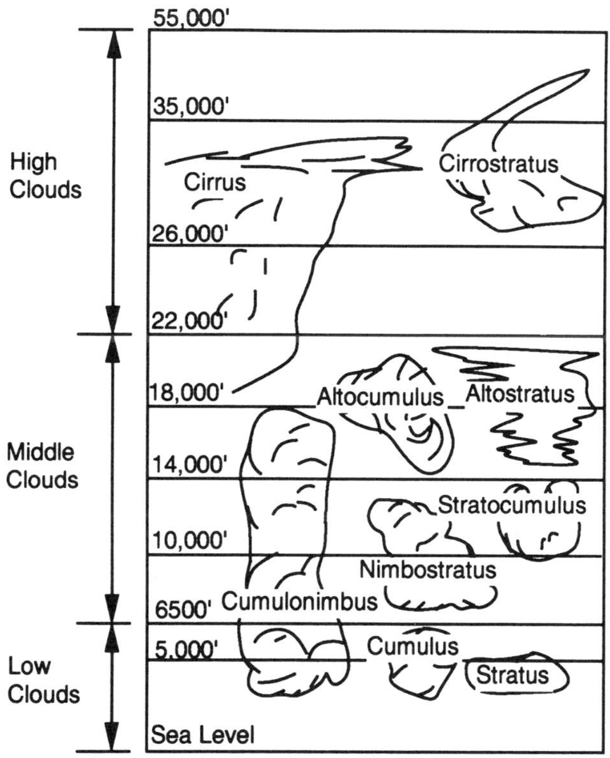

Figure 12.18 Cloud types. (After [6].)

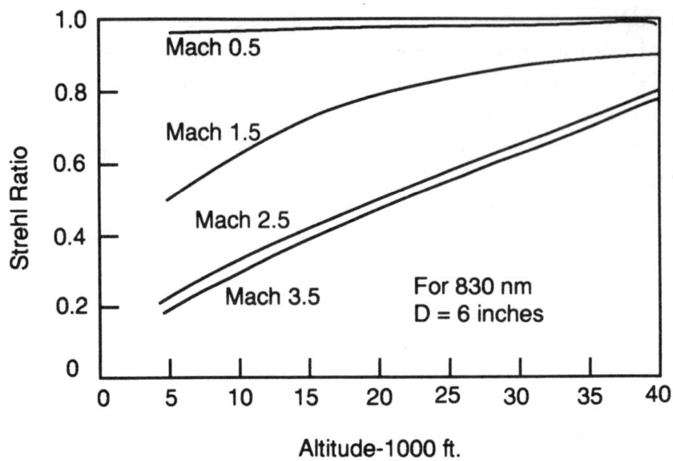

Figure 12.19 Strehl ratio versus altitude for several mach numbers. (After [6].)

12.2.1.4 Aircraft Boundary Layer Effects

A specialized case in propagating a laser beam through the atmosphere is where the transmitting terminal is in an aircraft. The fluctuations in density of the air that flows over the aircraft skin or window creates a thin turbulent boundary layer. This turbulent boundary layer causes the phase of the propagating beam to be distorted. This results in a loss of on-axis gain, a measure of which is the Strehl ratio. The Strehl ratio was discussed in Chapter 3 of this book.

A number of theories have been proposed to explain boundary layer effects. Two are the Fisher-Wolter [12, 13] and Kelsall-D'Amato [14, 15], and they approximate the boundary layer profile based on correlations for such boundary layer parameters as thickness, displacement thickness, and local Reynolds number. These two theories are valid and accurate only for the subsonic region (below the speed of sound at the pressure-altitude). Fivel [16] expanded the Kelsall-D'Amato theory to cover both the subsonic and supersonic regimes. The application of laser communications has been postulated for hypersonic vehicles, for example, the *National Aerospace Plane* (NASP) since optical frequencies do not suffer the communications "blackout" experienced with RF communications during hypersonic reentry of spacecraft. The effect of certain parameters on the Strehl ratio for a transmitted beam from an aircraft is shown in Figure 12.19 for the altitude effect and Figure 12.20 for the mach number effect for several window sizes.

12.2.1.5 Parametric Curves of Atmospheric Effects

This section presents the results of atmospheric calculations adapted from [6] for a number of parameters of satellite to aircraft and satellite to ground links. The

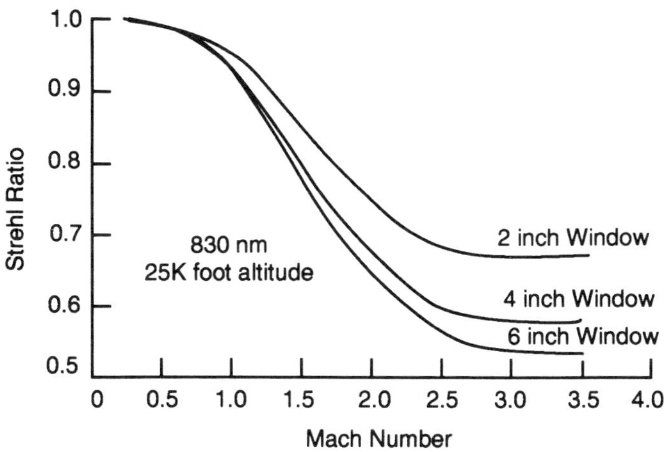

Figure 12.20 Strehl ratio versus mach number for several window sizes. (After [6].)

scenarios chosen are appropriate for the uplink/downlink missions, which complete the communications architecture that started with the crosslinks of section 12.1.

The altitude of the satellite does not affect the atmospheric losses since significant atmosphere does not extend beyond about 70 nmi. The altitude of the aircraft is assumed to be 30,000 ft, which is a normal cruising altitude. The atmosphere above the ground station is modeled the same, independent of its altitude above sea level. It should be noted that the curves shown here are probabilistic in nature and are highly scenario-dependent.

Trades of scintillation and absorption are shown. The common parameters are a 99% wind speed probability, a 10^{-6} associated communication bit error probability, and an 8-in receive aperture.

Figure 12.21 depicts atmospheric absorption for clear sunny day and hazy day conditions. Note the rapid increase in absorption for low-elevation angles where the signal has a long atmospheric path. Note also that a longer wavelength passes through the atmosphere with less loss than shorter wavelengths. Figure 12.22 is a set of curves for both ground station and aircraft path absorption with elevation angle. A hazy day is not shown for the aircraft since haze is a low-altitude effect. As may be expected, the loss for an aircraft is small. The aircraft link loss is less than 1 dB above 20-deg elevation angles while the ground station loss is greater than 2 dB.

The results of computing turbulence effects such as scintillation, beam spread, and spatial coherence loss are shown graphically in figures that follow. The required expectation margin, also known as frustration margin, is shown in Figure 12.23 as a function of elevation angle for a ground station and an aircraft terminal. At elevation angles greater than 20 deg, the frustration margin going to a ground

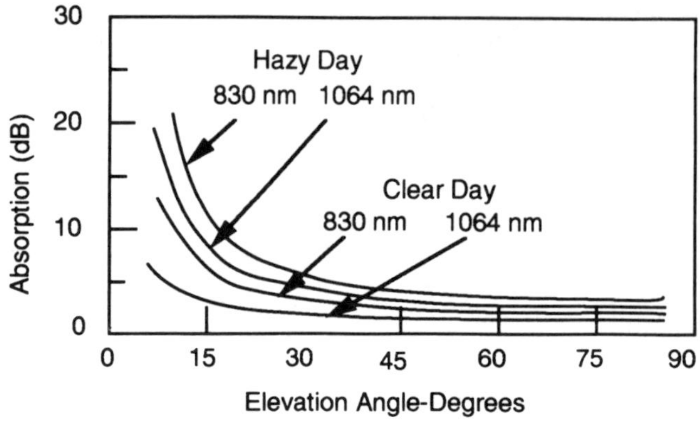

Figure 12.21 Clear / hazy day absorption for satellite to ground links. (After [6].)

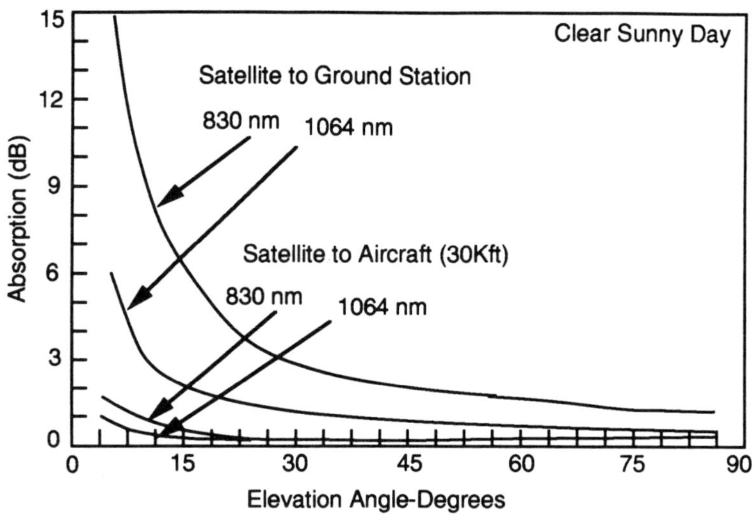

Figure 12.22 Absorption for satellite to air / ground links as a function of elevation angle. (After [6].)

station is much greater than that going to an aircraft from a satellite. For angles less than 20 deg, the frustration margin to an aircraft becomes larger than that to a ground station. At low-elevation angles, the beam propagating to an aircraft travels through the tropopause of the atmosphere farther than to a ground station. Therefore, the margin requirement is greater since the tropopause is a more turbulent region.

Examining the wavelength aspect of the curves for the ground station, one expects the order to be reversed with the longer wavelengths having smaller losses

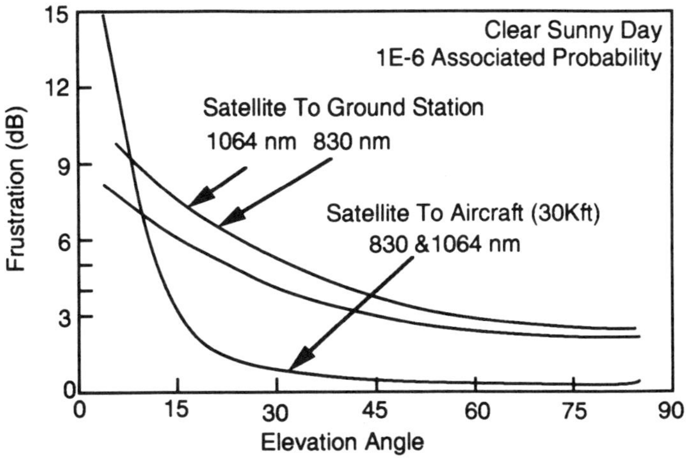

Figure 12.23 Frustration versus elevation angle for satellite to air / ground links. (After [6].)

than the shorter wavelengths. There are two effects to consider. First, limiting amplitude scintillation is when the variance of the log-normal distribution is a maximum. Therein, the variance is equal for any wavelength and the frustration margin requirement is the same. The second effect is aperture averaging. Aperture averaging reduces the variance for all wavelengths, but will have more of an effect at shorter wavelengths than at longer wavelengths, as may be seen in Figure 12.23. The satellite to aircraft frustration margin with respect to wavelength is not amplitude scintillation–limited at large elevation angles, but does become limited at elevation angles below 10 deg.

Another parameter is dynamic range, which is the sum of the exceedance surge height and the exceedance depth fade. This dynamic range is important since this is the range of the power levels on the receiver's detector caused by the dynamic temporal behavior of the atmosphere. Figure 12.24 is a set of curves of the dynamic range versus elevation angle for various wavelengths from a satellite to a ground station and to an aircraft.

For heterodyne and fiber receiver systems, the spatial coherence across the receiver aperture is extremely critical. As stated previously, making the receiver aperture much bigger than the coherence length does not improve the SNR for heterodyne systems.

12.2.2 Satellite to Ground Laser Communications

As mentioned earlier, as data rates increase into the multigigabit per second regime, RF communications technology is having difficulty coping with a number of problems. First, the high data rates force the spacecraft/ground terminal builder to

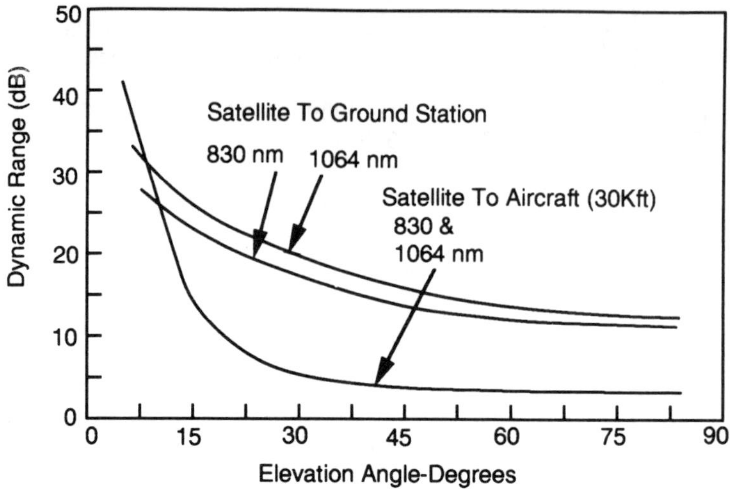

Figure 12.24 Dynamic range versus elevation angle for satellite to air / ground links. (After [6].)

use larger antennas (smaller beam divergence) and/or greater power in the transmitter to contend with the high-data-rate link demands. Another problem is frequency congestion, which is as serious as the demand for larger hardware and more power. The larger antenna/more power automatically implies larger spacecraft with larger power systems. The ground antennas are already large. As power increases, the sidelobes become very energetic and the intercept problem intensifies. Also, the RF uplink receiver can be jammed more readily than the downlink receiver, so this is another problem, particularly for a foreign ground station. Laser communications provides users with a viable solution for their high-data-rate congestion and security concerns. The laser communications solution works well also with the global military on the move scenario as will be discussed.

12.2.2.1 Global Military on the Move Scenario

This was briefly discussed in the context of the crosslinks needed for implementation. The application is long-haul communication trunks between CONUS and theater operations. The need was demonstrated in Desert Storm, where copious quantities of data were urgently needed by the theater commanders and by the troops in the field, on the move. In this type of scenario, high-quality video conferencing and sensor imagery transfer to and from airborne and terrestrial mobile command and control operations is needed. The data is perishable and of little use if not received in near real time due to the rapidly changing nature of the conflict. There is a real need for transmission of data at rates up to 25 Mbps to the troops on the move. Consideration has been given to mobile terminals where data is received by the terminal while moving.

The data would come via a GEO to a LEO satellite and be received by a mobile ground terminal. The downlink data rate would be 25 Mbps over a 300-nmi channel. The conceptual design for this application starts with the LEO terminal.

Both the uplink and the downlink are subject to atmospheric effects. Fades are experienced in the downlink, which may be somewhat ameliorated by an increase in aperture size (aperture averaging). One of the major effects is beam wander. It is not much of a problem on the downlink, but has a major effect on the uplink due to leveraging by the atmosphere that is close to the terminal. As a consequence, the uplink beam must be wide enough to encompass the expected beam wander (steering) while retaining enough energy for the LEO to track on.

With small-scale diversity, it is expected that an availability of over 90% can be realized. The telescope in the mobile terminal would view the LEO satellite through a window in the top of the truck. If a wide field of regard is required, a dual-window concept may be used. The baseplate is isolated. Moving receiver stabilization is a critical issue since the truck may be moving while receiving data.

To operate under these dynamic conditions, three degrees of isolation are required. The coarsest isolation is the truck mount, which uses a Turner (air-ride) suspension for removal of the major perturbations. The telescope gimbal is the next degree of isolation, acting to attenuate frequencies on the order of tens of hertz. A two-axis rate gyro such as that used on the M-1 tank for "shoot on the run" gun stabilization would sense the telescope motion and provide a correction signal for stabilizing the telescope. The final and fastest correction would be by sensing the tracked image motion and correction of this motion by a *fast steering mirror* (FSM) operating in two axes. Such a FSM can provide correction up to about 4 KHz. It appears that the moving receiver concept is valid for reasonable roads. It would not work in rough terrain, but the truck could be stopped and communications initiated in a few minutes. With the whole area socked in by clouds, the mobile command would revert to currently available low-data-rate RF communications.

Somewhat larger mobile terminals mounted on larger trucks would be required to increase the data rate beyond the 25-Mbps rate. Truck or trailer-mounted receivers are well within the current technology. Larger truck/trailer-mounted terminals are discussed further in the next subsection on high-data-rate communications.

12.2.2.2 High-Data-Rate Downlink

To continue what was started earlier in the chapter, describing a 4-Gbps crosslink design based on today's hardware, the downlink/uplink portion of the architecture will be discussed. A 4-Gbps downlink and a 100-Kbps command and control uplink is postulated. This asymmetric link is the user end of the architecture where data is gathered at high data rates by an aircraft or spacecraft, transferred through a GEO crosslink to another GEO and then to the ground. The command and control link is of lower data rate in the opposite direction.

The ground terminal is very similar to the GEO terminal for the 1-Gbps receive communication channels. The 4-Gbps is achieved by channelizing the four 1-Gbps channels. The ground telescope aperture is 24 to 48 in. Ground terminal size can be adjusted to compensate for the atmospheric losses, which will be on the order of 3 to 6 dB for midlatitude ground stations. The terminal can either be fixed,

Figure 12.25 Mobile ground receiver terminal. (Courtesy of Contraves, Inc.)

similar to a smaller astronomical observatory, or mobile as mounted on a flat-bed trailer. A mobile terminal may be used that carries a 36-in aperture telescope, which is exemplified by the five mobile ground terminals built by Contraves for NASA-GSFC several years ago and is illustrated in Figure 12.25.

12.2.2.3 Cloud Cover Mitigation

It is well-known that there is little correlation between cloud cover patterns across longer distances on a continent. This is evident from watching the weather on TV where moving cloud patterns are shown for major portions of the globe. This lack of correlation is global in its extent. For a high-data-rate mission downlinking of data, this effect can be used to an advantage. Studies performed early on for submarine laser communications established the viability of this approach. Since the high-data-rate data must be returned to CONUS, ground stations anywhere in CONUS could be implemented.

A study [17] indicates that of the 10 best of 289 CONUS sites, the probability of a *cloud free line of sight* (PCFLOS) does not exceed 0.752. A much higher PCFLOS, on the order of $> 95\%$ is required. Multiple ground terminals (diversity) increase the PCFLOS, enabling reception for a satellite with a high availability. Two types of diversity exist: small-scale diversity (local), and large-scale diversity (thousands of miles). For small-scale diversity, where the weather is somewhat correlated and distances between ground stations can be several hundred miles, three terminals in the southwest yield a PCFLOS of 0.921. The three terminals are located in Tucson, AZ, Roswell, NM, and Abilene, TX.

For larger scale diversity where terminals can be thousands of miles distant, the weather is uncorrelated and the availability for the best of seven NASA sites results in the following availabilities, as illustrated in Table 12.4. The performance of the best of 299 CONUS locations (not NASA locations) yields a high availability. The best four location diversity value for Phoenix, Arizona, Sacramento, California, Russel, Kansas, and Tampa, Florida is 0.989 [17].

Large-scale diversity produces the best availability, at over 98%. For this large-scale diversity and a centralized user at one location, it is necessary to collect

Table 12.4
Cloud Free Line Of Sight Availability

Two Terminals	*Three Terminals*	*Four Terminals*
Dagget, CA Roswell, NM	Dagget, CA Roswell, NM Vero Beach, FL	Dagget, CA Roswell, NM Vero Beach, FL Huntsville, AL
0.912	0.962	0.981

From [17].

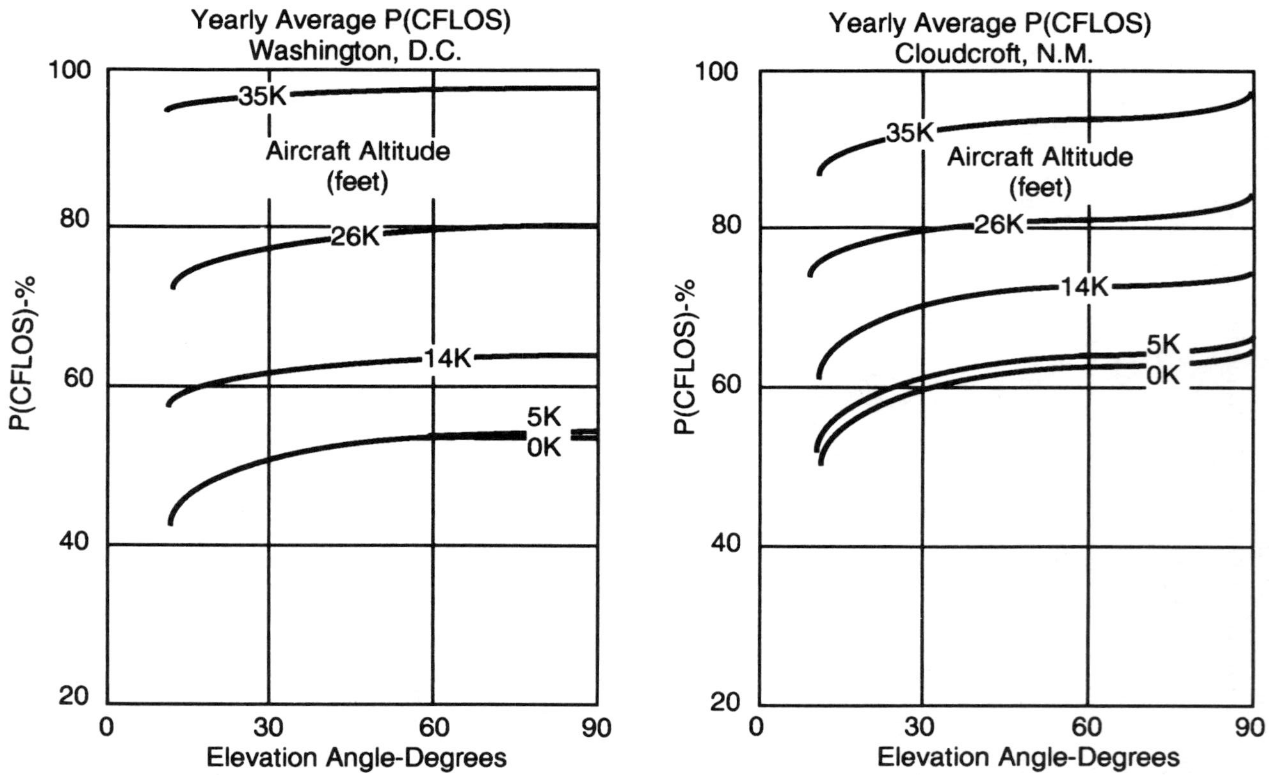

Figure 12.26 Probability of a cloud free line of sight as a function of elevation angle for several flight altitudes. (Courtesy of McDonnell Douglas Aerospace.)

the high-data-rate data at one of these stations and relay it to the user. To do this, a system connecting the ground terminals is necessary. The system must also handle the high data rates of the links. Distribution is practical via a *fiber-optic network* (FON). A commercial FON was considered by NASA (1990) for 600-Mbps data distribution. It is also a possibility that a long-term lease for such a network could be implemented.

Use of an aircraft receiver terminal is an appealing approach to cloud cover mitigation and also a low-cost option. An aircraft relay between the satellite downlink and the user on the ground provides extremely high availability. The curves of Figure 12.26 illustrate the average PCFLOS for an aircraft flying at various altitudes as a function of elevation angle for two locations, Cloudcroft, NM and Washington, DC. It should be noted from the curves that an aircraft flying at 35,000 ft has an availability of well over 95% for Washington and about 95% for Cloudcroft. This availability is obtained without active maneuvering to avoid high-altitude clouds.

The aircraft relay can offer advantages that include very little atmospheric effect. It can operate anywhere except in denied airspace and can be dispatched to intercept LEO ground tracks for LEO data dumps. Use of wide-body subsonic aircraft, already in military inventory, with simple terminal installation and the subsequent transmission to the ground at short range is easily implemented using current RF technology.

The implementation of an aircraft link was completed many years ago with NASA's Kuiper Observatory, which was a 36-in aperture telescope mounted in a C-141 aircraft. The telescope, built by Fecker (now Contraves, Inc.), viewed the sky through an open port to avoid use of an infrared window and tracked to an accuracy of an arc second. Large windows have been installed in military aircraft for years and they maintained high wavefront quality at subsonic speeds (also at supersonic speeds as was accomplished in photographic tests with the X-15 supersonic rocket-propelled aircraft). A 30-in window left over from a reentry signature program was installed in a C-135 in the AFTS program at the end of the 1970s [18] and it maintained near diffraction-limited quality in flight.

12.2.2.4 Other Satellite to Aircraft Applications

Some unusual applications for satellite to aircraft laser communications have surfaced in past years. One of the more interesting is Report Back where strategic airborne forces such as the strategic bomber fleet or a large group of cruise missiles would uplink their status at each leg of the flight. To accomplish this, a miniature terminal was conceptually designed that used a high peak power Nd:YAG Q-switched laser to advantage. The required data rate was low (hundreds to a few thousands of bits per second). The receiver on a GEO satellite received a hundred or more messages simultaneously from differing locations on Earth and decoded them.

A major feature of this system was the ability to reject jammers. This was accomplished by sophisticated electronics that rejects messages that do not possess the proper address. The conceptual GEO receiver was a gimbaled 12-in aperture wide-field folded Schmidt design. The detector was an array of $n \times n$ light pipes with an APD for each light pipe. The instantaneous FOV was adequate to view major portions of a continent.

Desert Storm, with the continuous SCUD missile launch site problem and the short response time necessary to eliminate them, initiated another interesting potential application for laser communications. When a missile launch was detected by an early warning satellite, the coordinates of the launch would be downlinked directly to a strike aircraft, which would proceed at high speed to the target area. The advantage of this approach is the very short response time. This approach required that the strike aircraft be positioned in sectors so that the travel time to the target would be on the order of a few minutes, short enough to prevent the enemy forces from tearing down the missile launcher and moving. The terminal required in the aircraft was extremely compact, the receive aperture requirement being less than 3 in. The aircraft receiver terminal would open-loop point based on aircraft attitude data and *global positioning system* (GPS) position coordinates. The instantaneous FOV of the receiver was adequate to encompass the error. The terminal would not transmit in the simplest version, but rather act on the received information. The terminal was small enough to be mounted under the canopy aft of the pilot.

The possibility of equipping ground troops with satellite to laser communicators for emergency purposes has been studied and is feasible, again subject to cloud cover restrictions. An individual unit would be compact and setup could be accomplished in minutes. There is an RF analogy currently in development and it is not too far a stretch to visualize a two-way laser terminal, particularly with the advent of the battery-powered miniature lasers currently available. Hand-held lasers are now available that output a Joule per pulse. For low data rates, little in the way of average power or a large collection aperture is required. The terminal would be analogous to the strike aircraft receiver discussed just previously and could be open-loop pointed well enough with a calculation based on GPS coordinates and compass/level information, just as a television installer aims a satellite dish.

12.3 SUBMARINE LASER COMMUNICATIONS

One of the more interesting potential applications of laser communications to appear is communications between a satellite and a submerged submarine operating at depth. Initial interest was shown in *submarine laser communications* (SLC) about 1970 and continued until 1993. The *Defense Advanced Research Projects*

Agency (DARPA), *Office of Naval Research* (ONR), and *Office of Naval Technologyt* (ONT) have funded SLC for nearly two decades.

12.3.1 SLC Background

A short history of SLC is appropriate to understand how the technology evolved. DARPA funded the work, which was done by the Naval Ocean Systems Center and several contractors. The first major effort was Project OSCAR (optical submarine communications by aerospace relay). The contractor was GTE Sylvania. The first aircraft to submarine tests were conducted in 1981. A frequency-doubled Nd:YAG laser (0.532 μm) mounted on an aircraft flying at 12,200m was aimed through clouds and ocean water to a submarine at depth. The submarine received the signal in a specially designed receiver and decoded the message. The results were encouraging and indicated further work would be appropriate. The frequency-doubled Nd:YAG approach was designated the "blue-green" approach.

In 1983, the effort was concentrated on two receivers, one based on quartz birefringence and the other on cadmium sulfide birefringence, both for the 0.532 μm frequency-doubled Nd:YAG laser wavelength. Study efforts included work on technologies, aircraft experiments, laser-satellite and mirror-satellite approaches. The use of ground-based laser beams reflected from mirror satellites was dropped in favor of a laser transmitter mounted in the satellite. An interesting fact of SLC, which may or may not be obvious, is that the laser wavelength and the filter wavelength must coincide. The designer of the SLC approach must come up with a very narrowband filter/laser combination, which also operates in a wavelength where seawater transmits. Not an easy task. DARPA and the Navy in recent years have solicited research into filter and laser combinations

In the mid-1980s, emphasis was placed on the development of a "blue" laser/filter combination. In certain types of seawater, a blue wavelength will penetrate deeper than the "blue-green" wavelength. The combination selected for development was a *xenon chloride* (XeCl) gas laser working with a cesium atomic resonance filter receiver at 755 nm. The receiver was referred to by the acronym ARFORM (*atomic resonance filter optical receiver module*). The XeCl laser underwent life tests with over a billion shots (pulses) demonstrated, but the drawback of this laser was its large size and weight. It needed a supply of gas to operate, was acoustically noisy, had a wear-out mechanism in the anode/cathode, and produced a very corrosive exhaust, all nontrivial design problems for a long-life satellite.

In the meantime, a blue-green transmitter and receiver was tested and in 1986 a P-3C aircraft communicated with an operational submarine operating under the icepack [19, 20]. The blue system was demonstrated in 1988 under Project Y-Blue, where a blue laser was installed in a Navy P-3 and transmitted messages to surface receivers and a submerged submarine, which carried a cesium ARFORM receiver.

A project, designated TALC, which stands for tactical airborne laser communications, was completed in 1991 and demonstrated two-way laser communications between an aircraft and a submarine. Tactical or attack submarines would be the submarine type for communications. The downlink for TALC was blue-green, a frequency-doubled Nd:YAG laser. The receiver was a previously demonstrated quartz birefringence filter receiver. The uplink used a solid-state blue laser and the receiver was equipped with a cesium atomic resonance filter. The test was a success [20].

The idea of using synchronous satellites was abandoned in favor of using aircraft or smaller LEO satellites, due primarily to cost. The technology for mounting a blue-green solid-state laser in a geosynchronous satellite was well-developed. Small satellites, although a number would be required, would still be more cost-effective than use of large geosynchronous satellites.

The most recent SLC activity, an advanced technology demonstration, centered around the demonstration of a blue-green downlink and blue uplink to a satellite in LEO. The project fell prey to deep budget cuts in 1993.

12.3.2 Operational Considerations

Submarines currently must surface to communicate. They must at least come to a depth that permits them to trail a wire enabling use of RF signals for communication, the most common being *very low frequency* (VLF), which is rapidly attenuated by sea water. Radio frequencies do not penetrate seawater to any depth excepting *extremely low frequencies* (ELF) where data rate is but a few bits per second. There are two types of antennas used with VLF, the floating wire type and the loop antenna. The loop may be carried on a retractable mast on the submarine's sail or it may be towed behind the submarine on a cable that also carries the received signal down to the submarine. There must be about 100m of floating wire at the surface; therefore, the deeper the submarine wishes to be, the more wire must be streamed. The wire must be lined up with the (great circle) azimuth to the transmitting station. The use of a mast-mounted loop antenna requires the submarine to be very near to the surface. With the mast extended and the loop just below the surface, the submarine is vulnerable to detection and might even collide with a surface ship. The near surfacing, of course, exposes the submarine to detection, which implies sub loss to a sub captain. The ballistic missile submarine fleet depends on stealth for survival and the original concept for SLC involved only the transmission of nuclear control orders to the force. Surfacing has been equated with detection and loss ever since World War II and this is not likely to change. The onshore transmission facilities are not expected to survive more than a few minutes in the event of a nuclear surprise attack.

The thought of two-way communications with a sub is appealing to fleet commanders since they can receive reports from the submarine and direct the submarine's actions. The submarine captain, on the other hand, cannot afford to compromise his or her position so he or she is very reluctant to implement two-way communications.

12.3.3 Technologies

12.3.3.1 Seawater Characteristics

Referring to Figure 12.27, it may be seen that there is a "window" of transmission through seawater in the blue-green (625 THz) of the visible region. The attenuation coefficient curve illustrated in Figure 12.27 shows that a minimum attenuation in this "window" that is about 80 dB lower attenuation than that which is only approached at the high and low extremes of the electromagnetic spectrum.

There are a number of types of seawater, varying considerably in clarity (about a factor of 3). The types, designated Jerlov followed by a numeral, are named after a scientist who did a great deal of work in studying seawater characteristics. The transmission of these types are illustrated in Figure 12.28 for blue-green wavelength at 0.532 micrometers. Note that the Type III is the most transmissive and is a type of very clear water found near the North and South

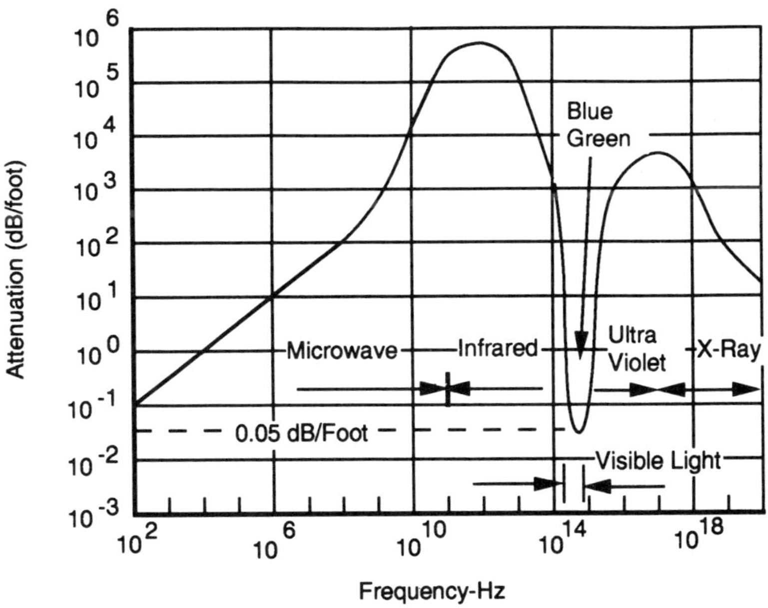

Figure 12.27 Blue-green seawater window.

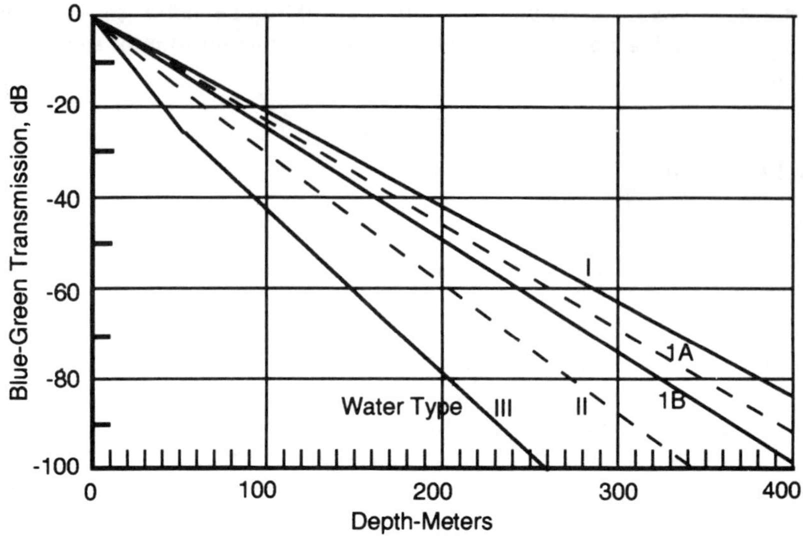

Figure 12.28 Seawater transmission of water types as a function of depth.

Poles. Its best transmission is more in the blue of the spectrum than in the blue-green. The blue wavelengths are best for cold water (North/Arctic) regions and provide deeper penetration in this type of water than the blue-green. The blue-green is good for temperate regions where plankton/microorganisms are abundant.

Another characteristic of seawater is the spreading of a laser pulse both temporally and spatially. When a short-pulse-length narrow laser beam enters the surface of the ocean, it immediately begins to spread in size as it penetrates the depths (multipath scattering). Something else happens also due to this multipath scattering. The short well-defined pulse starts to spread out also. What started as a 15-ns pulse length at the surface, gradually becomes a many-microsecond-long pulse at submarine depth. It should be intuitively obvious that with a many-microsecond-long pulse that signaling rate would be limited, which is the case. In order to use pulse interval modulation (PIM), for example, it may be remembered that a number of time slots must be created in which a pulse may located to represent the message characters. The requirement that the slot must be many microseconds long indicates that the maximum signaling rate will be on the order of hundreds of bits per second, which it is. If the submarine stayed within tens of meters of the surface, the maximum rate could be much higher. The spreading does occur in air also, but to a much lesser extent, so it is not a driver for SLC system design.

12.3.3.2 Clouds

A SLC system would be of little value if cloud presence prevented communications. Clouds are present everywhere in the world and are of many types. The attenuation of visible light is intuitively known to everyone since one cannot see though dense clouds. A laser beam cannot penetrate a cloud directly since the attenuation is extremely high. An interesting phenomenon occurs, however, when a laser beam from a SLC satellite or aircraft hits a cloud. Due to multipath scattering, the attenuation of the cloud is not the only factor contributing to the passage of the signal through the cloud. The apparent signal loss is much less than would be predicted from the attenuation alone. The multipath scattering spreads the beam greatly while preserving much of the signal so that not nearly as much link margin is required to penetrate clouds as might be expected based on the attenuation of a beam straight through the cloud This scattering phenomenon makes communications through clouds achievable. Cloud loss can range from about 4 to 14 dB [21] This effect is illustrated in Figure 12.29. Also to be noted is the additional spreading of the signal due to the seawater. Note that the submarine receiver must have a very broad FOV to pick up the signal.

12.3.3.3 Receiver Requirements

Since the Sun may also be shining in the same FOV and at the same wavelength (visible region), the only way to detect the laser signal out of the combined signal which includes sunlight is to assign a very narrow wavelength bandpass to the submarine receiver. This receiver requires a broad FOV, a steradian if possible; a very narrow bandpass, on the order of a couple of angstroms to reject the sunlight; a filter bandpass, which is centered on the laser wavelength; a filter, which responds rapidly enough for the data rate; and a detector, which is responsive at the required wavelength and speed

12.3.3.4 Receiver Developments

The detector for SLC receivers has been the photomultiplier due to its sensitivity at the blue and blue-green wavelengths and the fact that it has built into it a very nearly noiseless amplifier of high gain (10^5 to over 10^6). Attention has been paid to optimizing the quantum efficiency of the photocathode for the required SLC wavelength.

A major receiver development was the submarine laser communication receiver (SLCR), which operated in the blue-green portion of the spectrum at 0.532 μm. Figure 12.30 is a photo of one manufacturer's version of a SLCR module. The receiver used a quartz birefringence filter, which is made up of a number of quartz polarization plates that produce a narrowband filter with a large FOV. Several of

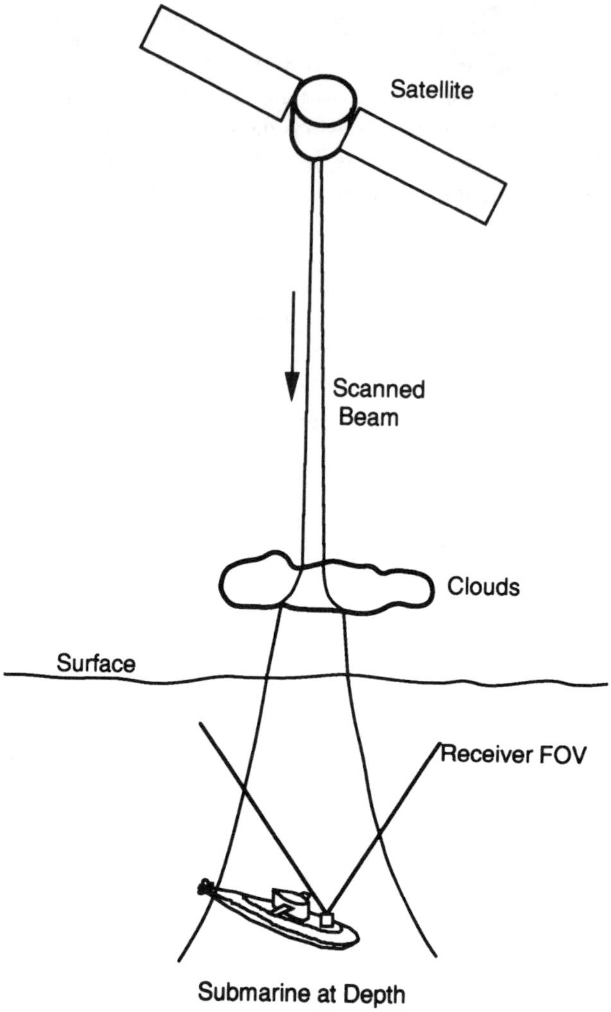

Figure 12.29 Propagation through water and clouds.

these receivers comprised a submarine shipset. The signal flow was arranged to sum the outputs of the receivers. Several receivers were used to collect enough signal to produce an adequate SNR. The shipset was designed to operate with a geosynchronous satellite-mounted transmitter. For a LEO satellite, a single receiver would suffice with a reasonable laser power due to the much shorter range of a LEO over a GEO satellite transmitter.

Figure 12.30 Submarine laser communications receiver module. (Courtesy of McDonnell Douglas Aerospace.)

Another contractor (Rockwell), produced another type of blue-green (0.532 μm) receiver that used a *cadmium sulfide* (CdS) *dispersive birefringence filter* (DBF) in conjunction with a *photomultiplier tube* (PMT) detector. The CdS filter shared difficulty of manufacture with the quartz birefringence filter, but toward the end of the filter development the process had been pretty well refined. The CdS filter was manufacturable in only a 1-in size so that many dozens of filter modules were required to achieve the collection area required for a GEO SLC receiver. Each filter module had its own PMT and the outputs of the entire array were summed to provide the signal. An optical filter module is shown in Figure 12.31. One of these filter modules was used in a test where a blue-green laser transmitted a signal

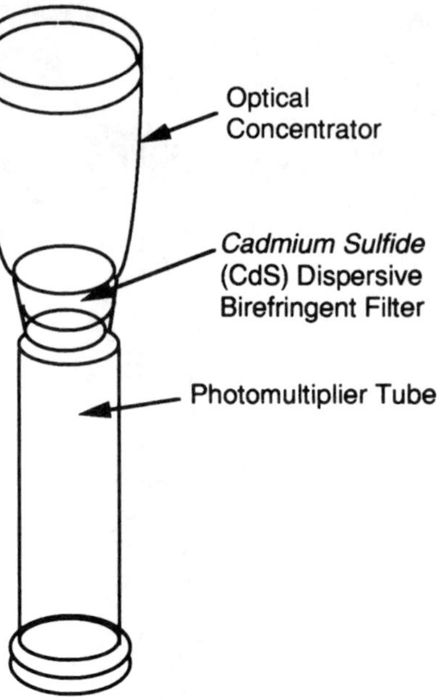

Figure 12.31 Dispersive birefringent optical module.

through clouds to a lower altitude aircraft to determine the pulse spreading characteristics of different cloud types.

The development of the atomic resonance filter for the blue (455 and 459 nm) approach was accomplished by several contractors. The operation of the ARFORM, which is based on absorption-fluorescence in cesium vapor, is illustrated in Figure 12.32. The blue light from the satellite passes into the cell after passing through a conventional filter that blocks red light while passing blue light. The blue light is absorbed by hot cesium vapor in a flat cell that is oven controlled. The cesium vapor then fluoresces in the red and the red light escapes the cell through a red filter which blocks the passage of the blue light. The red light that escapes the cell is collected by the photocathode of one or more photomultiplier tubes where it is converted to a summed electrical signal. The "integrating sphere" nature of the cell creates an ability to accept blue light from a very wide FOV. Subsequent to this earlier development, very small cesium cells were built, some as small as few inches in aperture, which would be useful for the receiver of a sub to aircraft or sub to satellite uplink. This type of filter was tested in the field several times, the latest being during the TALC exercise. There are other atomic resonance possibilities based on other materials including rubidium (421 nm), rubidium (532 nm), thallium-cesium (535 nm), potassium (532 nm), and magnesium (518 nm).

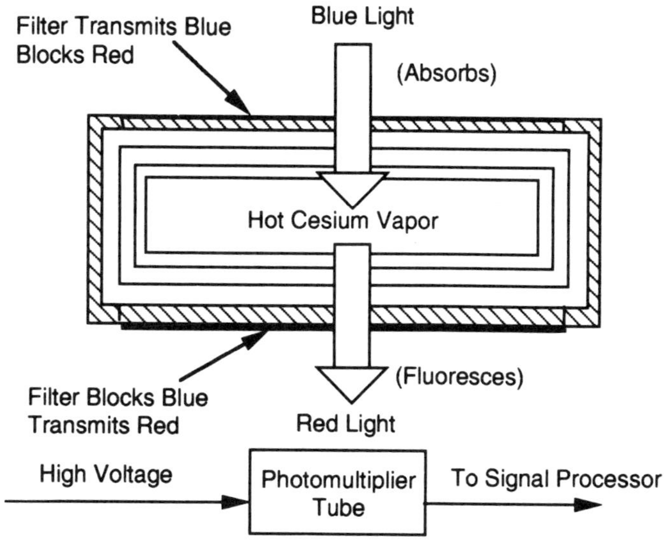

Figure 12.32 Cesium ARFORM filter operation.

Table 12.5
SLC Filter Characteristics

	Filter Type		
Characteristic	*Quartz Birefringent (Lyot)*	*CdS Dispersive Birefringent*	*Cesium Atomic Resonance*
Wavelength	532 nm	532 nm	455 or 459 nm
Bandpass-Angstroms	3.3	1.9	0.06 or 0.12
Field-of-view (H2O)	± 20 deg	±60 deg	> ±50 deg
Quantum efficiency	12%	12%	4.2% (15%*)
Transmission	10%	12%	50% or 20%

*With additional development.

A comparison of the filter technologies which have been used in receivers is shown in Table 12.5.

12.3.3.5 Laser Transmitters

As indicated earlier in the history of SLC, the blue-green (0.532 μm) frequency-doubled Nd:YAG laser approach was the first developed. This solid-state laser uses a Nd-doped YAG rod, pumped originally by flash lamps and then more recently by diode pumping. The laser possesses the right characteristics for a SLC transmitter. At the low data rates possible for SLC due to cloud/seawater pulse lengthening, the ability to Q-switch this laser results in a natural application for PIM where high-peak-power pulses are timed for positioning in time slots, each of which represents a discrete configuration of bits. The high peak power provides a high SNR with achievable lasers and detectors. The laser is also light and long-lived with diode pumping and is also capable of large energy pulses. An example of this potential is a 1-J/pulse, 20-pulse/sec, 15-ns pulse width diode-pumped laser, which is readily achievable at today's state of the art [22]. By using a PIM modulation format of 10 bits per pulse it is seen that a data rate of 10 bits per pulse times 10 pulses per second or 100 bits per second is available. It should be noted that the peak power of this laser would be 1 J/15 ns or 66.7 MW! The efficiency would be expected to be about 5%. The Nd:YAG lasing medium could be a rod or a slab. The laser could be built to survive in space and would weigh approximately 60 lb, which is light compared to a gas laser such as the XeCl laser, the shortcomings of which were mentioned previously.

With the advent of tunable lasers, the probability of lining up a laser wavelength with a good filter such as the cesium atomic resonance filter, or *atomic line filter* (ALF) as it is also called, greatly increased. The submarine transmitter is not very weight-limited so that a laser weighing thousands of pounds would be feasible to provide an uplink. A solid-state tunable laser such as an alexandrite laser could be tuned and frequency-doubled to make a wavelength match for a cesium filter. Again, for a submarine, efficiency is not all that important since plenty of power is available and providing coolant is not all that difficult. An interesting trade and design problem for a submarine uplink is whether to mount the entire laser head external to the hull or provide a window for the laser to penetrate the hull.

Other solid-state lasers have been considered to work with atomic resonance potential filters. These include rubidium (532 nm) with Nd:YAG, thallium-cesium (535 nm) with neodymium-doped beryllium lanthanate (Nd:BEL), potassium (532 nm) with Nd:YAG, magnesium (518 nm) with the second harmonic of titanium sapphire (Ti:Al2O3) pumped by diode lasers, and doubled Nd:YAG or, Cr:ZnWO4, which can be tuned so that it doubles directly into 518 nm.

12.3.4 Example System Characteristics

A spaceborne transmitter to communicate from a LEO satellite to a submarine is estimated to have the characteristics shown in Table 12.6.

The technologies for communications between an aircraft or satellite and a submarine at operational depth are in place and await a new start when the

Table 12.6
SLC Transmitter Characteristics

Equipment	*Weight (lbs)*	*Power (W)*	*Notes*
Gimbaled mirror	33	20	
Pointing/control electronics	6	16	
Transmit telescope	6		
Transmit optics	15		
Laser transmitter (includes cooling)	60/280	500/1.1 KW	1 to 20 J
Comm electronics	12	5	
Power regulator	50/180	140/2.6 KW	1 to 20 J
Command/control electronics	14	30	
Cables	25/40		1 to 20 J
Structure (baseplate)	20/40		1 to 20 J
Totals	214/626	731/3.8KW	1 to 20 J

operational need appears. The uplink is not as well developed as the downlink technology and additional work is needed.

12.4 DEEP SPACE APPLICATIONS

From Lesh [22], "Past missions have been starved for communications capacity and missions of the future will require even more of that capacity while taxing the spacecraft less in terms of mass, size, and power. The Voyager mission extracted more space-acquired data at the Saturn, Uranus, and Neptune encounters than was planned at launch due to improvements in the *deep space network* (DSN) during the interplanetary cruise period." The improvements included arraying of antennas within a given DSN complex, arraying of multiple DSN complexes, and the arraying of 27 antennas at the Very Large Array at Socorro, New Mexico.

The data return requirements for the Galileo mission required an antenna too large to launch erected, so a spare unfurlable TDRSS antenna was selected. There is always a potential failure with the deployment of a stowed antenna since a few have failed to deploy in the past.

NASA has made initial exploratory visits to the major bodies of the solar system, so future missions will require more detailed data and as a result, much higher communications bandwidths due to, for example, multispectral imagers and synthetic aperture radars (which are being considered for these missions). These considerations, coupled with the move toward smaller, lighter, less costly satellites that are launched more frequently create a trend toward very small, low-power communications systems. This is the appealing feature of laser communications for deep space. It is possible to design interplanetary spacecraft with high communications bandwidth in a small, low-power package using laser communications.

The enabling difference between RF and laser communications is due to the frequency (or wavelength). RF is at a big disadvantage due to frequency when it comes to antenna gain, which, of course, equates to beam divergence. Wavelengths of RF systems are thousands of times larger than those at optical. This is reflected in antenna size for a given gain at a given frequency. What this means to the designer of a high-data-rate link from Mars to Earth is illustrated in Figure 12.33. Note that the RF beam is 30 times wider than Earth while the laser beam illuminates only a portion of Earth. The appropriate optical technologies for use in deep space missions have been studied by NASA/JPL. NASA's Lewis Research Center has studied the technologies for a high-data-rate laser return link from Mars [23, 24].

Laser beams are very conserving of transmitted power. As mentioned previously, RF designers have gone all out in arraying Earth antennas, some which are nearly the size of a football field. One wonders how much more can be done with RF frequencies in future missions to deep space. Recognition of this has steered NASA/JPL to institute a long-term optical communications program for deep space

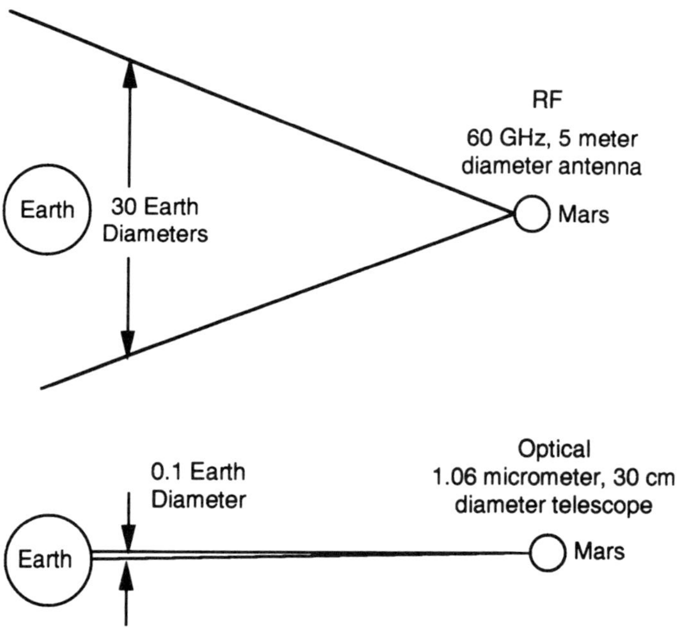

Figure 12.33 A comparison of RF and laser communication beams for communication from Mars to Earth. (After [23].)

that has been on-going for some 14 years. Both spacecraft and ground reception technologies are being developed, system level planning is in place, and a deep space optical communications demonstration was successfully performed late in 1992 [25].

12.4.1 Deep Space Communications Technologies

JPL has been operating under a long-range development and demonstration plan that was described in 1991 and updated in 1994 [26]. Key system elements include the spacecraft transmitter, acquisition/pointing/tracking, telescope and structures, detectors, filters, background mitigation techniques, and coherent-detection techniques. Also included is the definition of a direct-detection 10m diameter orbital optical receiver of "photon bucket" design, (the *deep space optical receiver antenna* or DSORA). This large, segmented, advanced composite receiver need only be as good as 100 times the diffraction limit for the aperture to operate.

For ground station technologies, the development of suitable uplink lasers, weather modeling and measurements, and optical network technology are included in the development plan. The progression has been orderly. Of course, the plan, as with all plans, must be modified somewhat based on funding constraints and the results obtained from technology developments. Many of these developments have been reported in the annual *Proceedings of The SPIE on Laser Communications*.

The earth receiver of the optical signals, whether one or more ground stations or an orbiting receiver in the form of a large telescope, must be included in the planning for future missions. Due to world political climate changes, large telescopes operated or supported by the military have become available for supporting civil space programs. An example of this cooperation was demonstrated on the Galileo Optical Experiment (GOPEX) program where the Air Force Phillips Lab Starfire Optical Range's 1.5m diameter telescope was used. A possible experimental future receiver for laser communications signals from space is the Air Force's Phillips Lab 3.5m diameter telescope shown in Figure 12.34.

12.4.2 The Galileo Optical Experiment

To receive signals from deep space, it is necessary to first establish a link between Earth and the spacecraft. This involves an uplink beacon laser, which is first transmitted to the spacecraft to provide a reference for the spacecraft transmitter. This type of beacon is best exemplified by the high peak power attainable in a solid-state laser such as a Nd:YAG laser, whether at 1064 nm or frequency doubled to 532 nm. The objectives of the experiment were to demonstrate that an uplink beacon laser could be directed to a distant spacecraft based only on the spacecraft's trajectory predictions and nearby reference stars, that the disturbance effects of atmospheric turbulence were well enough understood to be predicted and accommo-

Figure 12.34 A 35m telescope. (Courtesy of Contraves, Inc.)

dated, and lastly, to validate the communications link margin theoretical models used to predict performance. The GOPEX experiment [25] involved two ground station transmitters, one at Kirtland Air Force Base's Starfire Optical Range in Albuquerque, New Mexico and the other at Table Mountain Observatory near Wrightwood, California. Two stations were used to take advantage of diversity in avoiding the effects of cloud cover as was discussed previously in this chapter.

Frequency-doubled Nd:YAG lasers were used as beacons to illuminate the Galileo spacecraft's CCD detector, where transmission distance ranged from 600,000 km to 6,000,000 km. Pulsed laser signals were sent from both sites at pulse rates from 10 Hz to 30 Hz. The spacecraft's camera shutter was opened while the camera line of sight was slewed over the earth areas where the transmitters were located. The recorded pulse images on the CCD were then relayed back to Earth via the conventional RF link.

Nearly 50 confirmed detection images were obtained, which showed that the uplink beacon transmission is well understood and clears the way for further, more challenging experiments leading to a full optical communications capability for deep space beyond the year 2000.

References

[1] Chenoweth, A. J., Crofts, D. E., Dreisewerd, D. G., Green, S. I., Lambert, S. G., Morris, T. R., Simmons, S., and Casey, W. L. "The Maturity of Direct Detection Laser Crosslink Technology," *SPIE OE-LASE '92*, Vol. 1866, Los Angeles, CA, Jan. 1992.

[2] Dreisewerd, D., Lambert, S., Morris, T., Ulrich, G., and Casey, W. "A High Data Rate LEO to GEO Optical Crosslink Design," *IEEE Military Communications Conf. MILCOM, '92*, Vol. 3., Oct. 1992.

[3] Dreisewerd, D., Lambert, Ulrich, G., and Casey, W. "A Twice Synchronous Range GEO to GEO Optical Intersatellite Link," *IEEE Military Communications Conf. MILCOM, '92*, Vol. 1., Oct. 1992.

[4] Lambert, S., Ulrich, G., Chenoweth, A., Morris, T., and Casey, W. "Short-Range Multi-Terminal Satellite Crosslink Communications," *IEEE Military Communications Conf. MILCOM, '92*, Vol. 3, Oct. 1992.

[5] Hinrichs, C. A. 'The Engineering Aspects of Clear Air Laser Propagation," *IEEE Military Communications Conf. MILCOM, '85*, Oct. 1985.

[6] Ruggieri, N., and Price, T. *Laser Applications in Space Communications - Final Report*, Vol. 1, Appendix I, Atmospherics, Rome Air Development Center, May 29, 1987.

[7] Ochs, G., Lawrence, R. and Wang, T. "Stellar Scintillation Measurement of the Vertical Profile of Refractive Index Turbulence in the Atmosphere," *SPIE, Vol. 75*, 1976.

[8] Gage, K., VanZant, T., and Green, J. "Vertical Profiles of C_n^2 in the Free Atmosphere: Comparison of Model Calculations with Radar Observations," *18th Conf. on Radar Meteorology*, March 28–31, 1978.

[9] Balsey, B., and Peterson, V. "Doppler-Radar Measurements of Clear Air Atmospheric Turbulence at 1290 MHz," *Journal of Applied Meteorology*, March 1981.

[10] Hinrichs, C. A. "A Probabalistic Structure Constant Model," *McDonnell Douglas Internal Memo LC-0884-001*, Aug. 27, 1984.

[11] McClatchey, R., et al. *Optical Properties of the Atmosphere*, Third Edition, AFCRL-72-0497, Aug. 24, 1972.

[12] Fisher, R. "Effects of Flow Aerodynamics on Imaging Sensor Resolution," *SPIE, Vol. 79, Aerial Reconnaissance Systems*, 1976.

[13] Wolter, D. "Aerodynamics Effects on Airborne Optical Systems," *McDonnell Douglas Report No. A2582*, Dec. 1973.

[14] Kelsall, D., and D'Amato, R. "Optical Degradation By Aerodynamic Boundary Layers," *MIT Lincoln Laboratory Report ESD-TR-78-243*, Sept. 1977.

[15] Cebeci, T., et al. "A Finite -Difference Method for Calculating Compressible Laminar and Turbulent Boundary Layers: Part 1-General Description," *McDonnell Douglas Report DAC-67131*, March 1969.

[16] Fivel, H., and Hinrichs, C. "Effect of Turbulent Boundary Layer Flow on Optical Transmission," *AIAA 18th Thermophysics Conf.*, June 1983.

[17] Chapman, W., and Fitzmaurice, M. "Optical Space-to-Ground Link Availability Assessment and Diversity Requirements," *SPIE OE-LASE, '91*, Vol., 1417, Los Angeles, CA, Jan. 21–22, 1991.

[18] Katzman, M., Editor. *Laser Satellite Communications*, Englewood Cliffs, NJ: Prentice Hall, 1987.

[19] Painter, F. "Submarine Laser Communications," *Defense Electronics*, June 1989.

[20] "Submarine Laser Communications," *Forecast International/DMS Market Intelligence Report*, Aug. 1992.

[21] Weiner, T., and Karp, S. "The Role of Blue-Green Laser Systems in Strategic Submarine Communications," *IEEE Trans. On Communications*, Vol., COM-28, No. 9, Sept. 1980.

[22] Lesh, J. "Recent Progress in Deep Space Optical Communications," *SPIE OE-LASE '92*, Vol. 1866, Los Angeles, CA, Jan. 1992.

[23] Spence, R. "Comparison of Optical Technologies for a High Data Rate Mars Link," *SPIE OE-LASE 91*, Vol., 1417, Los Angeles, CA, Jan. 21–22, 1991.

[24] Spence, R. "Optical Communications and a Comparison of Optical Technologies for a High Data Rate Return Link From Mars," *NASA Technical Paper 3180*, Lewis Research Center, 1993.

[25] Lesh, J., Deutsch, L., and Weber, W. "A Plan for Development and Demonstration of Optical Communications for Deep Space," SPIE, Vol. 1522, Munich Germany, 11–12 June, 1991.

[26] Wilson, K., Lesh, J., and Yan, T.-Y. "GOPEX: A Laser Uplink to the Galileo Spacecraft on its Way to Jupiter," *SPIE OE-LASE '93*, Los Angeles, CA, Jan., 1993.

Appendix: Useful Formulas

Table A.1
FUNDAMENTAL WAVELENGTHS OF COMMON LASERS

Solid-State Lasers	
Nd:YLF (Frequency doubled)	523 nm
Nd:YAG (Frequency doubled)	532 nm
Nd:YLF	1,047 nm
	1,053 nm
	1,313 nm
	1,321 nm
Nd, Cr:GSGG	1,061 nm
Nd:YAG	1,064 nm
	1,319 nm
	1,320 nm
	1,335 nm
Nd:YAP	1,080 nm
Nd:YVO4	1,342 nm
Er,Yb:glass	1,535 nm
Tm:YAG	2,015 nm
Tm,Ho:YAG	2,091 nm
Er:YAG	2,900 nm
Tunable (Dye and Solid State)	
Ti:sapphire	670–1,070 nm
Organic Dyes	200–1,000 nm
Alexandrite	720–800 nm
Cr:LiSAF	780–1,010 nm
Forsterite	1,130–1,370 nm
Co:MgF2	1,750–2,500 nm
Semiconductor	
AlGaAs	780–860 nm
AlGaInP	630–700 nm
InGaAs	910–1,100 nm
InGaAsP	1,300 & 1,550 nm
Gas	
CO_2	10.6 μm

A.1 USEFUL PHYSICAL CONSTANTS

Planck's Constant	h	6.6254×10^{-34} J/sec
Speed of light in free space	c	2.998×10^{8} m/sec
Electronic quantum charge	q	1.602×10^{-19} coulombs
Boltzman's constant	k	1.381×10^{-23} J/K

A.2 CONVERSIONS FROM OPTICAL POWER TO CURRENT, PHOTOELECTRONS / SEC AND PHOTONS / SEC

Converting from received optical power levels on a photodiode to current levels at the preamplifier input

$$i_{\text{out-photodiode}} = P_{\text{in}}(M\eta q/h\nu) \tag{A.1}$$

where η is the responsivity of the photodiode, M is the gain of the detector, and $\nu = c/\lambda$.

Converting the current in (A.1) to photo-electrons per second

$$N_{\text{out-photodiode}} = P_{\text{in}}(M\eta/h\nu) \tag{A.2}$$

Converting to photoelectrons per second in (A.2) to photons per second yields

$$H_{\text{out-photodiode}} = P_{\text{in}}(M/h\nu) \tag{A.3}$$

A.3 EARTH AND ASTRONOMICAL CONSTANTS

Astronomical Unit (AU)	1.5×10^{8} kilometers (km)
Parsec	3.1×10^{13} km
Mean Earth Radius	3,441.66 nautical miles 6,373.95 km
Distance Moon to Earth	384,405 km
Moon Radius	1,738 km
Angular Diameter	
Sun From Earth	32′15″ or 9.38 mrad
Moon From Earth	31′07″ or 9.05 mrad

Table A.2
PLANETARY ORBITS

Planet	*Semimajor Axis of Orbit*	
	AU	$\times 10^6$ kms
Mercury	0.387	57.91
Venus	0.723	108.21
Earth	1.0	149.6
Mars	1.524	227.9
Jupiter	5.2	778.3
Saturn	9.546	1,428
Uranus	19.2	2,872
Neptune	30.0	4,498
Pluto	39.5	5,910

Table A.3
EARTH ORBITAL CHARACTERISTICS

Altitude Nautical Miles	*Period Days*	*Circular Orbit Velocity (ft / sec)*
100	0.06125	25,567
200	0.06386	25,213
300	0.06651	24,874
482	1/14th	24,290
906	1/12th	23,073
1,469	1/10th	21,713
2,256	1/8th	20,156
3,461	1/6th	18,313
5,592	1/4th	15,998
10,899	1/2	12,697
19,323	1 (synch)	10,088
32,740	2	7,999

A.4 USEFUL SIMPLE OPTICAL FORMULAS

Effective Focal Length / Field Angle

$$x = \alpha \cdot \text{EFL} \tag{A.4}$$

Note: Size of lens or f-number doesn't matter.

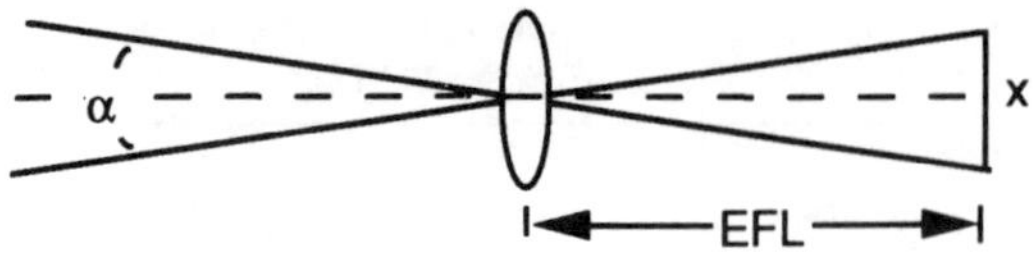

Figure A.1 Effective focal length / field angle.

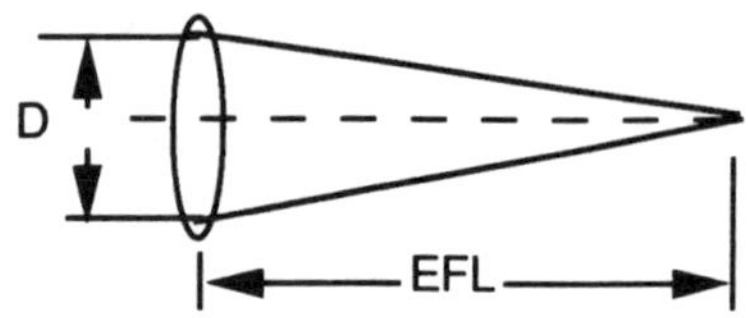

Figure A.2 F-number development scenario.

F-Number

$$f/\# = \frac{\mathrm{EFL}}{D} \tag{A.5}$$

Diffraction-Limited Spot Size

$$\theta_{\mathrm{spot}} = \frac{2.44\,\lambda}{D} \quad \text{angular spot diameter dimension} \tag{A.6}$$

$$d_{\mathrm{spot}} = \frac{2.44\,\lambda}{D}\mathrm{EFL} \quad \text{linear spot diameter dimension} \tag{A.7}$$

Gaussian Lens Formula

$$\frac{1}{\mathrm{EFL}} = \frac{1}{s} + \frac{1}{s''} \tag{A.8}$$

EFL is the focal length (+D for positive, − for negative). The values of s and s'' are conjugate distances.

Good for f-numbers greater than unity.

Magnification

$$m = \frac{s''}{s} \tag{A.9}$$

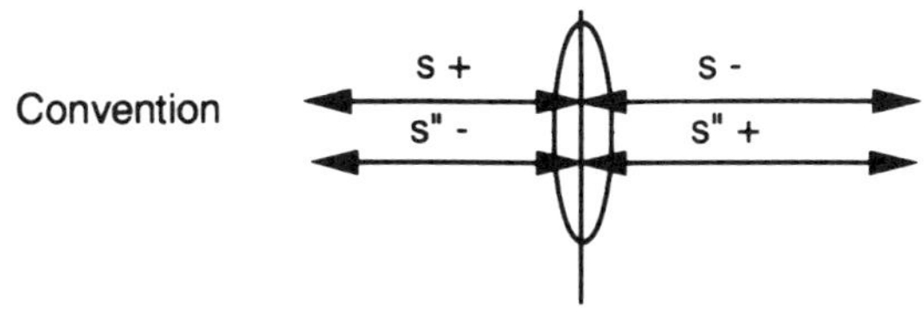

Figure A.3 Gaussian lens formula definition.

Numerical Aperture

In air, numerical aperture (N.A.)

$$\text{f}/\# = \frac{\text{N.A.}}{2} \tag{A.10}$$

Solid Angle

$$\Omega = 2\pi[1 - \cos(\theta)] \tag{A.11}$$

where θ is the half-angle cone.

For small values of θ, less than 20 deg, a good approximation is

$$\Omega \approx \pi\theta^2 \tag{A.12}$$

Far-Field Gain

For a Gaussian fed circular unobscured, untruncated aperture the gain is given as

$$G_{\text{trans}} = \frac{32}{\theta_t^2} \tag{A.13}$$

where θ_t is the transmit beam divergence at the $1/\text{e}^2$ point on the beam.

For a uniform fed circular unobscured, untruncated aperture, the gain is given as

$$G_{\text{trans}} = \frac{16}{\theta_t^2} \tag{A.14}$$

The off-axis far-field intensity can be approximated by

$$G(\beta) \approx e^{-8(\beta/\theta_t)^2} \tag{A.15}$$

where β is the off-axis angle.

Divergence For A Given Aperture Size

The $1/e^2$ divergence of the transmit beam for a given aperture size, D, can be approximated by

$$\theta_t \approx \frac{4\lambda}{\pi D} \tag{A.16}$$

where λ is the wavelength of the transmitted energy.

A.5 USEFUL MATHEMATICAL EXPRESSIONS

Sphere

$$\text{Area} = 4\pi r^2 \tag{A.17}$$

$$\text{Volume} = \frac{4}{3}\pi r^3 \tag{A.18}$$

where r is the radius of the sphere.

Cylinder

$$\text{Area} = 2\pi rh \tag{A.19}$$

$$\text{Volume} = \pi r^2 h \tag{A.20}$$

where r is the radius of the cylinder and h is the cylinder height.

Cone

$$\text{Area} = \pi r\sqrt{r^2 + h^2} \tag{A.21}$$

$$\text{Volume} = \frac{\pi r^2 h}{3} \tag{A.22}$$

Ellipse

$$\text{Area} = \pi ab \tag{A.23}$$

where a and b are the lengths of the semi-major and semi-minor axes, respectively

Triangle

$$\text{Area} = \frac{bh}{2} \tag{A.24}$$

where b is the base length and h is the height of the triangle.

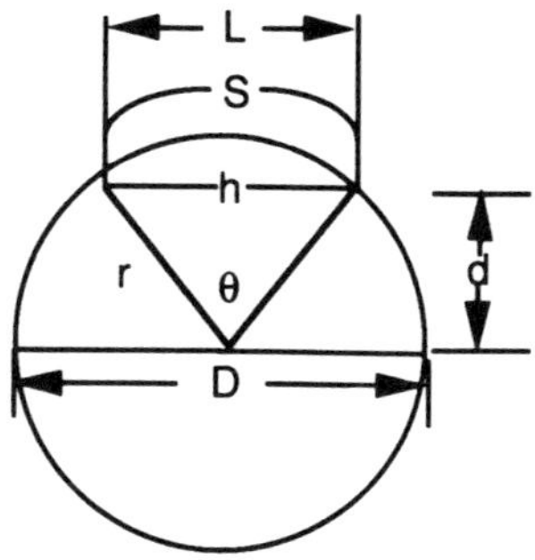

Figure A.4 A circle.

Circle

$$\text{Circumference} = 2\pi r = \pi D \tag{A.25}$$

$$L = 2r \cdot \sin\left(\frac{\theta}{2}\right) = 2\sqrt{r^2 - d^2} \tag{A.26}$$

$$d = r \cdot \cos\left(\frac{\theta}{2}\right) \tag{A.27}$$

$$h = r - d \tag{A.28}$$

$$\theta = \frac{S}{r} = \frac{2S}{D} = 2 \cdot \cos^{-1}\left(\frac{d}{r}\right) \tag{A.29}$$

A.6 GENERAL FACTOIDS

- 180 deg is equivalent to π rad.
- To convert radians to degrees multiply by 57.29.
- To convert degrees to radians multiply by 0.01745.
- 17.45 mrad equals one deg.
- One arc second is approximately 5 μrad (4.848 μrad).
- One arc minute is about 300 μrad (290.9 μrad).

Table A.4
ENCODER ACCURACY

Number of Encoder Bits	*Angular Resolution*
2	1.5708 rad
3	785.3 mrad
4	392.7 mrad
5	196.4 mrad
6	98.2 mrad
7	49.1 mrad
8	24.5 mrad
9	12.3 mrad
10	6.1 mrad
11	3.1 mrad
12	1.5 mrad
13	767 μrad
14	383 μrad
15	192 μrad
16	96 μrad
17	48 μrad
18	24 μrad
19	12 μrad
20	6 μrad
21	3 μrad
22	1.5 μrad
23	750 nrad
24	375 nrad

A.7 REFERENCE LIST OF SPIE VOLUMES ON LASER COMMUNICATIONS

Society of Photo Instrumentation Engineers (SPIE)
P. O. Box 10
Bellingham, WA 98227-0010
Telephone (206) 676-3290

"Selected Papers On Free-Space Laser Communications"
D. Begley, Editor, Volume MS30, 1991.

"Free-Space Laser Communication Technologies"
Volume 885, Los Angeles, CA, 1988.

"High Data Rate Atmospheric and Space Communications"
Volume 996, Boston, MA, September 1988.

"Space Sensing, Communication and Networking"
Volume 1059, Los Angeles, CA, 16-18 January 1989.

"Free-Space Laser Communications Technologies II"
Volume 1218, Los Angeles, CA, January 1990.

"Free-Space Laser Communications Technologies III"
Volume 1417, Los Angeles, CA, January 1991.

"Optical Space Communications II"
Volume 1522, Munich, Germany, June 1991.

"Free-Space Laser Communications Technologies"
Volume 1635, Los Angeles, CA, January 1992.

"Free-Space Laser Communications Technologies IV"
Volume 1866, Los Angeles, CA, January 1993.

"Free-Space Laser Communications Technologies VI"
Volume 2123, Los Angeles, CA, January 1994.

List of Acronyms

ACTS	advanced communication technology satellite
AFTS	airborne flight test system
AJ	anti-jam
ALF	atomic line filter
AM	amplitude modulation
APDs	avalanche photodiodes
AR	antireflection
ARFORM	atomic resonance filter optical receiver module
ARTEMIS	advanced relay and technology mission
ASE	amplified stimulated emission
ASIC	application specific integrated circuit
ASK	amplitude shift keying
ATM	asynchronous transfer mode
BE	brilliant eyes
BSTS	boost surveillance and tracking system
C^3	command, control, and communications
CCD	charge coupled device
CD	compact disc
CDM	code division multiplexing
CdS	cadmium sulfide
CID	charge injection device
CLS	crosslink set
CMOS	complementary metal oxide semiconductor
CO_2	carbon dioxide
CONUS	continental United States
CTD	charge transfer device
CW	continuous wave
DARPA	Defense Advanced Research Projects Agency

dB	decibels
DBF	dispersive birefringence filter
dc	direct current
DDLT	direct-detection laser transmitter
DPSK	differential PSK
DSCS	Defense Satellite Communication System
DSN	deep space network
DSORA	deep space optical receiver antenna
EDFAs	erbium-doped fiber amplifiers
EHF	extremely high frequency
ELF	extremely low frequencies
EO	electro-optic
ESA	European Space Agency
FDM	frequency division multiplexing
FET	field effect transistor
FEWS	follow-on early warning system
FM	frequency modulation
FON	fiber-optic network
FOV	field-of-view
FPN	fixed pattern noise
FSDD	full-scale development demonstration
FSK	frequency shift keyed
FSM	fast steering mirror
GaAlAs	gallium aluminum arsenide
GaAs	gallium arsenide
Gbps	gigabit per second
Ge	germanium
GEO	geosynchronous Earth orbit
Ghz	gigahertz
GOPEX	Galileo optical experiment
GPS	global positioning system
GSFC	Goddard Space Flight Center
HDRLT	high-data-rate laser transmitter
HDTV	high definition television
HeNe	helium neon
Hz	hertz
ICD	interface control document
IF	intermediate frequency
IMPATT	impact avalanche and transit time
InGaAs	indium gallium arsenide
InGaAsP	indium gallium arsenide phosphate
InP	indium phosphide

IO	imaging optics
JPL	Jet Propulsion Laboratory
LCC	lifecycle cost
LCE	laser communication experiment
LCS	laser crosslink subsystem
LDP	laser diode package
LED	light emitting diode
LEO	low Earth orbit
LO	local oscillator
LPD	lateral photo detector
LPI	low probability of intercept
Mbps	megabits per second
MgF_2	magnesium fluoride
MILSTAR	military strategic tracking and relay
MITLL	Massachussetts Institute of Technology's Lincoln Laboratory
MOPA	master oscillator, power amplifier
NASA	National Aeronautics and Space Administration
NASP	National Aerospace Plane
Nd:YAG	neodymium yttrium aluminum garnet
Nd:YAP	neodymium yttrium aluminum phosphate
Nd:YLF	neodymium yttrium lithium fluoride
NEA	noise equivalent angle
NRZ	nonreturn to zero
ONR	Office Of Naval Research
ONT	Office Of Naval Technology
P-HEMTs	power-high electron mobility transistor
PBM	pulse binary modulation
PC	personal computer
PCFLOS	probability of a cloud-free line of sight
PDBM	pulse delay binary modulation
PIM	pulse interval modulation
PIN	p-intrinsic
PLL	phase locked loop
PM	phase modulation
PMT	photomultiplier tubes
PPBM	pulse polarization binary modulation
PPM	pulse position modulation
PQM	pulse quaternary modulation
PSK	phase shift keyed
PSR	point source rejection
QAPDs	quadrant APDs

QPIN	quadrant p-intrinsic
QPPM	quaternary PPM
RF	radio frequency
RFP	request for proposal
rms	root mean square
ROM	rough order of magnitude
RS	Reed Solomon
RTV	room temperature vulcanizing
RZ	return to zero
SDI	strategic defense initiative
SDIO	Strategic Defense Initiative Organization
SDL	Spectra Diode Labs
SDM	spatial division multiplexing
SF	slope factor
SFTS	space flight test system
SILEX	Space Intersatellite Link Experiment
SLC	submarine laser communications
SLCR	submarine laser communication receiver
slik	special low "K"
SNR	signal to noise ratio
SONET	synchronous optical network
TDM	time division multiplexing
TDRSS	tracking and data relay satellite system
TTL	transistor-transistor logic
TWTA	traveling wave tube amplifier
ULE	ultralow expansion
U.S.	United States
UV	ultraviolet
VA	Viterbi algorithms
VCM	volatile condensable materials
VLF	very low frequency
WDM	wavelength division multiplexing
XeCl	xenon-chloride

About the Authors

Stephen G. Lambert has over fourteen years experience in aerospace communication system design of which thirteen years have been on space related laser communications programs. Mr. Lambert was involved in the initial development and testing of the optical receiver for the Laser Crosslink Subsystem program and has analyzed and specified performance requirements for numerous laser crosslink programs. He most recently was chief systems analyst for the Follow-on Early Warning System laser crosslink terminal. In addition to his program work, Mr. Lambert has been a communications system analyst and designer of many advanced concept laser communication payloads including aircraft, satellite and deep space laser communication terminals.

Early in his career, Mr. Lambert developed analytical tools including a model of the laser communication end to end free space link, which is still used today. For the past six years he has divided his efforts between program system engineering and advanced laser communication concept development.

Mr. Lambert received bachelor and masters degrees in electrical engineering from Purdue University, West Lafayette, IN, in 1980 and 1981, respectively. He has authored ten technical papers on laser communications and has served as a panel member at laser communications forums. Mr. Lambert is a member of the Institute of Electrical and Electronic Engineers and its Communications Society. He is currently employed by McDonnell Douglas Aerospace—St. Louis, MO.

William L. Casey has over thirty-six years experience in aerospace optical systems of which thirty-two have been on space-related programs. He was an optical payload specialist on numerous advanced space programs. Mr. Casey invented and built a Viewfinder Tracking System for Earth Resources used by three crews on Skylab and was the responsible engineer for the Space Shuttle Star Trackers.

His efforts for the last twenty years have been devoted exclusively to laser communications covering conceptual definition, design and development of space laser communications hardware. He was optical design task leader for McDonnell Douglas' laser communications activities for ten years and supervised optical design efforts at another aerospace company. Mr. Casey holds a patent for and built the first wavelength combined multiple diode laser used as a laser communications source. He has written two dozen technical papers on optics, sensors and laser communications.

Mr. Casey graduated Cum Laude in Mechanical Engineering from Lehigh University, Bethlehem, PA. He is a member of Tau Beta Pi, Pi Tau Sigma and is listed in Who's Who Worldwide, 1993–1994. He is currently consulting in the field of laser communications.

Index

The Artech House Optoelectronics Library

Brian Culshaw, Alan Rogers, and Henry Taylor, *Series Editors*

Acousto-Optic Signal Processing: Fundamentals and Applications, Pankaj Das

Amorphous and Microcrystalline Semiconductor Devices, Optoelectronic Devices, Jerzy Kanicki, editor

Bistabilities and Nonlinearities in Laser Diodes, Hitoshi Kawaguchi

Coherent Lightwave Communication Systems, Shiro Ryu

Electro-Optical Systems Performance Modeling, Gary Waldman and John Wootton

Elliptical Fiber Waveguides, R. B. Dyott

The Fiber-Optic Gyroscope, Hervé Lefèvre

Field Theory of Acousto-Optic Signal Processing Devices, Craig Scott

Frequency Stabilization of Semiconductor Laser Diodes, Tetsuhiko Ikegami, Shoichi Sudo, Yoshihisa Sakai

Fundamentals of Multiaccess Optical Fiber Networks, Denis J. G. Mestdagh

Germanate Glasses: Structure, Spectroscopy, and Properties, Alfred Margaryan and Michael A. Piliavin

High-Power Optically Activated Solid-State Switches, Arye Rosen and Fred Zutavern, editors

Highly Coherent Semiconductor Lasers, Motoichi Ohtsu

Iddq Testing for CMOS VLSI, Rochit Rajsuman

Integrated Optics: Design and Modeling, Reinhard März

Introduction to Electro-Optical Imaging and Tracking Systems, Khalil Seyrafi and S. A. Hovanessian

Introduction to Glass Integrated Optics, S. Iraj Najafi

Introduction to Radiometry and Photometry, William Ross McCluney

Laser Communications in Space, Stephen G. Lambert and William L. Casey

Optical Control of Microwave Devices, Rainee N. Simons

Optical Document Security, Rudolf L. van Renesse

Optical Fiber Amplifiers: Design and System Applications, Anders Bjarklev

Optical Fiber Sensors, Volume I: Principles and Components, John Dakin and Brian Culshaw, editors

Optical Fiber Sensors, Volume II: Systems and Applicatons, John Dakin and Brian Culshaw, editors

Optical Interconnection: Foundations and Applications, Christopher Tocci and H. John Caulfield

Optical Network Theory, Yitzhak Weissman

Optical Transmission for the Subscriber Loop, Norio Kashima

Reliability and Degradation of LEDs and Semiconductor Lasers, Mitsuo Fukuda

Semiconductor Raman Laser, Ken Suto and Jun-ichi Nishizawa

Semiconductors for Solar Cells, Hans Joachim Möller

Single-Mode Optical Fiber Measurements: Characterization and Sensing, Giovanni Cancellieri

Ultrafast Diode Lasers: Fundamentals and Applications, Peter Vasil'ev

For further information on these and other Artech House titles, contact:

Artech House
685 Canton Street
Norwood, MA 02062
617-769-9750
Fax: 617-769-6334
Telex: 951-659
email: artech@world.std.com

Artech House
Portland House, Stag Place
London SW1E 5XA England
+44 (0) 171-973-8077
Fax: +44 (0) 171-630-0166
Telex: 951-659
email: bookco@artech.demon.co.uk